Advances in
Integrated Optics

Advances in Integrated Optics

Edited by

S. Martellucci
The University of Rome "Tor Vergata"
Rome, Italy

A. N. Chester
G. M. Hughes Electronics Corporation
Los Angeles, California

and

M. Bertolotti
The University of Rome "La Sapienza"
Rome, Italy

Springer Science+Business Media, LLC

Library of Congress Cataloging-in-Publication Data

Advances in integrated optics / edited by S. Martellucci, A.N.
 Chester, and M. Bertolotti.
       p.   cm.
   Contains the proceedings of a conference held June 1-9, 1993, in
 Erice, Sicily.
   Includes bibliographical references and index.
   ISBN 978-0-306-44833-1   ISBN 978-1-4615-2566-0 (eBook)
   DOI 10.1007/978-1-4615-2566-0
   1. Integrated optics--Congresses.   I. Martellucci, S.
 II. Chester, A. N.   III. Bertolotti, Mario.
 TA1660.A38 1994
 621.36'93--dc20   94-38477
                                                                    CIP

Proceedings of the International School of Quantum Electronics 18th Course on
Advances in Integrated Optics, held June 1-9, 1993, in Erice, Italy

ISBN 978-0-306-44833-1

© 1994 Springer Science+Business Media New York
Originally published by Plenum Press, New York in 1994

All rights reserved

No part of this book may be reproduced, stored in a retrieval system, or transmitted in any form or by any means, electronic, mechanical, photocopying, microfilming, recording, or otherwise, without written permission from the Publisher

PREFACE

This volume contains the Proceedings of a two-week summer conference titled "Advances in Integrated Optics" held June 1-9, 1993, in Erice, Sicily. This was the 18th annual course organized by the International School of Quantum Electronics, under the auspices of the "Ettore Majorana" Centre for Scientific Culture.

The term Integrated Optics signifies guided-wave optical circuits consisting of two or more devices on a single substrate. Since its inception in the late 1960's, Integrated Optics has evolved from a specialized research topic into a broad field of work, ranging from basic research through commercial applications. Today many devices are available on market while a big effort is devolved to research on integrated nonlinear optical devices.

This conference was organized to provide a comprehensive survey of the frontiers of this technology, including fundamental concepts, nonlinear optical materials, devices both in the linear and nonlinear regimes, and selected applications. These Proceedings update and augment the material contained in a previous ISQE volume, *"Integrated Optics: Physics and Applications"*, S. Martellucci and A.N. Chester, Eds., NATO ASI Series B, Vol. 91 (Plenum, 1983). For some closely related technology, the reader many also wish to consult the ISQE volumes: *"Optical Fiber Sensors"*, A. N. Chester, S. Martellucci and A. M. Scheggi, Eds., NATO ASI Series E, Vol. 132 (Nijhoff, 1987) ; and , *"Nonlinear Optics and Optical Computing"*, S. Martellucci and A. N. Chester, Eds., E. Majorana Int'l Science Series, Vol. 49 (Plenum, 1990).

We have brought together some of the world's acknowledged experts in the field to summarize both the present state of I.O. technologies and their background. Most of the lecturers attended all the lectures and devoted their spare hours to stimulating discussions. We would like to thank them all for their admirable contributions. The conference also took advantage of a very active audience; most of the participants were active researchers in the field and contributed with discussions and seminars. Some of these seminars are also included in these Proceedings.

The Chapters in these Proceedings are not ordered exactly according to the chronology of the conference but they give a fairly complete accounting of the conference lectures with the exception of the informal panel discussions. We did not modify the original manuscripts in editing this book, except to assist in uniformity of style.

The volume begins with two broad tutorial treatments by the course Directors, G. C. Righini and M. Bertolotti. Chapters 3 through 11 survey optical materials and processing techniques, covering not only nonlinear optical materials but also the growing use of semiconductors, glasses, and polymers. Chapters 12 through 16 describe a variety of Integrated Optics devices and their characteristics. Chapters 17 and 18 describe experiments with solitons and an improved technique for measuring waveguide losses. Finally, Chapters 19 through 23 describe systems applications of Integrated Optics.

As described above, these Chapters span the theory, the techniques, and the applications of Integrated Optics technology today. References to additional articles will be found in the individual Chapters.

Before concluding, we acknowledge the invaluable help of Giancarlo Righini, from IROE-CNR, co-director of the conference. We also appreciate the excellent professional support of our editor at Plenum Publishing Company, Miss Joanna Lawrence. We would also like to appreciate the support of Eugenio Chiarati for much of the computer processing work. We wish to mention with sincere thanks Anna Mezzanotte, Margaret Kyoko Hayashi and Giorgia Nanula, assistants to the Editors (S.M., A.N.C., and M.B., respectively). Finally, we acknowledge the organizations who sponsored the conference, especially the generous financial support of the G.N.E.Q.P.-CNR.

The Editors:

Sergio Martellucci
Professor of Physics
The University of Rome "Tor Vergata"
Rome (Italy)

Arthur N. Chester
Senior Vice President, Research and Technology
G.M. Hughes Electronics Corporation
Los Angeles, California (USA)

Mario Bertolotti
Professor of Physics
The University of Rome "La Sapienza"
Rome (Italy)

May 15, 1994

CONTENTS

1. Introduction to Integrated Optics:
   Characterisation and Modelling of Optical Waveguides 1

   S. Pelli and G. Righini

2. Introduction to Non-Linear Guided Waves 21
   M. Bertolotti

3. Nonlinear Optical Materials 57
   C. Flytzanis

4. Integrated Optics on Lithium Niobate 79
   D. Delacourt

5. Propagation of Self-Trapped Optical Beams
   in Nonlinear Kerr Media and Photorefractive Crystals 95
   B. Crosignani

6. Advances in Semiconductor Integrated Optics 109
   A. Carenco

7. Silica on Silicon Integrated Optics 121
   R.R. Syms

8. Integrated Optics on Silicon: IOS Technologies 151
   S. Valette

9. Are Glasses Suitable for Optoelectronics? 165
   A. Montenero

10. Linear and Nonlinear Optical Properties
    of Polymer Waveguides 173
    F. Michelotti

11. Fabrication and Characterization
    of Conjugated Polymer Waveguides 185
    S. Sottini

12. Linearized Optical Modulators
    for High Performance Analogs 199
    G. Tangonan, J.F. Lam and J.H. Schaffner

13. An Example of Ti:LiNbO$_3$ Device Fabrication:
    The Mach-Zehnder Electrooptical Modulator 207
    P. Cusumano and G. Lullo

14. All-Optical Switching in AlGaAs Semiconductor
    Waveguide Devices 213
    J.A. Aitchison

15. Integrated Optics Sensors 227
    O. Parriaux

16. Multi-Quantum Well Integrated Stacks
    for Detection in Mid Infra-Red 243
    I. Gravè, A. Shakouri, N. Kuze and A. Yariv

17. Spatial Optical Solitons-Experiments 259
    Y. Silberberg

18. Optical Losses Characterization of Channel
    Waveguide through Photodeflection Method 273
    **R. Li Voti, M. Bertolotti, L. Fabbri, G. Liakhou, A. Matera,
    C. Sibilia and M. Valentino**

19. Integrated Optics Applications for Telecommunications 279
    H.-P. Nolting

20. Optical Signal Processors and Applications 303
    M.N. Armenise and V.M.N. Passaro

21. Progress of High-Speed Optoelectronics 313
    H.-W. Yen

22. Optoelectronic Switching Applied to Radar Steering 323
    G. Tangonan, R.Y. Loo and W.W. Ng

23. Application of Optical Links in High Energy Physics Experiments 331
    C. Da Via, M. Glick and J. Söderqvist

Index 337

Chapter 1

INTRODUCTION TO INTEGRATED OPTICS: CHARACTERISATION AND MODELLING OF OPTICAL WAVEGUIDES

S. PELLI and G. C. RIGHINI

1. Introduction

The first demonstration of the laser in 1960 opened the way to the development of lightwave technology; then the production of low-loss optical fibers in the 70s made guided-wave optical communication systems become a reality. One of the problems associated with the development of long-haul systems was obviously related to the introduction in the transmission line of a number of repeaters, able to recondition and to reamplify the optical signal. The solution offered by conventional optics was unsatisfactory, due to the size and electrical power consumption, as well as to the critical dependence on temperature variations, mechanical vibrations, and the presence of moisture. The alternative first suggested by S.E. Miller, a researcher at Bell Laboratories, was to miniaturise the repeater, integrating all the components onto a single chip and interconnecting them via optical waveguides: the concept of integrated optics was born. More than two decades have passed since then: innovative research has been carried out on a vast spectrum of waveguide devices, and in recent years the goal of performing useful optoelectronic functions in a number of commercial applications has eventually come to fruition.

The bases of linear integrated optics, as concerns both propagation theory and the most common manufacturing technologies, are generally well established, and we can refer the interested readers to a number of books on the subject.[1-5] There is, however, a lot of activity still going on and, as a consequence, an increasing need of fixing some standardisation, especially concerning the definition and measure of the operational parameters of integrated optical waveguides, components and devices. The knowledge of waveguide characteristics such as propagation constant, chromatic dispersion, propagation loss, in-plane and out-of-plane scattering, is fundamental for assessing waveguide quality; providing a feedback of the characterisation results to the designers and to the people in charge of the fabrication process is the key to improving both manufacturing throughput and device performance.

The scope of the present paper is to review the basic measurement techniques which are employed for waveguide characterisation and to discuss an example of a numerical modelling approach, in the case of gradient-index waveguides.

S. Pelli and G. C. Righini - Research Institute on Electromagnetic Waves (IROE), CNR via Panciatichi, 64 - 50127 Firenze, Italy

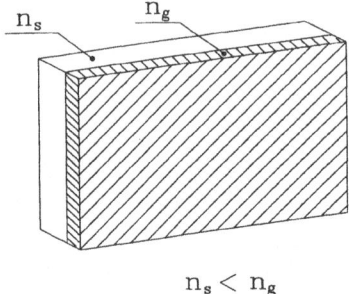

Fig. 1. The simplest structure of a planar waveguide; the guiding film has refractive index n_g higher than the substrate index n_s and the overclad index (here air, $n_c = 1$).

2. Waveguide Characterisation Using the Prism Coupler

The principle of optical confinement in a layer of dielectric material is based on the phenomenon of total internal reflection. The simplest planar waveguide, which is sketched in Fig. 1, is constituted by a three-layer structure: the guiding layer, with refractive index n_g, is supported by the substrate, having index n_s, and is covered by a cladding (which in this case is air, $n_c = 1$). Provided that the refractive index of the film n_g is greater than n_s and n_c, any light ray entering the film in such a way that the angle Θ formed by the ray in the film with the normal to the film surface is greater than the critical angle for the interface with the smallest index difference, namely $\Theta > \arcsin(n_g/n_s)$, experiences total reflection at both the upper and lower interfaces: the light beam is thus trapped within the film. Those rays which, in addition, fulfill the phase condition:

$$2\,k\,n_g\,d\,\cos\Theta - 2\,\phi_a - 2\,\phi_s = 2\,m\,\pi \tag{1}$$

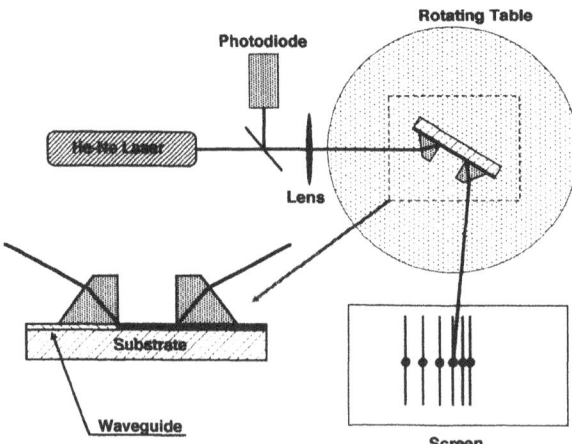

Fig. 2. Sketch of the two-prism experimental setup for the measurement of the effective index of guided modes from their optimum coupling angles (often referred to as m-line setup).

(where $k=2\pi/\lambda$, d is the film thickness, and m an integer which designates the mode number) correspond to stationary solutions and represent optical guided modes supported by the film. This is essentially a resonance condition; $-2\phi_a$ and $-2\phi_s$ are the phase changes suffered by the light beam at the film-air and film-substrate interfaces, respectively. Their values are given by the Fresnel coefficients and depend on the polarisation of the beam.[1b]

The practical excitation of a guided wave is one of the fundamental experimental procedures, both for the testing and for the real operation of an integrated optical device; the most straightforward methods are those of transverse coupling, in which the laser beam enters directly through an exposed cross-section of the waveguide (end-fire and end-butt coupling pertain to this class). When only the surface of the waveguide is accessible it is necessary to use longitudinal couplers such as prism, grating and taper couplers, in which the beam is incident obliquely onto the guide through a structure which provides the phase matching between the incident wave and a guided mode.

The prism-coupling technique[2b] is the most commonly used, because of some inherent advantages: high coupling efficiency (up to 80% for a Gaussian beam), applicability both to planar and to channel waveguides, and selective excitation of any of the guided modes. The main drawback is related to the critical effect on the coupling efficiency of a few factors: the form and position of the incident beam, and the adjustment of the air gap between the prism bottom base and the waveguide surface (gap thickness has usually to be less than 1 µm). The experimental setup includes a sample holder that allows the experimenter to regulate the pressure of the waveguide against the base of the prism in order to change the air gap thickness, and a precision rotating stage that allows the angle of the incident beam Θ' to be varied with respect to the waveguide.

In most cases, the excitation of the mode(s) of the waveguide is made evident by a visible streak along the propagation path, which is due to the guided light scattered out of the plane of the waveguide itself. Only in very low-loss guides, with attenuation below 0.2 dB/cm, will the streak be not so easily observed: a confirmation of the guided-wave excitation can be obtained by placing a second prism to outcouple light and by observing the light pattern onto a screen, as sketched in Fig. 2. If all the modes are excited at the same time (as occurs, for instance, when the incident beam is focused onto the prism), the pattern consists of *m-lines*, each one of them appearing as a brighter spot superimposed onto a weak line which extends along the direction parallel to the plane of the waveguide. Such a weak line is produced by in-plane scattering of the guided waves. By rotating the support table, one can preferentially excite one mode, so that the corresponding spot becomes the brightest one. If we do not use a focusing optical system, and the incident beam is collimated, we obtain the excitation of one mode at a time, and correspondingly a single line should be visible on the screen. Even in that case, due to scattering from topographical and/or index inhomogeneities, a portion of the power in the excited mode is coupled into the other modes and thus the entire set of m-lines is generally visible: the excited mode, however, produces a line on the screen which is much brighter than the other ones.

If the waveguide material is highly absorbing, the propagation length can be very short and it is not possible to use two prisms: a single symmetric, tent-shaped, prism is therefore used. In this case (see Fig. 3) we observe on the screen a bright spot corresponding to the reflected laser beam; when a waveguide mode is excited, part of the beam energy is coupled into the guide itself and absorbed so that a *dark line* appears in the centre of the spot.

Both of these experimental arrangements can be used to measure the propagation constants of the guided modes, since the propagation constant of the m-th mode is given by $\beta_m = k\, n_g \sin\Theta_m$, where Θ_m is defined according to Eq.(1) and it is correlated in a simple way to

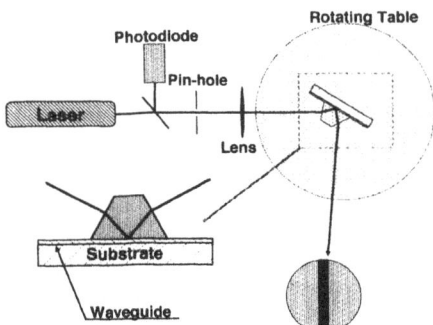

Fig. 3. Sketch of the single (tent-shaped) prism experimental setup for the measurement of the effective index of guided modes from their optimum coupling angles (often referred to as dark-line setup).

the external incidence angle Θ'. Thus, we can measure the angles Θ'_m corresponding to the excitation of the different modes and from these values calculate the respective propagation constants (or the effective indices $n_m = n_g \sin\Theta_m$). To have an idea of the measurement accuracy, the error in n_m is of the order of 10^{-4} when the synchronous coupling angles are determined with an accuracy of about 10^{-4} rad (20 seconds of arc).

In the case of a step-index waveguide, i.e. having a constant refractive index n_g, the refractive index and thickness of the guiding layer can be easily determined by measuring the effective indices of at least two modes and by using Eq.(1). When a waveguide is single-mode, one can solve the system of two equations in two variables by measuring either the TE and TM fundamental mode at the same wavelength, or the TE fundamental mode at two wavelengths. When the available measurements are more than two, iterative numerical procedures are generally used to increase the accuracy of the computation.

The same straightforward approach cannot be used for graded-index waveguides, namely those guides which are produced by diffusion processes (ion-exchange in glass, Ti-indiffusion in lithium niobate, ...) and have a refractive index distribution along the direction normal to the waveguide. This case is treated in Section 5.

3. Measurements of Other Waveguide Characteristics

A variety of specialised test and characterisation techniques can be employed in the evaluation of waveguides and components, especially when one is concerned not only with dielectric materials like glasses but with semiconductors as well. In the latter case the electronic properties are also very important, and measurements of the dopant profile, resistivity, carrier mobility etc. may be necessary. The techniques to be used include all the structural analysis developed for the characterisation of surfaces and solid-state materials (e.g. scanning and transmission electron microscopy, electron microprobe, ESCA, SIMS) plus optical techniques such as photoluminescence, Raman spectroscopy and so on.

Limiting ourselves to the most general problems, two simple and widespread measurements which are complementary to the prism-coupler characterisation are the direct observation of the mode profile and the evaluation of waveguide performance as a function of the wavelength. The setup for the former measurement is sketched in Fig. 4: the laser light is

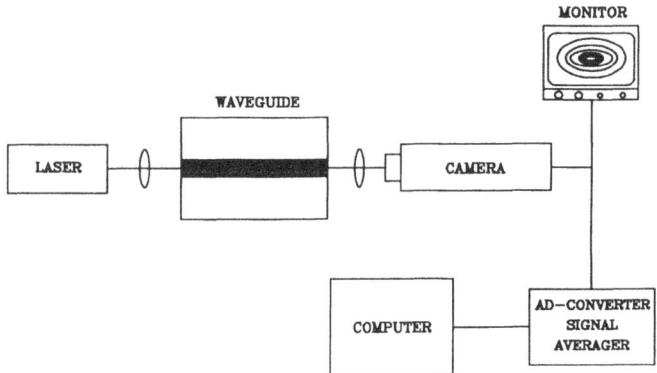

Fig. 4. Measurement of the modal field distribution in a channel waveguide.

end-fire coupled to the waveguide and the output face is imaged by an optical system onto a vidicon or CCD camera; a computer allows a detailed analysis of the image and therefore of the intensity profile of the mode. This knowledge is quite important to check the quality of the waveguide (this is especially true for channel waveguides) and also to properly model devices such as directional couplers and Mach-Zehnder interferometers; the mode profile is also a critical parameter for efficient coupling of the waveguide to external optical fibers. Moreover, from observation of the profile of the fundamental mode it is also possible to calculate the refractive index profile using the scalar wave equation.[6]

The experimental arrangement shown in Fig. 5 refers to the measurement of the transmission of the waveguide as a function of the propagating wavelength, using a white-light source and a monochromator in front of the detector. This measurement allows one to determine the range of wavelengths in which the given material structure has acceptable losses and can therefore be employed; it can also be important for analyzing whether the film has the same properties as the bulk material or if changes have been produced during the waveguide fabrication process. By the same technique, the spectra of species adsorbed on the film or of a fluid in contact with the film can be obtained.

Both these techniques have the common disadvantage that the waveguide end faces have to be accurately polished in order to permit efficient end-fire in/out coupling: the processing of the sample edges is often one of the most critical and time-consuming operations in the manufacture of waveguides and simple components.

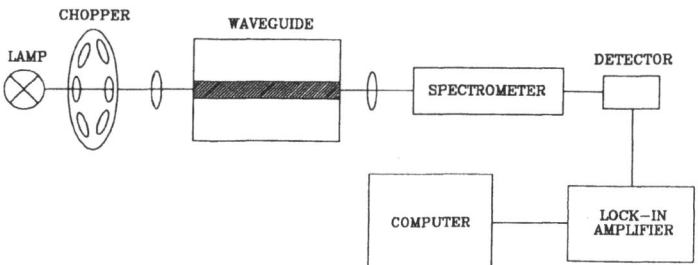

Fig. 5. Measurement of the spectral transmittance of a channel waveguide.

4. Measurements of Transmission Losses and Scattering

Visual observation of the light streak in the waveguide can give a quick but non-quantitative measure of the propagation loss coefficient a. As has already been said, the streak is very weak and difficult to observe in films with loss below 0.2 dB/cm, while in high-loss films ($\alpha > 5$ dB/cm) an approximate value of the loss can be inferred from the measurement of the length L (cm) of the streak itself. Since the dynamic range of the eye is of the order of 27 dB, the loss α in dB/cm can be evaluated as $\alpha = 27/L$. In the intermediate range of propagation losses it is necessary to perform a direct measurement; moreover, in order to provide feedback to the manufacturers it is often very useful to separate the contribution of the absorption from that of bulk or surface scattering. Most of the techniques briefly described in the following measure the total losses, i.e. due to both absorption and scattering. A very accurate measurement of absorption-only losses is possible by calorimetric techniques,[7] in which the temperature rise of a sample during laser irradiation is detected. The results of measurements in out-diffused lithium niobate waveguides show that absorption losses are of the order of 0.02 dB/cm, almost two orders of magnitude smaller than the total losses, thus demonstrating that, at least in the samples considered, scattering is the limiting mechanism in waveguide transmission. The authors have estimated that the minimum waveguide absorption coefficient measurable for samples with optimum geometry is on the order of 10^{-4} dB/cm.

4.1. Prism-Coupler Loss Measurements

One of the simplest and at the same time very effective methods of measuring the total loss is based on the use of two prisms,[8] according to the sketch depicted in Fig. 6. The intensity measured by the detector is plotted as a function of the propagation length Z, which can be varied by keeping prism 1 fixed and moving prism 2. The most critical factor of this nondestructive method is that the outcoupling efficiency should remain constant at all the different positions of prism 2: this aim is sometimes reached more easily by using between the prism and the waveguide a liquid with refractive index slightly lower than that of the guide. The limit of accuracy of the method is approached when the film has losses lower than 0.2 dB/cm, although some authors claim that the error can be kept down to ± 0.01 dB/cm. To achieve such results, however, the complexity of the measurement increases, and a good deal of operator skill is necessary.

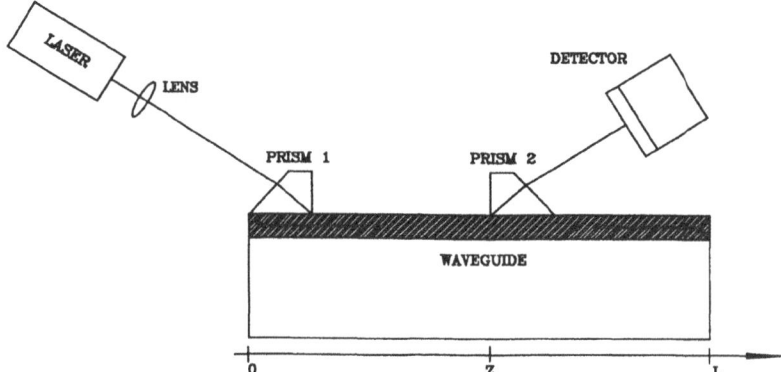

Fig. 6. Measurement of the transmission losses in a planar waveguide by the two-prism sliding method.

Introduction to Integrated Optics

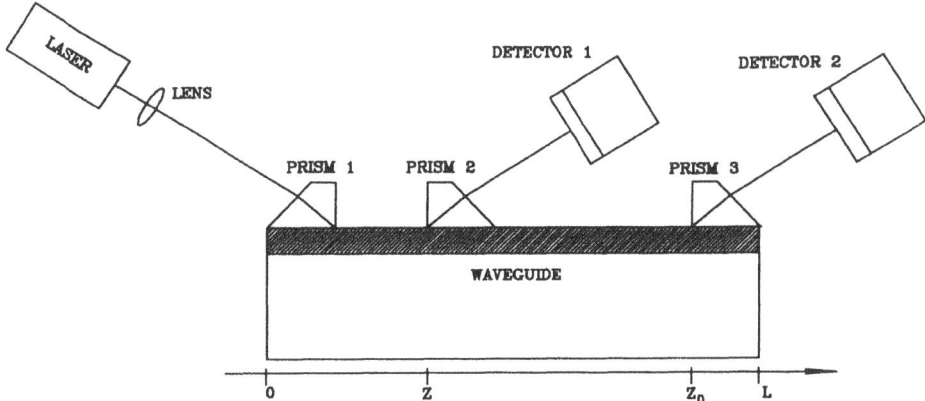

Fig. 7. Accurate measurement of the transmission losses in a planar waveguide by the three-prism sliding method.

A substantial relaxation of the constraints on the stability of the efficiency of the outcoupling prism is obtained when one uses the setup shown in Fig. 7, where three prisms are independently clamped to the guide. In this case both the prisms 1 and 3 are kept fixed, and prism 2 is moved along the guide. It has been shown[9] that, by measuring the intensity at detector 2 when prism 2 is removed and subsequently the intensities at detectors 2 and 1 for the various positions of the prism 2, an expression for the intensity of the guided light as a function of propagation length can be derived which is independent of the outcoupling efficiency of both prisms 2 and 3. Thus, even if in principle this technique is not any more accurate than the two-prism technique, it is more practical and rapid. An accuracy of the order of ± 0.01 dB/cm can now be achieved without taking particular care with the prism clamping pressure and without the use of matching fluids.

Another experimental arrangement, which is equivalent to the three-prism setup, is shown in Fig. 8; here, the light which was not coupled out by prism 2 and continued to

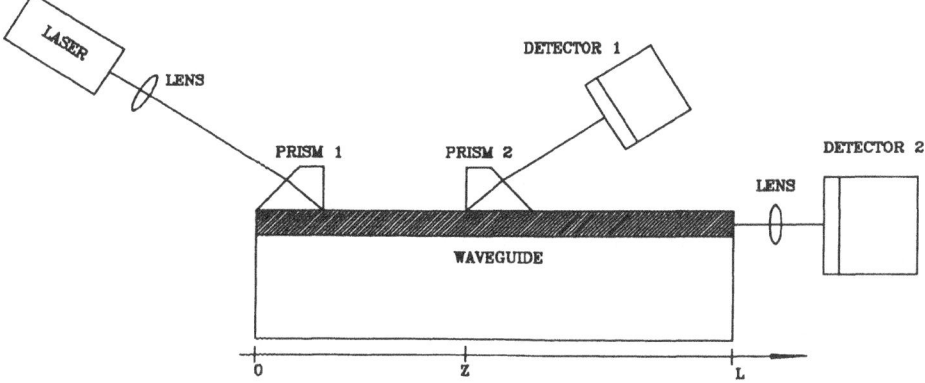

Fig. 8. Combination of the prism-sliding and end-fire techniques to measure the transmission losses of a planar waveguide.

propagate inside the waveguide is collected by a lens and measured by detector 2. This arrangement, which requires polishing of the output face of the guide, can be useful for short-length samples, where the positioning of three prisms and the movement of the intermediate one can be quite difficult.

4.2. Scattering-Detection Techniques

A simpler technique for measuring losses is one using a fiber optic probe,[10] according to the setup in Fig. 9. If we assume that the scattering centres in the waveguide are uniformly distributed and that the intensity of the scattered light in the transverse direction is proportional to the number of scattering centres, we can take the intensity of the scattered light along the guide (i.e. the brightness of the light streak) as proportional to the guided-light intensity at each point. Thus, to derive the plot of transmitted intensity versus propagation length it is sufficient to detect the scattered light by means of an optical fiber which is accurately moved along the streak, while keeping constant both its angular position and its distance from the waveguide. The use of the fiber probe in combination with a microscope allows better control of the distance, thereby improving the overall accuracy.

A variation of the method in which the distance does not need to remain constant, consists in moving the fiber probe normally to the guide until it just touches the guide surface and then withdrawing it until it just loses contact; the distal end of the fiber is connected to a detector, whose output goes to a chart recorder. By repeating this kind of measurement at several points along the streak and by analysing the resulting trace on the recorder chart, it is possible to measure attenuations as low as 0.3 dB/cm.[11]

The assumptions given above concerning waveguide uniformity are correct for most of the waveguides actually employed in device fabrication; an indication of the accuracy of a specific measurement can be given by the extent of the scatter of data points around the best-fit straight line (in a logarithmic-scale plot): the larger the extent, the lower the accuracy. By this method losses down to 0.2 dB/cm can be measured, with an error of ± 0.1 dB/cm.

All the previous techniques cannot be easily employed to measure losses of devices, be they Y-branches, directional couplers, or bidimensional components such as waveguide lenses. To overcome this difficulty, the scattering-detection method can be modified by using, instead of the fiber probe, a video camera onto which the image of the entire circuit is projected.[12]

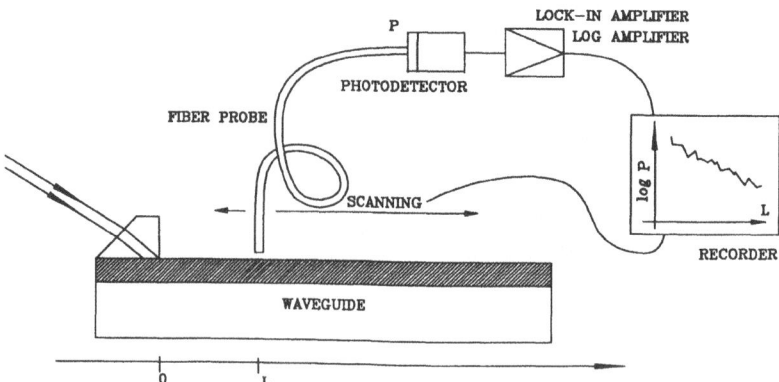

Fig. 9. Measurement of the waveguide transmission losses by the scattering detection method: here an optical fiber is used to collect the light scattered out of the guide.

Introduction to Integrated Optics

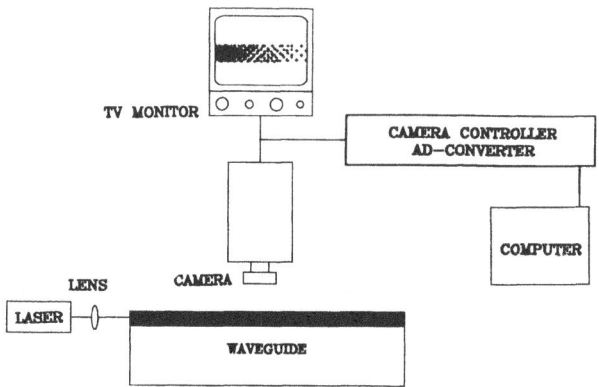

Fig. 10 This measurement technique is analogous to that sketched in Fig. 9, except that here a TV camera is used to collect scattered light instead of an optical fiber.

Easily available software allows the analysis of the image, as well as the evaluation of the propagation loss and the insertion loss of a component, if desired. In the setup of Fig. 10, end-fire coupling is assumed but the camera-detection technique can obviously also be used together with an input prism coupler. To improve the sensitivity of this technique, and to measure losses as low as 0.1 dB/cm, the use of a fluorescent overlayer has been suggested[13]; fluorescent light is emitted by the Nile Blue A perchlorate dye through the linear Stokes process, and therefore the fluorescent light intensity is proportional to the guided intensity. This method, however, requires special preparation of the sample and is not applicable to high-index materials; moreover, its use for measurements at near-infrared wavelengths would require the search for other suitable dyes.

A common advantage of all the scattering-detection methods is that they are not only nondestructive but also noncontacting; moreover, the optimum conditions for loss measurements have also been determined through an extensive set of measurements.[14]

Since scattering losses are usually dominant in dielectric waveguides, the assumption that absorption and radiation losses are negligible is acceptable, and in this approximation the above measurements of the scattered loss can be taken as equivalent to the measurement of the total loss. If, however, one is interested in deriving some specific information on the scattering characteristics of the waveguide, the fiber-probe or camera measurements can be repeated as a function of the mode and the collecting angle; in other words, for a given propagating mode one can move the detector in order to make an angular scanning in the plane containing the light streak and perpendicular to the guide surface: the position and size of the scattering centres, as well as the correlation length, can be thus evaluated.[15]

Additional useful information for the design of specific devices (e.g. for the design of an integrated optical spectrum analyser) is represented by the amount and the angular distribution of the in-plane scattering, namely of the light which is scattered by the waveguide inhomogeneities inside the plane at angles φ with respect to the propagation direction (see Fig. 11). For this measurement, the output beam is spatially Fourier-transformed by a lens, and the light distribution in the Fourier plane is measured by scanning it with a slit and photodetector (an image sensor can also be used to get the entire transform distribution): a plot of the scattered intensity vs the angle φ can be obtained, since the scanning coordinate y is related in a simple way to the angle φ.[16] Measurements on high-quality lithium niobate waveguides have shown that the scattering level can be as low as -55 dB at an angle $\varphi = 0.5°$ for a propagation length of 20 mm.

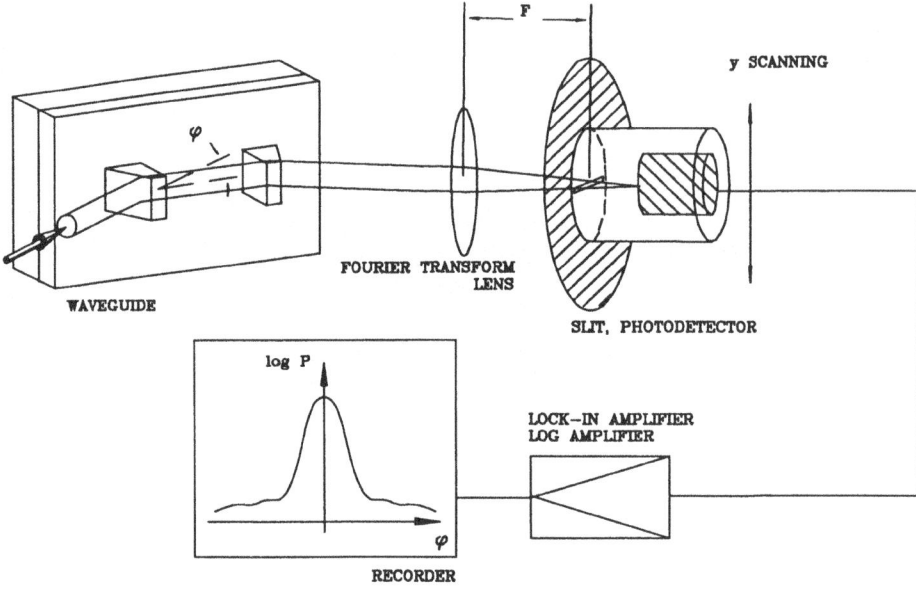

Fig. 11. Measurement of the in-plane scattering characteristics of a planar waveguide.

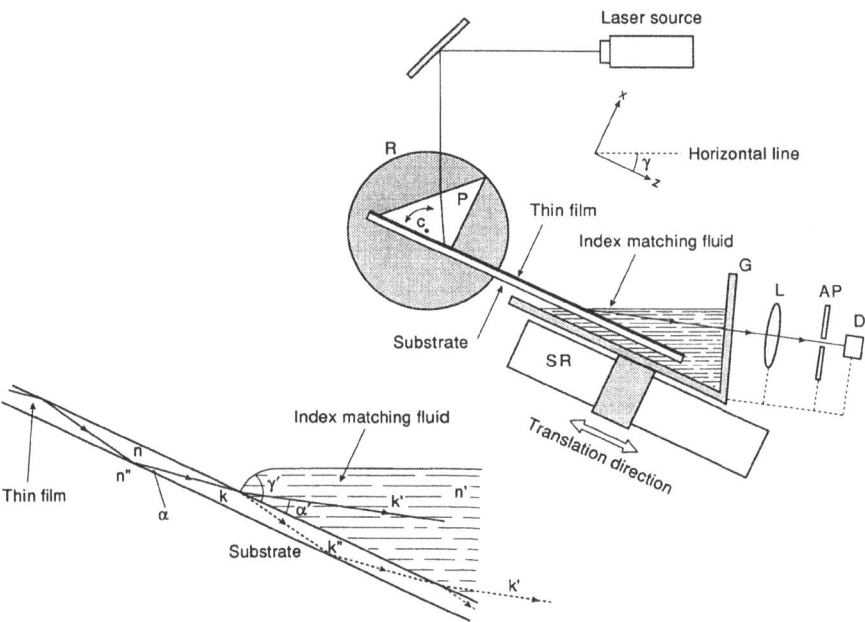

Fig. 12. Experimental setup for precision measurements of the optical attenuation in a waveguide, based on the outcoupling of guided light into a liquid with refractive index n' slightly higher than (n) of the guiding film. R, rotation stage; P, prism coupler; G, glass cell; L, focusing lens; Ap, aperture; D, photodiode; SR, sliding rail.

4.3. Precision Measurements of the Waveguide Losses

Because the quality of the materials and the fabrication processes have continued to improve, the propagation loss in a passive waveguide may well be quite small, below 0.1 dB/cm; it is therefore necessary to use more accurate measuring methods, without increasing their complexity. In a very recent paper an experimental technique has been reported, which is claimed to offer repeatability and accuracy of the measured attenuation typically better than 5%, even for measurements of losses below 0.1 dB/cm with less than 1cm-long guiding paths.[17] The experimental setup is shown in Fig. 12. A collimated laser beam is coupled into the thin-film waveguide by a prism coupler P which is mounted onto a rotation stage R. A V-shaped glass cell G, filled with a fluid having refractive index n' slightly higher than that of the guiding film, is mounted onto the sliding rail SR and can be moved in a direction parallel to the guide. As the film is immersed into the fluid, the guided beam does not undergo anymore total reflection at the upper interface and all the guided light is eventually outcoupled into the fluid, in the same direction (corresponding to the wave vector **k'**). The light emerging from the glass cell is focused by the lens L through the aperture AP and collected by the detector D. By translating the glass cell with respect to the sample, i.e. by immersing the waveguide for different lengths, it is possible to measure the guided-light intensity as a function of the propagation distance. Values of loss as low as 0.05 dB/cm have been measured, with standard deviation of the data below 0.01 dB.

4.4. Photothermal Deflection Technique

Another class of noncontacting loss measurement is that based on the photothermal deflection (PTD) effect. When some energy of a laser *pump* beam is absorbed by a material, a thermal gradient is produced, which in turn produces a refractive-index gradient in the absorbing and surrounding media. The PTD technique is based on the measurement of the refraction of a second laser beam, the *probe* beam, induced by such an index gradient.[18] Both crossed-beam and collinear-beam configurations are possible; in the former,[19] which is sketched in Fig. 13, the probe beam almost perpendicularly crosses the waveguide that is heated by the pump laser coupled into the waveguide itself. In the collinear configuration[20,21] the probe beam is parallel to the film surface and is refracted by the refractive index gradient induced in the gas (air) region close to the surface of the sample. The crossed-beam configuration was used[19] to measure the propagation loss of K-exchanged glass waveguides, which turned out to be 1.2 ± 0.2 dB/cm. Both the pump and probe beams were from He-Ne lasers at 633 nm: the probe beam was focused to a spot on the waveguide surface, and its deflection was detected by a bicell photodetector placed below the sample; the differential voltage of the bicell was amplified and separated from the noise by using a chopper and the lock-in detection. It was calculated that using 3 mW of pump power the induced index change was less than 10^{-5} : this gives an idea of the sensitivity of the method, which, according to some authors, in optimum conditions should be able to measure losses as low as 10^{-3} dB/cm.

5. Modelling of Graded-Index Optical Waveguides

Efficient numerical tools are necessary to simulate the operation of guided-wave components and devices, in order to properly design them and avoid "trial and error" steps towards their optimisation.

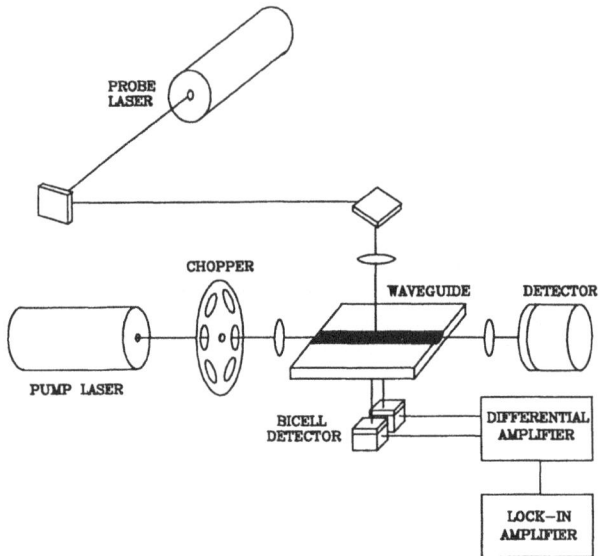

Fig. 13. Experimental setup for the photothermal deflection measurement in the crossed-beam configuration.

An example of a well known and widely used numerical algorithm for the modelling of integrated optical circuits is the one designated as the Beam Propagation Method,[22] which can be applied to both two- and one-dimensional guiding structures. In order to use these simulation tools, however, one of the most important parameters required is the refractive index profile of the guiding structure, which strongly affects the form of the guided field distribution; it can be of crucial importance especially in the design of nonlinear devices. It is therefore very important to determine the refractive index profile as a function of the process parameters used in the realisation of the waveguiding device. As mentioned in Section 2, the case of step-index waveguides permits a rather simple treatment; in the following discussion, we will consider some of the problems associated with the fabrication process of a very widely used class of graded-index waveguides, namely those produced by ion-exchange in glass, and with the reconstruction of the index profile from the experimental measurements of effective indices.

5.1. Characteristics of the Diffusion Process in Glass

Ion-exchange is the most widely used technique to obtain optical waveguides in various substrates such as glasses or lithium niobate.[23,24] Therefore it can be worthwhile to briefly recall the fundamental aspects of the dynamics of ion-exchange which can allow one, by assuming that the index distribution is proportional to the concentration of the exchanged ions in the substrate, to forecast the form of the refractive index profile.

The interdiffusion process between the metallic ions in the substrate and those in the salt melt is described by the Nernst-Planck equation[25]

$$F = -D\left(\nabla C - zC\frac{f}{RT}E\right) \qquad (2)$$

where F is the ion flux, C is the concentration of any of the mobile ions normalised to the total concentration of the exchanged ions, E is the local electric field generated by the concentration gradient, D is the diffusion coefficient (assumed to be independent of the concentration), z is the ion valence, f is the Faraday constant, R is the gas constant and T is the temperature.

In one dimension, considering a one to one exchange of two ion species a and b, with $F_a = -F_b$ it can be proved that

$$\frac{\partial C_a}{\partial t} = \frac{\partial}{\partial x}\left(\frac{D_a}{1-\alpha C_a}\frac{\partial C_a}{\partial x}\right) \tag{3}$$

with

$$\alpha = 1 - \frac{C_a}{D_a} \tag{4}$$

Under these assumptions the solution of Eq. (3) can be written as

$$C_a(x,t) = C_\infty \mathrm{erfc}\left(\frac{x}{2\sqrt{Dt}}\right) \tag{5}$$

where t is the diffusion time and D is defined by

$$D = \frac{D_a D_b}{D_a C_a + D_b C_b} = \frac{D_a}{1-\alpha C_a} \tag{6}$$

Hence, we would expect to find in most cases refractive index profiles of the form

$$n(x) = n_s + (n_0 - n_s)\mathrm{erfc}\left(\frac{x}{d}\right) \tag{7}$$

where n_s is the substrate index, n_o is the surface index and d is the exchange depth parameter. However, it can be shown that this holds true only when the ion depletion of the salt melt near the substrate surface during the exchange process can be neglected; in any other case the expected profile form would be a half-gaussian profile. We will see in the next sections how these models can be compared with practical cases.

5.2. WKB Method

Analysis of the guided wave propagation can be carried out in the most complete way by starting from Maxwell equations, with the proper boundary conditions.
The resulting wave equation:

$$\frac{\partial^2 E_y}{\partial x^2} + \left[k_0^2(n_f^2 - n_e^2) - k_0^2(n_f^2 - n^2(x))\right]E_y = 0 \tag{8}$$

however, can be explicitly solved only for few special systems (among which are step-index waveguides); an approximate approach has to be followed for graded-index structures.

In particular, the WKB approximation (*Wentzel-Kramers-Brillouin*, developed first in quantum mechanics)[26,27] can be a useful tool to handle this kind of problem, provided that the term $k_0^2(n_f^2 - n^2(x))$ varies slowly over a distance λ.
Under the WKB approximation the modal dispersion equation becomes

$$k_0 \int_0^{x_t} \sqrt{n^2(x) - n_e^2(m)}\, dx = m\pi + \phi_a + \frac{\pi}{4} \qquad (9)$$

$$n(x_t(m)) = n_e(m) \qquad (10)$$

$$\phi_a = \sqrt{\frac{n_e^2 - 1}{n_f^2 - n_e^2}} \qquad (11)$$

$n_e(m)$ being the effective index of the mode of order m, $n(x_t(m))$ being the so-called "turning points" (from their physical meaning of maximum penetration depth of the guided light in a ray optics description) and ϕ_a is half the phase shift caused by the reflection at the air-waveguide boundary.

In order to obtain the parameters that describe the refractive index profile of the waveguide, namely its mathematical form, the surface index n_o and the diffusion depth d, it is therefore necessary first to obtain the effective indices of the guided modes of the waveguide, given the substrate refractive index and the operational wavelength.

The effective indices can be measured by means of the experimental setup shown in Figs. 2 and 3, as already explained in Section 2.

Unfortunately, neither Eq. (9) cannot be solved analytically, so many numerical methods[28] have been proposed; here we will concentrate on two of them.

5.3. Dispersion Curves Method

This method, proposed by Yip and Albert,[29] requires the choice of a definite profile form, whose parameters must be fitted to the experimental data.
With the change of variable:

$$x' = \frac{x}{d} \qquad (12)$$

Eq. (8) gives:

$$d = \frac{k\left(m\pi + \phi_a + \frac{\pi}{4}\right)}{\int_0^{x_t} \sqrt{n^2(x) - n_e^2(m)}\, dx} \qquad (13)$$

This expression allows one to directly calculate the exchange depth, once the effective indices of the waveguide and all parameters of the index profile are given. Hence, in order to obtain the form of the index profile, it is sufficient to make the change of variable (12) and then, starting from a value slightly higher than that of the 0th-order effective index, increase the value of the surface index, until the sum of the squares of the differences between the

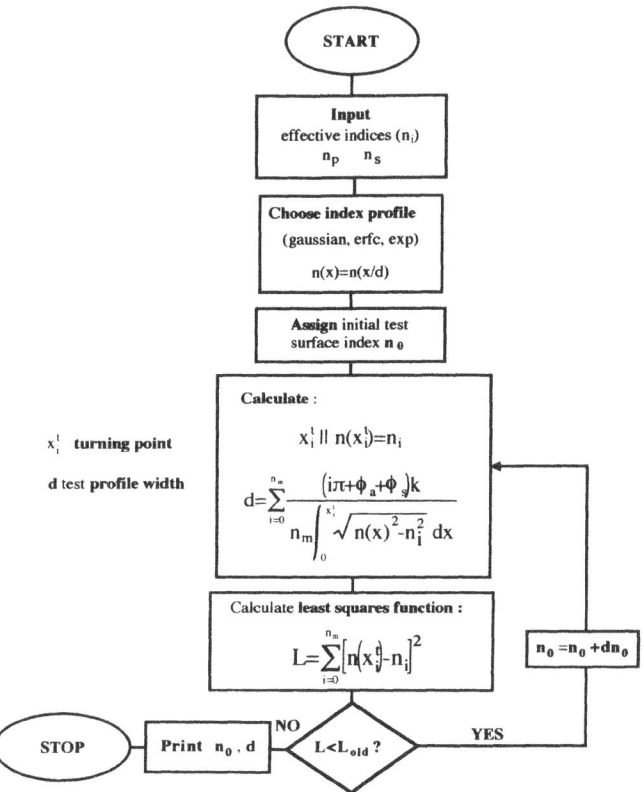

Fig. 14. Flow-chart of a program capable of computing the refractive index profile of a graded-index waveguide following the method proposed in Ref. 29.

experimental effective indices and the ones calculated by means of Eqs. (9-13) reaches a minimum. In Fig. 14 the flow-chart of a program capable of such a task is shown.

As trial profiles exponential, half-gaussian, erfc or second-order polynomial functions can be chosen. The most appropriate profile is usually the one which minimises the square sum mentioned above.

After this step, the modal dispersion curves are computed for a definite set of waveguides fabricated under the same conditions, using as surface index the mean of the surface indices of the whole set of waveguides considered. Agreement between the experimental data and the average dispersion curves is then checked as a final control on both the homogeneity of the set of waveguides and of the goodness of the fit process as a whole.

These dispersion curves are very important in the design of IO components, as they define the relationship between the effective indices and the thickness of the waveguides.
Another relation which can be verified is the one between exchange depth and time, as shown in Fig. 15. As can be seen in the figure, the behaviour of the depth as a function of the exchange time follows quite well the theoretical square root function of Eqs. (5, 7), even for relatively long exchange times. The knowledge of this relation allows one to produce a waveguide of predetermined thickness and hence, by using the dispersion curves, with a priori well-defined propagation constants.

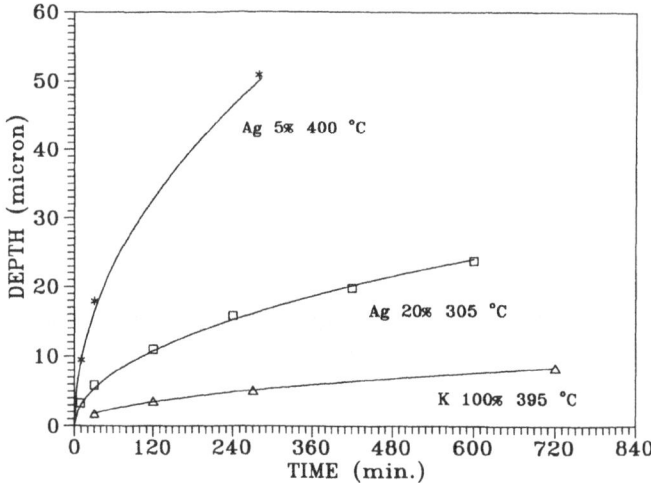

Fig. 15. Exchange depth vs. exchange time for various ion-exchange processes in a soda-lime glass.

5.4. Chiang Method

In this case, no profile is chosen at the beginning of the computation. The aim of this method, developed by K.S. Chiang,[30] is actually not to find any particular mathematical function as refractive index profile, but to determine through a recursive process an "empirical" index profile.

First of all, the experimental effective indices are interpolated with a polynomial function. This is a somewhat crucial step, because great care must be taken in the choice of the polynomial degree; too high a degree could for example lead to the enhancement of oscillation in the interpolated polynomial, caused by random errors in the experimental effective indices.

It is easy to show that by putting $x_t(m)=0$ in Eq. (8) the surface index is obtained as the effective index of the non-physical -0.75 order mode; thus, it is sufficient to extrapolate the interpolated polynomial to immediately obtain the surface index. Of course, as this parameter is the result of an extrapolation, it is very sensitive to the accuracy of the interpolation, but as this process is very straightforward this method becomes very attractive. In order to determine the entire index profile, the effective index profile is then approximated by a staircase function, which, once substituted in the integral of Eq. (9), transforms the integral into a series of sums, which are easier to deal with. In Fig. 16, where the flow chart of the whole process is shown, N_i's are successive sampling points of the interpolated effective index profile, whereas $\overline{N}_i$'s are the means between two successive sampling points. The recursive formula shown in Fig. 16 can then be derived and the whole profile computed. As a final step, the empirical profile can be fitted to a mathematical analytical function like the ones mentioned in the previous section, in order to have a result which is easier to treat in the following design processes.

5.5. Experimental Results

In order to provide an assessment of the agreement between the two numerical algorithms, the data for various ion-exchanged waveguides were processed using both methods, and the results are shown in Figs. 17 and 18.

The two figures correspond to different ion-exchange processes in the same soda-lime glass substrates, Fig. 17 referring to a 20% Ag$^+$/Na$^+$ exchange and Fig. 18 to a 100% K$^+$/Na$^+$ exchange.

In both figures the experimental data, the profiles obtained by the dispersion curve method, the "empirical" profile obtained by Chiang's method and the fit of an analytical function to the profile resulting from Chiang's algorithm are shown.

As it can be seen, the agreement between the two methods is fairly good, especially when considering the two fitted analytical profiles. In both cases, a gaussian profile seems to fit the experimental data best. This is not always the case, however; for example in Fig. 19, where the case of a Cs$^+$/K$^+$ exchange waveguide is considered, the proper choice consisted of a erfc type profile, as suggested by Eq. (5).

We compared the results in terms of surface index and exchange depth obtained by means of the two methods for a set of 20%, 5% Ag$^+$/Na$^+$ and K$^+$/Na$^+$ exchanged waveguides. Apart from a few exceptions the agreement is quite good.

In order to gain more confidence in these numerical methods, however, it is useful and sometimes necessary to make direct measurements either of the concentration of exchanged

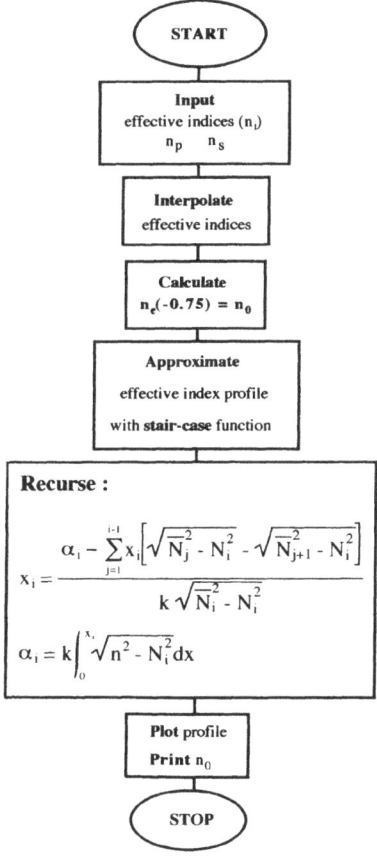

Fig. 16. Flow-chart of a program capable of computing the refractive index profile of a guided-index waveguide following the method proposed in Ref. 30.

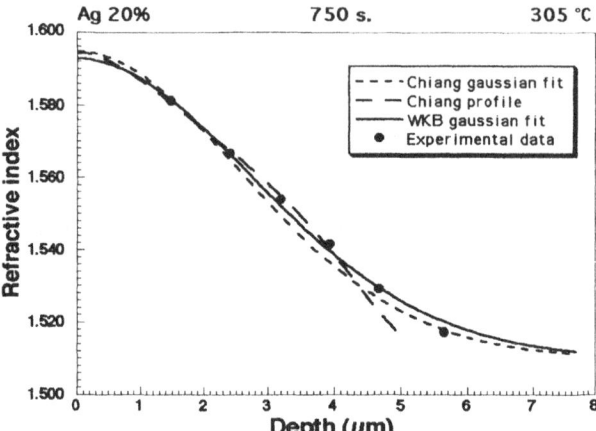

Fig. 17. Comparison of the reconstructed refractive index profiles of a waveguide produced by 20% Ag$^+$/Na$^+$ ion-exchange, as obtained by using the two numerical methods described in the text.

ions as a function of depth (which can be accomplished by such techniques as Rutherford Back Scattering or SIMS)[31, 32] or of the refractive index itself by interferometric techniques.[33] Unfortunately, these techniques are generally difficult of implementation; the limited number of tests we have been able to perform, however, tend to support quite well the simulation results.

6. Conclusions

A number of valuable characterisation techniques have been developed for the accurate measurement of the various parameters of an optical waveguide, from the thickness and refractive index to absorption and scattering losses. Most of the experimental techniques briefly

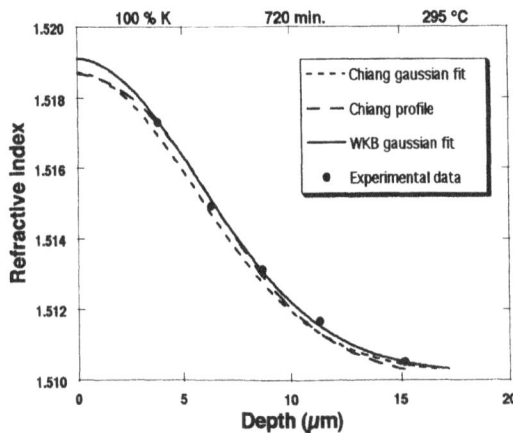

Fig. 18. Comparison of the reconstructed refractive index profiles: here the waveguide was obtained by 100% Cs$^+$/K$^+$ ion-exchange.

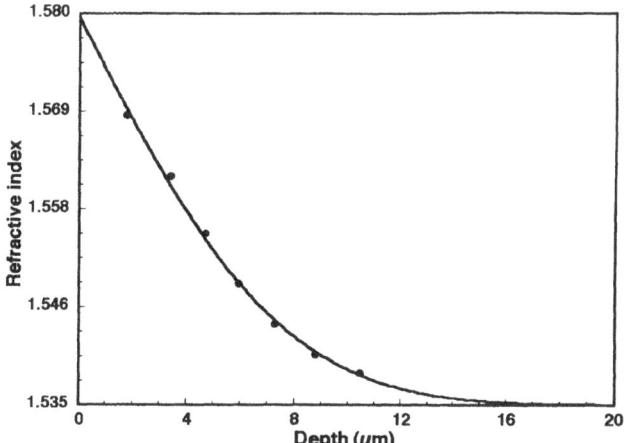

Fig. 19. Reconstructed refractive index profile of a 100% Cs$^+$/K$^+$ ion-exchanged waveguide, obtained by means of the dispersion curves method; in this case an erfc profile is used.

mentioned in the previous sections apply to any kind of waveguide structure, but it may be worthwhile to assess the best technique for the specific sample to be investigated.

It also should be emphasized that this summary is not at all inclusive, as the ingenuity of researchers has led to the development of a much larger number of measurement techniques, some of which, however, are suitable or cost-effective only in very specific cases.

Similar observations can be made concerning to the proposed modelling techniques. Here only two simple numerical techniques, easily implemented on a personal computer, have been summarised, which allow one to reconstruct the refractive index profile of graded-index waveguides. The results obtained by these methods, however, are accurate enough to give a good description of the realised waveguides and to be used as input data to more sophisticated modelling programs, e.g. based on the *Beam Propagation Method* or finite differences, which in turn allow one to optimise the design of discrete integrated optical components as well as to simulate the operation of more complex guided-wave circuits.

ACKNOWLEDGEMENTS. The financial support of the CNR Special Project TEO (Electro-Optic Technologies) is gratefully acknowledged.

References

1. T. Tamir, Editor, "Integrated Optics", Springer-Verlag, Berlin (1975); b) see Chapter 2 (Theory of dielectric waveguides).
2. R.G. Hunsperger, "Integrated Optics: theory and technology", Springer-Verlag, Berlin (1982); b) see Chapter 6 (Waveguide input and output couplers).
3. H. Nishihara, M. Haruna and T. Suhara, "Optical integrated circuits", McGraw-Hill, New York (1989).
4. M.A. Mentzer, "Principles of Optical Circuit Engineering", Marcel Dekker, New York (1990).
5. S.I. Najafi, "Introduction to glass integrated optics", Artech House, Norwood (1992).
6. L. McCaughan and E.E. Bergmann, J. Lightwave Techn. LT-1: 241-244 (1983).
7. S.D. Allen, E. Garmire, M. Bass and B. Packer, Appl. Phys. Lett. 34: 435-437 (1979).

8. H.P. Weber, F.A. Dunn and W.N. Leibolt, Appl. Opt. 12:755-757 (1973).
9. Y.H. Won, P.C. Jaussaud and G.H. Chartier, Appl. Phys. Lett. 37:269-271 (1980).
10. S. Dutta, H.E. Jackson, J.T. Boyd, R.L. Davis and F.S. Hickernell, IEEE J. Quantum Electron. QE-18:800-805 (1982).
11. N. Nourshargh, E.M. Starr, N.I. Fox and S.G. Jones, Electron. Lett. 21:818-820 (1985).
12. Y.Okamura, S. Yoshinaka and S. Yamamoto, Appl. Opt. 22:3892-3894 (1983).
13. Y.Okamura, S. Sato and S. Yamamoto, Appl. Opt. 24:57-60 (1985).
14. C. De Bernardi, A. Loffredo and S. Morasca, Proc. SPIE 651:259-262 (1986).
15. M. Imai, Y. Ohtsuka and M. Koseki, IEEE J. Quantum Electron. QE-18:789 (1982).
16. D.W. Vahey, Proc. SPIE 176:62 (1979).
17. C.-C. Teng, Appl. Opt. 32:1051-1054 (1993).
18. A.C. Boccara, D. Fournier, W. Jackson and N.M. Amer, Opt. Lett. 5:377 (1980).
19. R.K. Hickernell, D.R. Larson, R.J. Phelan, Jr. and L.E. Larson, Appl. Opt. 27:2636-2638 (1988).
20. H. Xue-bo and C. Wen-bin, Measurements of the absorption distribution of optical thin films by scanning photothermal microscopy, in "Photoacoustic and Photothermal Phenomena", P.Hess and J.Pelzl, Eds., Springer-Verlag, Berlin (1988).
21. C.Sibilia, M. Bertolotti, L. Fabry, G. Liakhou and R. Li Voti, Thermal diffusivity measurements in multilayers through photodeflection method: theory and experiments, in "Quantum Electronics and Plasma Physics: 6th Italian Conference", G.C.Righini, ed., Editrice Compositori, Bologna (1991).
22. J. van Roey, J. van der Donk and P. E. Lagasse, J. Opt. Soc. Am., 71:803 (1981).
23. T. Findakly, Opt. Eng., 24:780 (1985).
24. G. Stewart, C. A. Miller, P.J.R. Laybourn, C.D.W. Wilkinson and R. M. de la Rue, IEEE Jour. Quan. El. 13:192 (1977).
25. J. Breton and P. Laborde,Proc. SPIE, 1128:80 (1989).
26. H.Nishihara, M. Haruna and T. Suhara, Optical waveguide theory, in: "Optical Integrated Circuits", McGraw Hill, USA (1985).
27. L. I. Schiff, Approximation methods for bound states, in: "Quantum Mechanics", Mc Graw-Hill, Singapore (1985).
28. J. M. White and P. F. Heidrich, Appl. Opt., 15:151 (1976).
29. J. Albert and G. L. Yip, Appl. Opt., 24:3692 (1985).
30. K.S. Chiang, Jour. Light. Tech., 3:385 (1985).
31. G. C. Righini, S. Pelli, R. Saracini, G. Battaglin and A. Scaglione, Proc. SPIE, 1513:418 (1991).
32. R. A. Betts, C. W. Pitt, K. R. Riddle and L. M. Walpita, Appl. Phys., A 31:29 (1983).
33. R. A. Betts, F. Lui and T. W. Whitbread, Appl. Opt., 30:4384 (1991).

Chapter 2

INTRODUCTION TO NONLINEAR GUIDED WAVES

M. BERTOLOTTI

1. Introduction

The aim of these lectures is to discuss basic phenomena in planar waveguides in which some or even all of the constituent layers are fabricated from optical nonlinear materials. Nonlinear effects and nonlinear propagation in waveguides have been studied since the 1970s both theoretically and experimentally, and a number of review papers have detailed the various phenomena[1-3].

In a waveguide, light can be confined to a very small cross section area for a comparatively long distance, so that pump sources of moderate or low power levels can produce effectively nonlinear phenomena.

A crucial issue for frequency conversion is phase matching between the interacting waves. This is certainly one of the main advantages offered by waveguides in nonlinear processes.

Moreover the possibility of controlling the refractive index by changing the intensity of the light propagating in the waveguide makes available a number of devices in which light is controlled by light. An example of this is the directional coupler which will be briefly discussed below.

The effects which we will discuss, as examples of nonlinear processes which can take place in waveguides, are: second harmonic generation; parametric generation; nonlinear propagation effects in Kerr-like media; nonlinear coupling; and, spatial solitons.

The applications of these effects are well known. Second harmonic and parametric generation are valuable for producing new wavelengths in integrated optics, while nonlinear propagation and coupling effects in nonlinear materials with third order nonlinearities may pave the way for all optical devices. Unfortunately, the always present absorption, saturation of the nonlinear refractive index coefficient, and the unavoidable presence of thermal effects have prevented obtaining in most cases the hoped for results. Very recently the use of second-order effects, mimicking refractive index changes with light intensity, is giving hope for improving the situation, and also the possibility of generating spatial solitons seems to be very promising.

In the following we will give a short review of the above mentioned effects.

M. Bertolotti - Dipartimento di Energetica, Università di Roma "La Sapienza", Via A. Scarpa 16, 00161 Roma, Italy

2. Effects in Media with Second-Order Nonlinearities

2.1. Second Harmonic Generation

Second order nonlinear phenomena in planar structures have some unique features compared with the corresponding effects in bulk media: these features are connected with the necessity of using the actual power distribution in the waveguide resulting from the mode structure of both the pumping and the generated signal. This distribution is described by an overlapping integral which can dramatically reduce mode conversion effects, unless all the interacting modes have the same mode number. The control of phase matching also involves the guided-wave vectors of the structure.

Second harmonic generation in a waveguide can be described using coupled mode theory. The wave equation for the guided electric field at frequency ω can be written as

$$\nabla^2 E(\vec{r},t) + \frac{\omega^2}{c^2} n_i^2(z) E(\vec{r},t) = -\mu_0 \omega^2 P^{NL}(\vec{r},t), \tag{1}$$

where $n_i(z)$ is the refractive index of the i-th medium and P^{NL} is the nonlinear polarization. The z-direction is assumed to be normal to the guiding surfaces.

The nonlinear polarization for second-order phenomena can be written

$$P^{NL}(\vec{r},t) = P^{NL}(z,\omega_s) \cdot e^{i(\omega_s t - \vec{\beta}_p \vec{r})} + c.c. \tag{2}$$

where

$$P^{NL}(z,\omega_s) = \varepsilon_0 \chi^{(2)}(-\omega_s; \omega_a, \pm \omega_b) : E^{(m)}(\omega_a, z) E^{(m')}(\omega_b, z) a^{(m)}(\omega_a, x) a^{(m')}(\omega_b, x), \tag{3}$$

with $\omega_a = k_a c$, $\omega_b = k_b c$, $\chi^{(2)}$ being the second-order susceptibility (the minus sign for a frequency corresponds to taking the complex conjugation of the appropriate field, and m, m' are mode indices). The wavevector associated with the nonlinear polarization source field is

$$\vec{\beta}_p = \vec{\beta}^{(m,a)} \pm \vec{\beta}^{(m',b)}, \tag{4}$$

which is not necessarily equal to the wavevector $\beta^{(n)}(\omega_s)$ of the propagating field of frequency ω_s. The case

$$\beta_p = \beta^{(n)} \tag{5}$$

corresponds to phase-matching.

Eq. 1 for a planar waveguide infinitely extended along the y-direction, with z normal to the guiding surfaces (see Fig.1), in the case of isotropic media with $P^{NL}(r,t)=0$, has normal mode solutions for fields propagating along the x-direction which can be separated into TE waves with field components E_y, H_x and H_z and TM waves with field components H_y, E_x and E_z. For TE modes the m-th order guided mode can be written as:

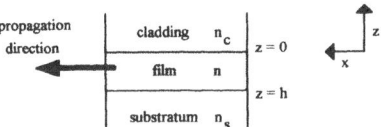

Fig. 1. Waveguide geometry.

$$E_y^{(m)}(\vec{r},t) = \frac{1}{2}E_y^{(m)}(z) \cdot a^{(m)}(x) \cdot e^{i(\omega t - \beta^{(m)} x)} + c.c. \tag{6}$$

The mode structure along the z-direction is contained in the $E_y^{(m)}(z)$ term. The term $a^{(m)}(x)$ takes into account changes in amplitude occurring along propagation direction due to interaction. $E_y^{(m)}$ is normalized so that $|a^{(m)}(x)|^2$ is the guided wave power in watts per meter of wavefront along the y-axis.

The generation of the nonlinearly produced signal waves at frequency ω can be usually treated in the slow varying envelope approximation (SVEA) using the well-known coupled-mode theory[4].

In the case of TE modes, the Eq. 6 is substituted into Eq.1. Assuming SVEA one can take

$$\frac{d^2 a^{(n')}(\omega_s, x)}{dx^2} \ll \beta^{(n')}(\omega_s) \cdot \frac{da^{(n')}(\omega_s, x)}{dx} \tag{7}$$

and thus obtain

$$\sum_{n'} 2i\beta^{(n')}(\omega_s) \cdot \frac{da^{(n')}(\omega_s, x)}{dx} E_y^{(n')}(\omega_s, z) e^{-i\beta^{(n')}(\omega_s)x} = -\omega_s^2 \mu_o P_y^{NL}(z, \omega_s) e^{-i\beta_p x} . \tag{8}$$

Using the orthogonality relations between modes, by multiplying both sides of Eq. 8 by $E_y^{*(n)}(\omega_s, z)$ and integrating over z one obtains

$$\frac{da^{(n)}(\omega_s, x)}{dx} = i \frac{\omega_s}{4} \int_{-\infty}^{\infty} P^{NL}(\omega_s, z) E^{*(n)}(\omega_s, z) \cdot e^{-i[\beta_p - \beta^{(n)}(\omega_s)]x} dz . \tag{9}$$

Two cases of second-harmonic generation can now be obtained according whether mixing of two co-directional guided waves of different polarization or mixing of two oppositely propagating guided waves is considered.

The nonlinear polarization fields are

$$P^{NL}(2\omega, z) = \frac{1}{2}\varepsilon_o \chi_{i,j,k}^{(2)} E_j^{(m)}(\omega, z) E_k^{(m)}(\omega, z) \left[a^{(m)}(\omega)\right]^2 \tag{10a}$$

$$P^{NL}(\omega, z) = \varepsilon_o \chi_{i,j,k}^{(2)} E_j^{*(m)}(\omega, z) E_k^{(n)}(2\omega, z) \cdot a^{(n)}(2\omega) \cdot a^{*(m)}(\omega) , \tag{10b}$$

where the factor 1/2 in Eq. 10a has been introduced to conserve energy. Substituting into Eq.9 for the case of co-propagating waves gives

$$\frac{da^{(n)}(2\omega,x)}{dx} = i\frac{\omega\varepsilon_o}{8}\int_{-\infty}^{\infty}\chi_{i,j,k}^{(2)}E_j^{*(n)}(2\omega,z)E_j^{(m)}(\omega,z)E_k^{*(n)}(\omega,z)\left[a^{(m)}(\omega,x)\right]^2 \cdot e^{-2i\Delta\beta x}dz$$

$$\frac{da^{(m)}(\omega,x)}{dx} = i\frac{\omega\varepsilon_o}{4}\int_{-\infty}^{\infty}\chi_{i,j,k}^{(2)}E_i^{*(m)}(\omega,z)E_j^{(n)}(2\omega,z)E_k^{*(m)}(\omega,z)\cdot a^{(n)}(2\omega,x)\cdot a^{*(m)}(\omega,x)\cdot e^{2i\Delta\beta x}dz$$
(11)

with

$$\Delta\beta = \left[\beta^{(m)}(\omega) - \beta^{(n)}(2\omega)\right].$$

If the fundamental beam is undepleted and conversion to the harmonic power after an interaction distance L is small, one has

$$\left|a^{(n)}(2\omega,L)\right|^2 = (\beta L)^2 |K|^2 \frac{\sin^2(\Delta\beta L)}{(\Delta\beta L)^2}\left|a^{(m)}(\omega,0)\right|^4 \tag{12a}$$

with

$$K = \frac{c\varepsilon_o}{4}\int_{-\infty}^{\infty}\chi_{i,j,k}^{(2)}E_j^{(m)}(\omega,z)E_k^{(m)}(\omega,z)E_i^{*(n)}(2\omega,z) \tag{12b}$$

The generation is essentially governed by the same laws as it is in the bulk except for the presence of the overlap integral K (Eq. 12b).

Two conditions must therefore be fulfilled. The first one is the usual phase-matching condition $\Delta\beta = 0$ which now can be fulfilled using β's corresponding to different modes but also considering that the β corresponding to one mode can be changed by changing the thickness of the waveguide. The second condition is to maximize the K-integral.

The dispersion in β with film thickness makes

$$\beta^{(n)}(2\omega) > \beta^{(n)}(\omega)$$

and reinforces the usual dispersion of the refractive index so that SHG with n = m is not possible with modes of the same polarization. It is however possible if n>m. This leads to a greatly reduced cross-section because the overlap integral involves interference effects when products of fields are integrated over the z-coordinate.

Several realizations have been made using different materials and different phase-matching conditions[5-31]. Among the most used materials are lithium niobate and semiconductors. Poled nonlinear organic polymers have in some cases very high nonlinear coefficients[13,32-39], and are at present considered very interesting materials. The crucial issue for frequency conversion, represented by phase-matching between the interacting waves, has been achieved by many different ingenious geometries[5-39]. Birefringence is most commonly utilized, combined with temperature adjustment for fine tuning. The so-called Cherenkov-scheme[5,40,41] (see Fig.2) is an alternative phase-matching method, where the SH light radiates down into the substrate at a certain angle. This technique can be advantageous in specific cases and has been used for generation of blue light.

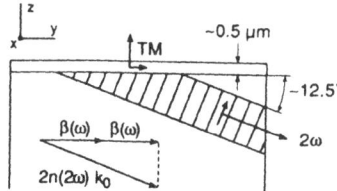

Fig. 2. Cherenkov scheme.

Quasi-phase-matching is a versatile method for phase matching using a periodic structure (Fig. 3) with the period chosen so that the phase mismatch is compensated for. For an efficient interaction a periodic modulation of the optical nonlinearity is required. Such a modulation can be obtained by periodically alternating the crystal orientation, so that the effective nonlinearity alternates between $+\chi_{eff}$ and $-\chi_{eff}$. The technique was first demonstrated by using a stack of differently oriented GaAs plates for frequency doubling of 10.6 μm radiation[42]. However this implementation is not very practical and has seldom been used. Ferroelectric crystals such as $LiNbO_3$, $LiTaO_3$ and KTP exist as multidomain crystals which can be alternately oriented (see chapter 7, this volume) and several experimental results have been obtained[23,43-46]. Very recently[47] second harmonic generation in ion-implanted waveguides in $KTiOPO_4$ with a conversion efficiency near to 25% for 1 μJ pulse excitation has been reported. The KTP crystal was specifically cut for type 2 $[n_z(\omega)+n_{xy}(\omega)]/2 = n_{xy}(2\omega)$ bulk phase-matching SHG at 1.064 μm (see Fig.4). SHG was obtained between $TE_0(n_z)$ and $TM_0(n_{xy})$ modes at 1.0741 μm. SHG has also been obtained in channel glass waveguides[48] where the second-order bulk susceptibility is zero for symmetry reasons. The effect has been obtained by the gradient and surface nonlinear polarization terms and can be used to study the surface properties of the waveguide. SHG was also obtained[49] in two-dimensional glass waveguiding films prepared by simultaneous illumination with 1064 and 532 nm laser light through a mechanism similar to that producing harmonic effects in optical fibers[50]. The case of contradirectional waves allows one to produce a nonlinear polarization in cw operation

$$P_y^{NL}(2\omega,z) = \chi_{yyy}^{(2)} E_y^{2(m)}(\omega,z) a_+^{(m)}(\omega,x) \cdot a_-^{(m)}(\omega,x) \qquad (13)$$

where a_+ and a_- describe guided waves propagating along the ±x axis respectively. The nonlinear polarization has no spatial periodicity parallel to the surface and hence the harmonic fields are radiated in directions normal to the waveguide surfaces[51,59]. In the pulsed regime the second harmonic field is only produced when the pulses overlap, and the second harmonic intensity is proportional to the convolution integral of the two fundamental pulses, an effect which has been proposed for pulse characterization[52].

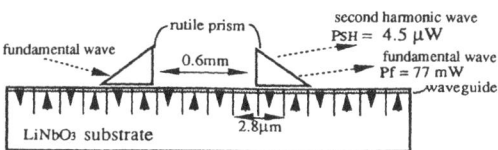

Fig. 3. Periodic structure.

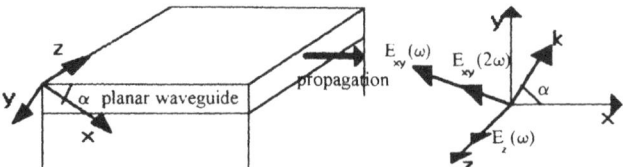

Fig. 4. Waveguide geometry for SHG in KTiOPO$_4$.

2.2 Parametric Processes

A number of other second-order nonlinear processes are possible including sum and difference frequency generation, parametric amplification, and parametric oscillation. Sum and difference frequency generation is essentially a generalization of second harmonic generation. Parametric amplification and oscillation have also been obtained in a number of cases[58-61].

Up to now the second order phenomena have been discussed with respect to one single nonlinear layer. This is not the only possibility. As an example, let us consider a more complex structure consisting of three layers[62] (see Fig.5), each one supporting a mode at a different frequency. We will show that this structure can serve as a nonlinear coupler.

Let us write the total field in the structure as

$$E = \sum_{j=1}^{3} \left\{ \sum_m C_j^{(m)} e^{-i\omega t} + c.c. \right\} \tag{14}$$

where the index j corresponds to a frequency ω_j, index m is the mode index (for each frequency), and

$$C_j^{(m)} = A_j^{(m)}(x,t) \cdot u_j^{(m)}(z) \cdot e^{i\alpha_j x} + B_j^{(m)}(x,t) \cdot v_j^{(m)}(z) \cdot e^{i\beta_j z} \tag{15}$$

$A_j^{(m)}(x,t)$ and $B_j^{(m)}(x,t)$ being the amplitudes of the fields that, for fixed frequency, are guided by layer a) and b) respectively, $u_j^{(m)}(z)$ and $v_j^{(m)}(z)$ the modal functions and α_j and β_j the guided mode wave vectors.

If we suppose that each layer can support only one TE mode (m=1) we may put $A_j^{(m)}(x,t)=a_j$, $B_j^{(m)}(x,t)=b_j$. In this way, all possible configurations of the refractive index of the structure are described by the following set of equations

$$\begin{cases} -i\left(\dfrac{4\omega_j}{x}\right)a_j + G_{jR} b_j e^{-i(\delta_j - k_{jL} + k_{jR})x} = e^{[-i(\beta_j - k_{jR})x]} \int u_j^* R_j dz \\ -i\left(\dfrac{4\omega_j}{x}\right)b_j + G_{jL} a_j e^{-i(\delta_j - k_{jL} + k_{jR})x} = e^{[-i(\beta_j - k_{jL})x]} \int v_j^* R_j dz \end{cases} \tag{16}$$

where we assume that

Nonlinear Guided Waves

$$\omega_3 = \omega_1 + \omega_2 \qquad (\vec{k}_3 = \vec{k}_1 + \vec{k}_2)$$

and

$$R_j = \frac{\omega_j}{c} C_i C_k^* \chi^{(2)}(\omega_j) \qquad k_j = n(\omega_j)\frac{\omega_j}{c}$$

$n_j(\omega_j)$ being the refractive index of the j-th layer, and

$$k_{jR} = n_j(\omega_{j-1})\omega_j/c \qquad k_{jL} = n_j(\omega_{j+1})\omega_j/c$$

$$G_{jR} = \int u_j^*(k_{jR} - k_j) v_j dz$$

$$G_{jL} = \int v_j^*(k_{jR} - k_j) u_j dz.$$

The set of Eq. 16 can be specialized to the simpler case of a structure in which only two frequencies are parametrically involved in the interaction: i.e. a three layer structure in which only the central layer is a nonlinear one, in which the frequency ω_2 propagates while the two external ones are linear and can support a single mode at frequency ω_1, so that $\omega_1 = \omega_2 - \omega_1$.

When the pump at frequency ω_2 is undepleted the solution of the field, for this special case of nonlinear coupler, can be easily obtained. The output powers from the external layers, if a power $P_a(0)$ is sent initially in channel a, turns out to read

$$\begin{cases} P_a(L) = P_a(0)\cos^2\left[(\chi_{ab} + \chi_{abnl}P_c)x\right] \\ P_b(L) = P_a(0)\sin^2\left[(\chi_{ab} + \chi_{abnl}P_c)x\right] \end{cases} \qquad (17)$$

where P_c is the pump power of the central layer c, χ_{ab} contains the overlap integral of the external mode a and b, and χ_{abnl} is proportional to the second order nonlinearity of the central layer. If $P_c = 0$ a complete power transfer from channel a to b can occur at coupling lengths L_c such that

$$L_c^{(n)} = (2n+1)/2\chi_{ab} \qquad n=0,1,2... \qquad (18)$$

Fixing the coupler length $L_c^{(n)}$, we can find a critical control power P_{crit2} which switches the power from a to b

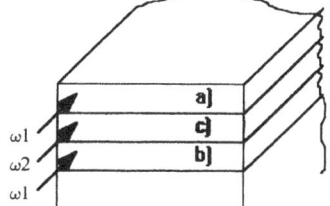

Fig. 5. Multi layer waveguide.

$$P_{crit2}(n) = (\chi_{ab})^2 / (\chi_{abnl})^2 (2n+1) . \tag{19}$$

The critical switching power can then be reduced by increasing the number of coupling periods n (for a nonlinear coupler driven in a conventional mode this opportunity does not exist).

3. Effects in Media with Third-Order Nonlinearities

3.1. Nonlinear Wave Propagation in Planar Structures

The key concept on which all nonlinear guided wave optical devices are based is that the local intensity of the guided wave controls the propagation wavevector; that is, the field profile and propagation constant can become power dependent when one or more of the layered media is characterized by an intensity dependent refractive index. This behaviour is characteristic of media possessing third-order nonlinearities. These media need not be crystalline because third order nonlinearities exist also in media possessing an inversion center, and this widens the possibility of choice of structures and materials. Two cases can occur. In the first case, the nonlinear change in the refractive index is small in comparison with the refractive index difference between the guiding media. In this case the dependence of the propagation wavevector on the power flow can be evaluated from coupled mode theory, and the guided wave field distribution (i.e., the field profile) can be approximated by linear guided modes. In the second case, the optically induced change in refractive index is comparable with, or larger than, the index difference between the guiding media. In this case both the propagation wavevector and the field distribution are power dependent, and this dependence can be evaluated from a more exact approach in which the nonlinear wave equation is solved subject to continuity of tangential electric and magnetic fields across all the interfaces. Several reviews exist on this subject[1-3] and we will only give a brief outline here.

The third-order nonlinear polarization vector is in general

$$P_i^{NL}(\omega) = \varepsilon_o \chi^{(3)}_{i,j,k,l} E_j(\omega) E_k^*(\omega) E_l(\omega) , \tag{20}$$

where i = x,y,z, and $\chi^{(3)}$ is the third-order susceptibility. Note that it is necessary to take the conjugate of one of the mixing optical fields so that the output signal is at the same frequency as the input signal.

If the optical field associated with a plane or a guided wave is large enough, it can change the refractive index of the medium. For a plane wave in an isotropic material the Fourier component of the polarization field at the frequency ω is

$$P_i(\omega) = \varepsilon_o \left[\chi^{(1)}_{ii} + 3\chi^{(3)}_{eff} |E_j(\omega)|^2 \right] E_i(\omega) , \tag{21}$$

where $\chi^{(1)}_{ii} = n_o^2 - 1$, n_o being the linear part of the refractive index. Expressing $|E_j(\omega)|^2$ in term of the local intensity

$$I = \frac{1}{2} c\varepsilon_o n_o |E_j(\omega)|^2$$

the intensity-dependent refractive index can be expressed in the form

$$n = n_o + n_2 I, \tag{22}$$

with

$$n_2 = \frac{3\chi_{eff}^{(3)}}{c\varepsilon_o n_o}, \tag{23}$$

where $n_2 > 0$ for self-focusing Kerr-like nonlinearities and $n_2 < 0$ for self-defocusing Kerr-like nonlinearities. For guided waves the expression for the nonlinear polarization vector is

$$P_i^{NL}(z) = c\varepsilon_o^2 n_o^2 n_2 \left[\frac{2}{3} E_i(z) \sum_j E_j(z) E_j^*(z) + \frac{1}{3} E_i^*(z) \sum_j E_j(z) E_j(z) \right]. \tag{24}$$

For TE polarized waves the only nonzero component of P^{NL} is

$$P_y^{NL}(z) = c\varepsilon_o^2 n_o^2 n_2 |E_y(z)|^2 E_y(z). \tag{25}$$

In this case only the ε_{yy} component of the nonlinear dielectric tensor enters into Maxwell's equations, and this will be written as

$$\varepsilon_{yy} = n_o^2 + \alpha |E_y|^2, \tag{26}$$

with

$$\alpha = c\varepsilon_o^2 n_o^2 n_2. \tag{27}$$

The wave equation, which takes the form

$$\frac{d^2 E_y}{dz^2} - (\beta^2 - n_o^2 k_o^2) E_y + \alpha k_o^2 |E_y|^2 E_y = 0, \tag{28}$$

can be applied to study a number of different situations.

3.1.1. Surface-guided Waves. Transverse electric (TE) polarized electromagnetic waves cannot be guided by the interface between two dielectric media whose refractive indices do not depend on intensity. However, when at least one of the two dielectric media exhibits a power-dependent refractive index, surface guided waves can exist, as has been studied by many authors[63-66]. TM polarized surface plasmon polaritons can be guided by the interface between a dielectric and a metal by virtue of the negative dielectric constant of the metal. However, they have very high attenuation coefficients.

If we consider the TE case for a nonlinear interface between an optically linear semi-infinite dielectric medium (substrate) with dielectric constant ε_s in the region I ($z<0$) and a semi-infinite Kerr-law nonlinear medium (cladding) with dielectric function

$$\varepsilon = \varepsilon_c + \alpha |E|^2 ,$$

in the region II (z >0) the nonlinear Maxwell equations read

$$\begin{cases} \dfrac{d^2 E_y^{(I)}}{dz^2} - q_s^2 E_y^{(I)} = 0 & z < 0 \\ \dfrac{d^2 E_y^{(II)}}{dz^2} - q_c^2 E_y^{(II)} + \alpha_c k_o^2 \left(E_y^{(II)} \right)^3 = 0 & z > 0 \end{cases} \qquad (29)$$

where $q_s^2 = \beta^2 - \varepsilon_s k_o^2$, $q_c^2 = \beta^2 - \varepsilon_c k_o^2$, $\alpha_c = c \, \varepsilon_o n_o^2 n_{2c}$. For waves guided by a single interface that are characterized by $E_y(z) \to 0$ as $|z| \to \infty$ (that is, the fields decay exponentially away from the boundary), the solutions of Eqs. 29 are well known[63-69]:

$$\begin{cases} E_y^{(I)}(z) = E_o e^{q_s z} & z < 0 \\ E_y^{(II)}(z) = \sqrt{\dfrac{2}{\alpha_c}} q_c \left\{ \cosh \left[q_c (z - z_c) \right] \right\}^{-1} & z > 0 \end{cases} \qquad (30)$$

where the positive root is taken for q_s and q_c, and assuming a self-focusing nonlinearity. For TE polarized waves both the field E_y and its derivative dE_y/dz are continuous across the interface z=0 between the nonlinear and linear medium. This leads directly to the eigenvalue equation

$$\varepsilon_s = \varepsilon_c + \frac{1}{2} \alpha_c E_o^2 , \qquad (31)$$

where E_o is the surface field. We see from Eq. 31 that the field amplitude is fixed at the boundary because ε_s and ε_c are constants, and if the limit $\alpha_c \to 0$ is taken in Eq. 31, then $E_o \to +\infty$, that is TE polarized electromagnetic waves do not exist in the linear limit for a single interface. The guided wave power in watts per meter of wavefront is given in terms of the Poynting vector as

$$P = \frac{1}{2} \int_{-\infty}^{\infty} \text{Re}(\vec{E} \times \vec{H}^*)_x dz = \frac{\beta}{2c\mu_o} \int_{-\infty}^{\infty} E_y^2(z) dz . \qquad (32)$$

For Kerr-law media this expression can be evaluated analytically[67]

$$P(\beta) = P_o \beta \left[\frac{\varepsilon_s - \varepsilon_c}{q_s} + 2(q_s + q_c) \right] , \qquad (33)$$

where

$$P_o = \sqrt{\frac{\varepsilon_o}{\mu_o}} (2\alpha_c k_o)^{-1} .$$

The β-power Eq. 33 can be viewed as the nonlinear dispersion equation $\omega = \omega(k,P)$, that is the frequency-wave number relationship for a given power level.

In real materials it is not possible to optically change the refractive index indefinitely, and a saturation effect sets in. The values of the saturated change Δn_{sat} of the refractive index vary from 10^{-1} to 10^{-4}. For a nonlinear interface the saturation effect is important because the interesting flux-dependent surface-guided wave properties occur when the optically induced change in the refractive index Δn_{sat} is comparable with, or larger than, the refractive index difference $n_s - n_c$ which exists at low powers between the substrate and the cladding.

If we write in general

$$\varepsilon_{xx} = \varepsilon_{yy} = \varepsilon_{zz} = \varepsilon_c + \varepsilon_c^{NL}\left(E_y^2\right) , \qquad (34)$$

the dispersion relation Eq. 31 becomes

$$\varepsilon_s = \varepsilon_c + \varepsilon_c \frac{1}{E_0^2} \int_0^{E_0^2} \varepsilon_c^{NL}\left(E_y^2\right) d\left(E_y^2\right) , \qquad (35)$$

from which the surface field E_0 can be determined. For a Kerr-law nonlinear cladding Eq. 35 reduces to Eq. 31. From Eq. 35 we obtain the important result that TE polarized surface-guided waves can be supported by a single nonlinear interface if, and only if, $\varepsilon_s > \varepsilon_c$. The permitted beta region for a nonlinear surface-guided wave is

$$k_o \sqrt{\varepsilon_s} < \beta < k_o \sqrt{\varepsilon_c + \varepsilon_{sat}} \qquad (36)$$

Numerical results for TE polarized surface-guided waves are shown in Figs.6 a,b,c for Kerr-law claddings (r = 2), power-law claddings (r ≠ 0), and saturable cladding respectively. We note that the minimum power required for the excitation of nonlinear TE polarized surface-guided waves increases with decreasing ε_{sat}. In the curves of Fig.6a two branches are visible[70].

The question now arises whether the excited waves are stable, i.e., propagate without changing the transverse profile of the electric field or are unstable on one or even both branches of the curve. The analysis of this case[2] shows that the curves are stable whenever $dP/d\beta > 0$. Figs. 6b and 6c show the evolution of the field distribution with propagation distance for the two cases of $\beta = 1.5607$ on the negative-sloped branch ($dP/d\beta < 0$) of Fig. 6a and of $\beta = 1.574$ on the positive-sloped branch.

Simple inspection shows that in the first case the profile changes drastically after a short propagation distance and the nonlinear surface wave is ejected into the linear substrate, while in the second case the profile remains constant after propagation.

Similar considerations as the ones reported above can be made for TM waves. In this case, however, TM polarized surface polaritons can exist also in linear media if $\varepsilon_c > 0$ and $\varepsilon_s < 0$, and $\varepsilon_c < |\varepsilon_s|$. This condition can be satisfied, e.g., if the substrate is a metal. The analysis of the TM case is complicated greatly by the existence of two field components, both of which can contribute to field-dependent dielectric constants and refractive indices, and for this reason we will not discuss this case here.

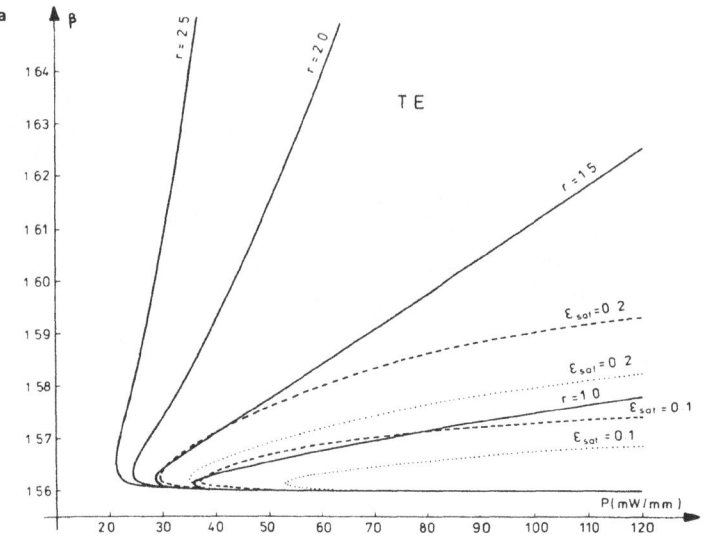

Fig. 6a. The dependence of the effective index β on the power flux for TE waves guided by the interface between a self-focusing cladding (n_c = 1.55, n_{2c} = 10^{-9} m²/W) and a linear substrate (ns = 1.56). Different curves refer to a Kell-law cladding (r = 2), power-law claddings (r ≠ 2) and saturadle claddings.

3.1.2 Nonlinear Waves Guided in Thin Films. Thin film waveguides in which some combination of the film, the cladding, or the substrate are nonlinear, exhibit many interesting power-dependent characteristics. The problem is solved also in this case using Eq. 28 for

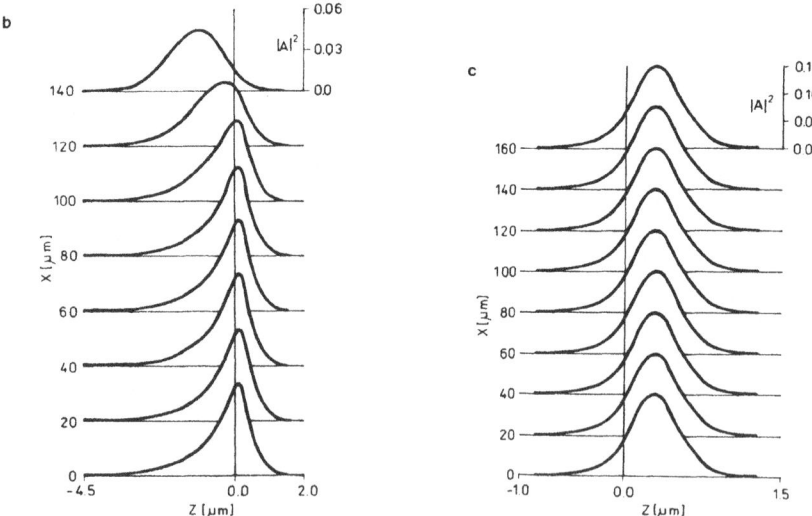

Fig. 6b. and c Evolution of nonlinear surface wave field distribution with propagation distance. Same values of refractive indices as Fig. 6a : b) β = 1.5607; c) β = 1.574.

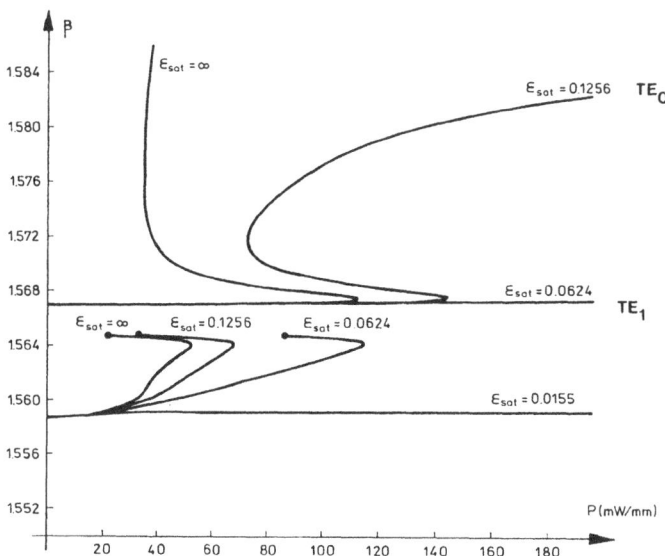

Fig. 7. The propagation constant β versus power P for a self-focusing cladding. Here $n_f = 1.57$, $n_c = n_s = 1.55$, $n_{2c} = 10^{-9}$ m²/W, d = 2 μm, and λ = 0.515 μm.

each of the media involved in the structure and applying the usual boundary conditions. Several cases can be considered according to which medium or media (cladding, substrate, guiding film) is or are nonlinear, and according to whether positive or negative nonlinearity is considered, and solutions in closed form using Jacobi integrals can be found[2]. In all cases complex curves describing the behaviour of β as a function of power are found. Although in some cases solutions in closed form can be given for Kerr-type nonlinearities, the most common way of discussing results is to look at computer generated simulations. If saturation is taken into account the shape of these curves tends to be smoothed. As an example, Fig.7 shows the propagation constant beta versus power for a self-focusing cladding for different values of saturation[72]. The unique features of the TE_0 solutions are the existence of wave propagation for $β > n_f$ and the local maximum in the guided wave power. Moreover the TE_1 branch terminates at some value of $β < n_f$. Compared with the Kerr-law case ($ε_{sat} = ∞$) the characteristic behaviour of TE_0 and TE_1 nonlinear wave solutions is preserved provided that the saturable value $n_{sat} = \sqrt{ε_c + ε_{sat}} - n_c$ is not too small. The absolute maximum in the TE_1 guided wave power depends strongly on the saturation value, and below a certain value of n_{sat} no change with power is predicted. The important conclusion is that if n_{sat} is too small, the most interesting power-dependent features of nonlinear guided waves can be altered and in some cases eliminated.

To take into account also the stability of solutions[73-83] we will consider the different possible cases and present some examples without going through calculations:

1. Nonlinear bounding, both cases of: a) single nonlinear bounding (a1- self-focusing cladding, and a2- self-defocusing cladding); and, b) two nonlinear bounding media (b1- both media self-focusing, and b2- both media self-defocusing).

2. Nonlinear film. This is the more complex case, as many possibilities can occur. Perhaps the most spectacular propagation in this case is via solitons, which we will discuss briefly in Section 3.4.

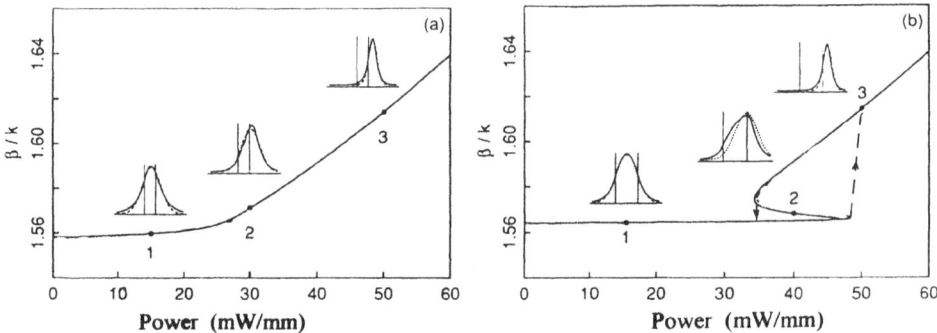

Fig. 8. Effective index as a function of power for a waveguide with linear core and substrate and nonlinear cladding. Core widths are: (a) 0.6 μm and (b) 1.2 μm.

a1 - *Single nonlinear bounding: self-focusing* cladding. For the case of self-focusing bounding media, both analytical and numerical stability analyses have been done[71]. Fig.8 gives[84] β as a function of power for a waveguide with linear core and substrate and nonlinear cladding (see Fig.9). This is the case considered in Fig.7 but in Fig.8 saturation is not taken into account. Curve a is for a film width 0.6 μm, and b for 1.2 μm. The inset figures show the field distributions at the labeled points. For the small value of the film thickness of Fig.8 the P-β curve increases monotonically. In the case of Fig. 8b the arrows indicate that the actual form of the solution has a bistable form[75]. As power is increased from zero, the field remains centered in the waveguide film until a critical power is reached (in this case P = 48mW/mm). At this point, there is a discontinuous jump and the field moves into the cladding and narrows, as indicated by the inset at point 3. As power is decreased again, the field remains concentrated outside the core boundary until a second critical power is reached (in this case P = 34 mW/mm), and again there is a discontinuous jump back to the low-power form of solution.

a2 - *Single nonlinear bounding: self-defocusing cladding.* For a self-defocusing medium, β decreases monotonically with guided wave power[83bis,85]. If $n_s > n_c$, cutoff occurs governed by the linear substrate refractive index value. However, for $n_c > n_s$, cutoff occurs at a finite power and this phenomenon has been proposed for building upper threshold devices[85]. The interesting point is that the TE_0 nonlinear guided wave is always stable in spite of the fact that its nonlinear dispersion curve always has a negative slope dP/dβ <0. Fig. 10 shows[83bis] the typical form of the TE_0 nonlinear dispersion curves (P versus β) for both symmetric (Fig. 10a) and asymmetric (Fig. 10b) linear waveguides with a self-defocusing cladding medium.

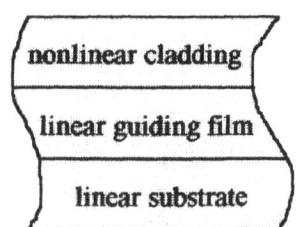

Fig. 9. Nonlinear cladding waveguide geometry.

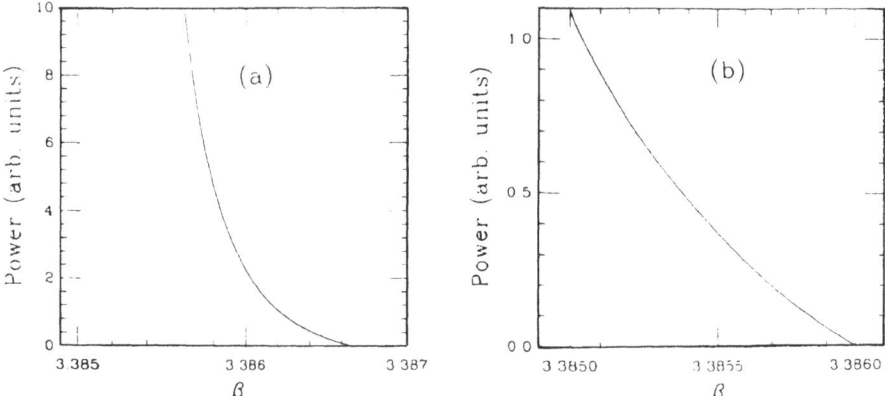

Fig. 10. The effective index of TE_0 mode versus guided wave power. The parameter values used are d = 1.07 µm, λ = 0.82 µm, n_c = 3.385, n_f = 3.39 and (a) n_s = 3.385, (b) n_s = 3.38.

This calculation is obtained solving Eq. 28 for the nonlinear guided waves as a function of effective index and calculating the power according to Eq. 32. The notable feature of these dispersion curves is that the slope is always negative.

b1 - *Two nonlinear bounding media: both media are self-focusing.* Two possibilities can be considered: the media have equal nonlinearity or have different nonlinearities. For a nonlinear cladding and substrate two kinds of solutions are found (see Fig. 11): symmetric (S) and asymmetric (A). The branch of solution (S) which is obtained above a critical power has the field which moves symmetrically into both the cladding and the substrate. It is known that this symmetric branch is unstable[75,86].

b2 - *Two nonlinear bounding media: both media are self-defocusing.* Using direct linear analysis with exact solutions, in a waveguide with a linear film bounded symmetrically by a Kerr self-defocusing medium, TE_0 and at least TE_1 guided waves are stable[83,83bis]. The calculation of possible even and odd modes supported by this system is elementary. These are constructed from eigenmodes that are expressed in terms of cos or sin functions in the linear film and cosech functions in the self-defocusing medium. Some dispersion curves and mode

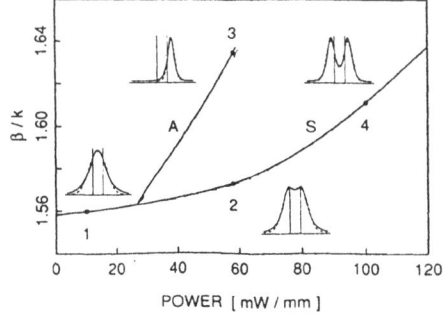

Fig. 11. Effective index as a function of power for a waveguide with linear core and symmetrical nonlinear claddings. The core width is 0.6 µm. The inset figures represent the fields at the labeled points.

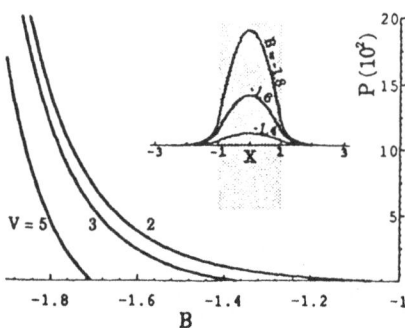

Fig. 12. Dispersion curves for TE_0 mode for several values of V. Inset: mode profiles with V = 3.

profiles are shown in Figs.12 and 13. All dispersion curves in this system have negative slope $dP/d\beta < 0$. Mode profiles of these waves, shown in the insets, always have peaks inside the film because the self- defocusing effect only causes the difference $n_f - n_c$ to be even sharper and the guiding effect to be enhanced. Furthermore, the induced index profile of the whole waveguide has a W shape, which is an interesting and potentially useful feature. A larger guided power P has a larger fraction of itself contained within the film. For even modes P is infinite at values of β corresponding to $B = d^2(\beta^2 - k_o^2 n_\omega^2) = -4k^2\pi^2$ (d is the film thickness) for k = 1,3,5..., while odd modes have infinite power at points corresponding to k = 2,4,6.... Combining this fact with the condition $-V^2 < B < 0$ ($V = kd\sqrt{n_\omega^2 - n_c^2}$ is the waveguide parameter) leads to an interesting conclusion. To have an upper power threshold, the TE_0 mode requires that $V < \pi/2$ (i.e. the waveguide to be single mode), and the TE_1 mode requires that $V < \pi$ (i.e., the waveguide is two moded).

3.1.3. Propagation in nonlinear waveguides in case of low intensity. When the nonlinear waveguide media are excited at a power levels not high enough to produce power-dependent field distributions, it can be seen from the previous examples that the variation in effective index with power is linear. This result can also be obtained from first-order perturbation theory or coupled mode theory[89]. One can then write

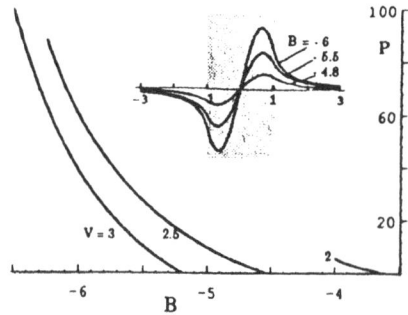

Fig. 13. Dispersion curves for TE_1. Inset: some mode with V = 2.5.

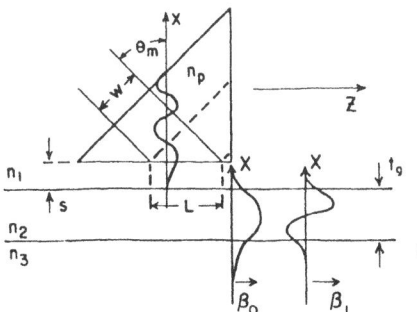

Fig. 14. Prism coupling.

$$\beta^{(m)} = \beta_o^{(m)} + \Delta\beta_o^{(m)}|a^{(m)}(x)|^2 \tag{37}$$

Such a power-dependent wavevector can affect the synchronous coupling condition for a dual-channel coherent coupler or prism or grating couplers.

3.2. Nonlinear Coupling of Radiation to the Waveguide

3.2.1 Prism Coupling. Prism coupling (see Fig.14) is the most common way of exciting modes in a waveguide[90]. When the direction of the incident laser beam is such that the parallel component of the plane wave wavevector of the incident light in the prism is equal to the guided wave wavevector, coupling becomes effective and the optical energy can be transferred from the prism to the waveguiding film and vice versa. In the presence of a nonlinear medium, when the guided-wave power increases as the guided wave grows under the base of the prism, β changes, synchronism with the external field is lost, and the coupling efficiency is reduced. If the coupling is optimized at low excitation levels, an increase in the incident power will cause a reduction in the efficiency with distortion in the angular profile, a shift of the optimum coupling angle, and changes in the beam profile (see Fig.15). The original coupling conditions can be recovered in some cases by readjusting the coupling angle. In the case of pulsed excitation, the above effects become time dependent and the temporal pulse profiles are also distorted. The first experiments were performed by using a liquid crystal between the prism and a linear glass waveguide[91], a CdS_xSe_{1-x} doped glass[92], ZnO[93], ZnS[94], and organic waveguides[95]. Previously the nonlinear properties of prism and grating couplers were used to measure the nonlinearities in waveguides via the power-dependent shift in the optimum coupling angle[96-98]. The earliest work actually utilized surface plasmon modes[96] and only later were integrated optics waveguides used[97-98]. Subsequently the optical limiting properties were studied for both cw and pulsed excitation, as well as pulse distortion[91,93,99].

The origin of the nonlinearity is very important in the coupling process. A thermal nonlinearity (refractive index changes with temperature, which in turn changes because of the absorption of light) due to thermal diffusion can lead to an equilibrium uniform temperature (and hence index change) in the coupling region. The coupling process can then be treated in terms of a resonantly coupled field[100,101] whose amplitude is constant along the coupling

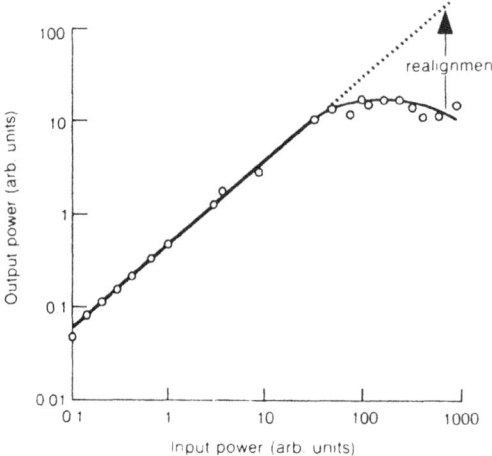

Fig. 15. Experimentally determined nonlinear response of a prism-coupled doped polymer waveguide.

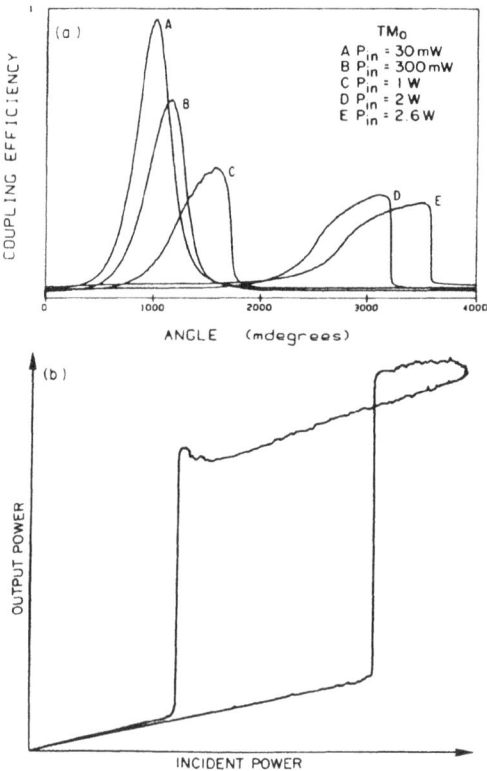

Fig. 16. The angular dependence of the coupling efficiency with: a) increasing incident cw power; and, b) bistability obtained by increasing the incident power at a fixed positive angular detuning.

region. It has been shown that a nonlocal nonlinearity of this kind is needed for bistability[102,103] which is the result of longitudinal feedback caused by diffusion of the nonlinear change of refractive index. On the contrary if the nonlinearity is purely local (due for example to electronic fast processes) a spatial variation in the refractive index is maintained across the whole coupling region and a detailed analysis of a guided-wave mode growing in the coupling region is necessary[104,105].

The theory for the coupling of high-power laser pulses into waveguides embracing both Kerr-law nonlinearities (instantaneous turn-on and turn-off times) and integrating nonlinearities (a turn-off longer than the pulse width) has been developed [106]. The distortions are symmetric for Kerr nonlinearity with an instantaneous relaxation time, and asymmetric for integrating nonlinearities.

An example of prism coupling is shown in Fig.16a for cw operation. The angular dependence of the coupling efficiency with increasing incident cw power at $\lambda = 0.515$ μm for a ZnS waveguide is shown. The excited nonlinearity was a thermal one. There is both a shift in the peak coupling angle and a decrease of efficiency, with a marked distortion of the curve with increasing power[107]. Eventually bistability in the output versus input response is obtained (Fig.16b).

3.2.2. Nonlinear Grating Coupling.
In the case of grating coupling the resonant input coupling angle must satisfy the phase matching condition for the wave vectors of the input radiation[108]

$$\beta_o^{(m)} = k_o \cos\theta_o \pm \frac{2\pi}{d} m, \tag{38}$$

where $\beta_o^{(m)}$ is the propagation constant of the guided wave, d is the grating period, $k_o = 2\pi/\lambda$, θ_o is the resonant input coupling angle with respect to the plane of the waveguide and m = 1,2.... . Actually losses are introduced by the grating and the propagation constant of the mode is

$$\beta^{(m)} = \beta_o^{(m)} + j\alpha_L \tag{39}$$

The term α_L is due to the leakage of the energy into the diffracted orders scattered by the grating[108].

Many of the considerations given for nonlinear prism coupling are valid also in the case of nonlinear grating coupling[109]. A number of experiments on the use of grating coupling in nonlinear waveguides exist[110]. The geometry shown in Fig. 17 is particularly instructive because it shows the role of the different elements involved in the coupling. The coupling is realized with a grating placed between a linear guide and its linear substrate and a very thin semiconductor (CdS) layer (20 nm) is deposited on the top of the guiding film. This layer is

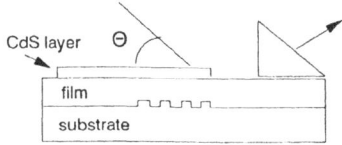

Fig. 17. Geometry of the grating coupler.

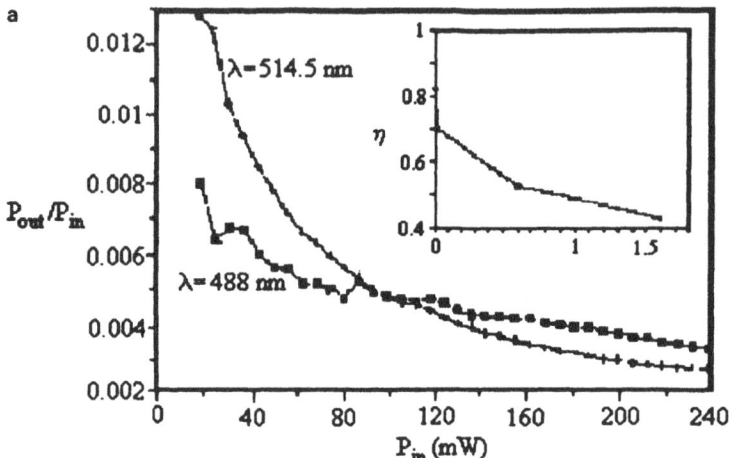

Fig. 18a. Efficiency of the grating coupled waveguide. In the inset, the theoretical prediction.

the nonlinear element. The reduction of the coupling efficiency is shown for cw operation in Fig.18a for two wavelengths of 514.5 and 488 nm. In this case the decrease in coupling efficiency is due to the combined effect of a thermal change of the refractive index of the linear film due to absorption of radiation by the CdS film which produces a loss of phase matching and of a thermal change in absorption of the CdS itself (see Fig.18b). In fact the absorption coefficient of CdS increases as the light power increases. The thermal change of refractive index can be verified by readjusting the angle for perfect synchronism.

In the pulsed regime the change of refractive index was too small and only the change of absorption of the CdS layer was important. Fig.19 shows the results at 584 and 571.5 nm changing the repetition frequency of the pulses (each pulse 4 ps long). At high frequency (9.5MHz) the coupling efficiency increases with power because transmission of the thin CdS layer increases at these wavelengths, due to heating (at variance with the behaviour at shorter wavelengths shown in Fig.18b). At low repetition rate (147 KHz) the coupling efficiency decreases. At this low repetition rate the thermal change of CdS absorption is negligible and the electronic contribution of the refractive index change can be seen to produce a decrease of coupling efficiency. The results of a numerical simulation based on these assumptions are shown in the insets and exhibit good agreement with the experimental findings.

3.3. An Example of Application: The Nonlinear Coherent Coupler

A directional linear coupler in its simpler form consists of two waveguides that are both adjacent and parallel (see Fig.20), constructed from dielectric materials that have large nonlinear susceptibilities. Given that the two waveguide channels are sufficiently close together, light directed into one of the channels can penetrate into the other channel and (at low light intensities) the resultant field overlap can cause a periodic power exchange between the two channels. In this case the minimum distance along the coupler after which a maximum amount of power is transferred from one channel to the other is referred to as the coupling length.

The coupler may be studied by using the coupled-mode theory in which a perturbation polarization responsible for the coupling contains the refractive index of the waveguides.

Nonlinear Guided Waves

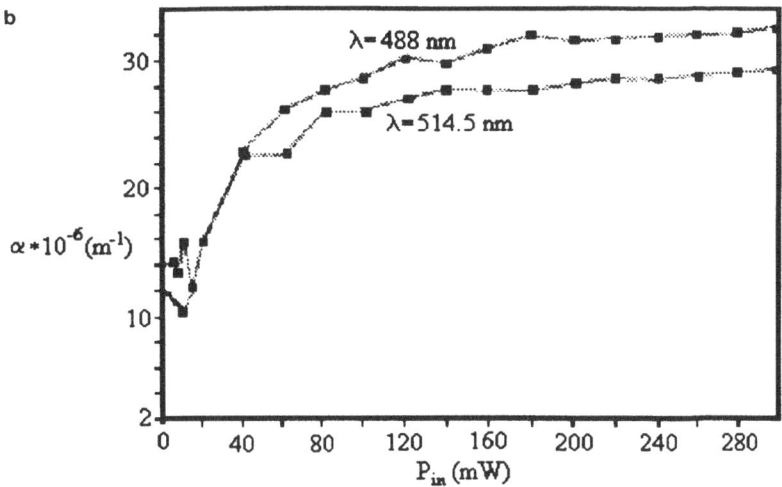

Fig. 18b. Absorption coefficient α of CdS as a function of light power.

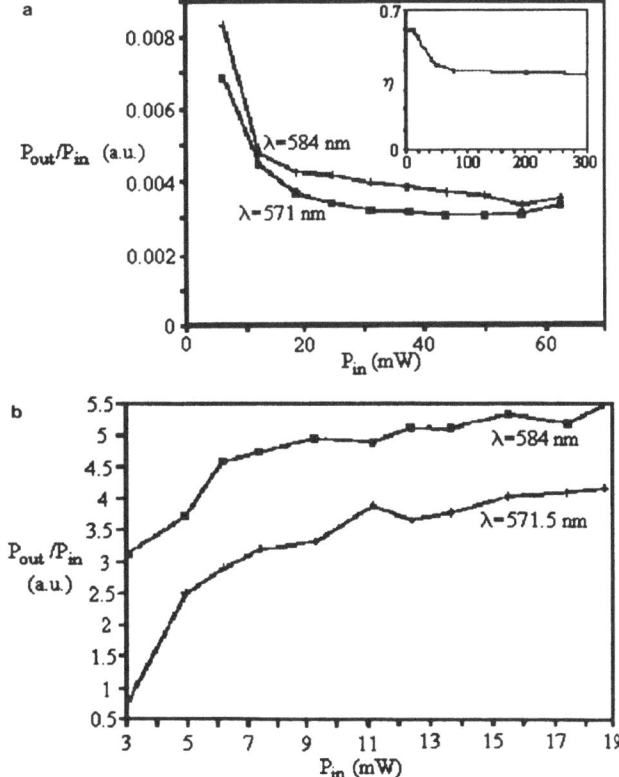

Fig. 19. Coupling efficiency as Figs. 18 but in pulsed regime, a) repetition frequency 147 kHz, b) repetition frequency 9.5 MHz.

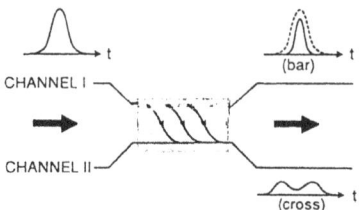

Fig. 20. Operation of the nonlinear coherent coupler.

From the difference between the refractive indices, we obtain the coupling constant K of the structure. Complete power transfer occurs in a distance $L_b = \pi/2K$ if the detuning parameter δ is zero, that is in the case of complete phase matching (since $\delta = (1/2)(\beta_1 - \beta_2)$, where β_1, β_2 are the wave-vectors of two modes propagating inside the structure). If δ is not zero, the maximum fraction of power that can be transferred is proportional to

$$\frac{K^2}{K^2 + \delta^2} \qquad (40)$$

In a nonlinear coupler constructed from a Kerr-type medium the dependence of refractive index n is given by Eq. 22. Varying the input light intensity can cause the light, that at low intensity comes out one channel, to exit from the other channel.

The nonlinear coupler was first studied by Jensen[111]. The interaction typically involves two coupled co-directional or contra-directional guided wave modes which exchange power. They are governed by the nonlinear coupled mode equations

$$\begin{cases} -i\dfrac{da_1}{dz} = \left[\dfrac{\delta\beta}{2} + \Delta\beta_1\left(|a_1|^2\right)\right]a_1 + Ca_2 \\ -i\dfrac{da_2}{dz} = \left[-\dfrac{\delta\beta}{2} + \Delta\beta_2\left(|a_2|^2\right)\right]a_2 + Ca_1 \end{cases} \qquad (41)$$

where a_1, a_2 are the amplitudes of mode in individual waveguides 1 and 2, $\delta\beta = \beta_1 - \beta_2$ is the linear propagation constant difference of the two guides, C is the linear coupling coefficient and $\Delta\beta_i(|a_i|^2)$ is the power-dependent refractive index change in channel i for arbitrary nonlinearity. For saturable nonlinearity, $\Delta\beta_i$ can be given by the form[89]

$$\Delta\beta_i = 2Cw_i\left[1 - \exp(-2P_i/P_{ci}w_i)\right] \qquad (42)$$

where w_i is the normalized saturation parameter related to the actual saturation value $\Delta\beta_{sat}$

$$w_i = \Delta\beta_{isat}/2C \qquad (43)$$

and P_{ci} is the critical power defined for the Kerr effect, and $P_i = |a_i|^2$ is the power in waveguide i. If the two channels are identical, $\delta\beta = 0$ and the coupling occurs via field overlap quantified by

$$\Gamma = \omega\varepsilon_o \int_{-\infty}^{\infty}dx \int_{-\infty}^{\infty}dy \; E_1^*(r)E_2(r) \tag{44}$$

As a result, when only one of the channels is excited with low powers at the input, the power oscillates between the two channels with a beat length L_b. As the input power is increased, a mismatch is induced in the wavevectors ($\delta\beta \neq 0$) which decreases the rate of power transfer (see Eq. 40). This leads to an increase in the effective beat length. There is a critical power associated with the solutions for which an infinitely long, lossless nonlinear directional coupler acts as a 50:50 splitter. For higher input powers, the initial wavevector mismatch is too large to overcome and the power effectively stays in the input channel. Therefore if the device length is terminated at $L_b/2$, then at low powers the signal comes out of channel 2 and at high powers out of channel 1. Unfortunately due to the saturation of the refractive index change, the fraction of switched power reaches a saturation value, that is increasing the power past a certain point does not lead to an increase in the fraction of the light switched from one output channel to the other. The effect of saturation (together with linear loss) was first shown by Stegeman et al.[89,112,113]. Figs. 21 show the results[114] for a saturable nonlinear coupler with mismatched arms of length $2L_b/\sqrt{1+\delta_n^2}$, where $\delta_n = \delta\beta/2C$ is a function of the normalized

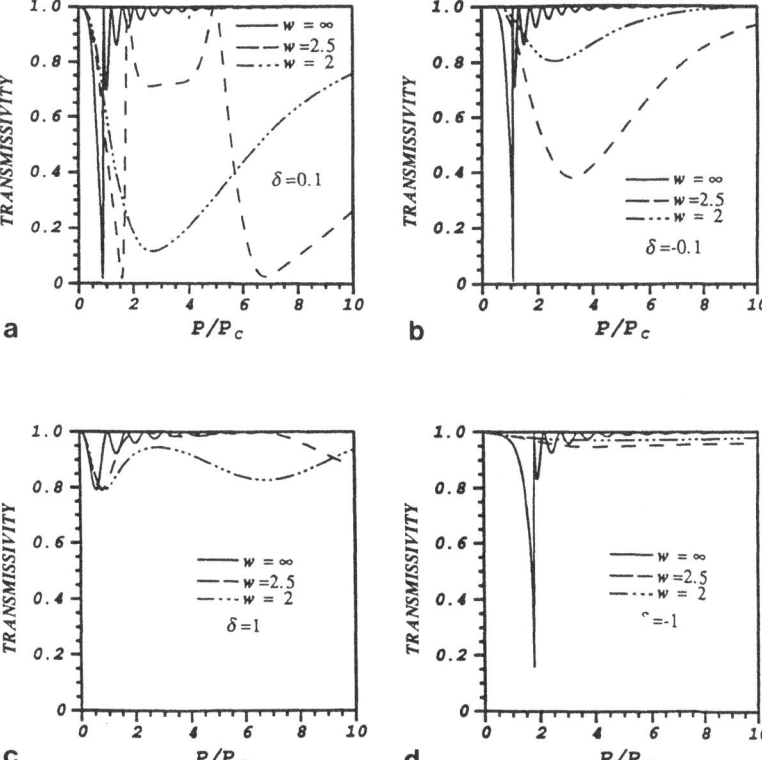

Fig. 21. Trasmissivity of channel 1 (or power in guide 1) versus normalized power for a one-beat mismatched nonlinear coupler when guide 1 is initially excited for various saturations w: a) $\delta = 0.1$; b) $\delta = -0.1$; c) $\delta = 1$; d) $\delta = -1$.

input power incident in one channel. Figs.21a,b give the transmissivity of channel 1 as a function of the normalized power incident in one channel for linear mismatches $\delta_n = \pm 0.1$, when guide 1 is initially excited. The most interesting characteristic is the switching on or off from channel 1 to channel 2 by a slight change in the input power. In the absence of saturation ($w \to \infty$), a sharp switching occurs at a particular power that can be larger or smaller than P_c, depending on the sign of δ. As the magnitude of saturation w decreases, the transmission notch broadens considerably. For w = 2.5 almost all of the power can be switched at two power values for $\delta = 0.1$ (Fig. 21a) whereas the switch notch is broadened significantly with only 60% of total power switched at one power value for $\delta = -0.1$ (Fig.21b), indicating a dramatic nonreciprocity of the device. As the magnitude of the mismatch increases, the switching characteristic is deteriorated quickly by saturation, as Fig. 21 shows.

The very stringent requirements on material properties derived from the above analysis has limited the number of successful implementations. Most experiments have been performed in the near infrared with GaAs-AlGaAs multi-quantum well structures[115]. Operation at photon energies below half bandgap in AlGaAs has also proved successful[116]. The use of two different saturable nonlinear channels has been used in semiconductor-doped glass waveguides in which the two channels both saturated in absorption, but at different rates[117]. In general, by combining one self-focusing channel with another channel of different nonlinearity, which can be self-focusing, linear or self-defocusing, the detrimental effect of nonlinear saturation is greatly reduced compared with that of the conventional coupler composed of two identical self-focusing channels[118].

Fig. 22[118] shows the switching characteristic for channels as described in the insets. Although this Figure refers to a fiber coupler, the main results apply in general, and show that the best results are obtained for a coupler constructed by combining one self-focusing with one self-defocusing guide. Other switching geometries can be implemented, as the Y or X waveguide junction switches[119], which however we will not discuss here.

The crucial parameter to be used to characterize the operation of the coupler can be derived through the following reasoning. The variation in the fraction of light switched in the nonlinear coupler can be written as

$$w = \Delta n_{sat} \frac{L}{\lambda} \cong \Delta \beta_{sar} \frac{k_o L}{2\pi} \qquad (45)$$

where w is the figure of merit. A change in w from 1 to 2 corresponds essentially to a change from zero to 100% switching. A phase change $\Delta \Phi^{NL} = 2\pi w$, with w=2, is needed to swicth. If saturation is present one might expect to obtain the desired phase change by increasing w by an increment of the device length. This is in fact true if the waveguide attenuation coefficient satisfies $L \ll 1/\alpha$. A material figure of merit can therefore be defined as

$$W = \frac{w}{\alpha L} = \frac{\Delta n_{sat}}{\alpha \lambda} \qquad (46)$$

One may expect that in materials with low attenuation good switching can be obtained also in the presence of saturation. This is effectively the case. Directional couplers with two equal channels have been implemented successfully in AlGaAs channel waveguides operated with photon energies below one-half the bandgap[112,113].

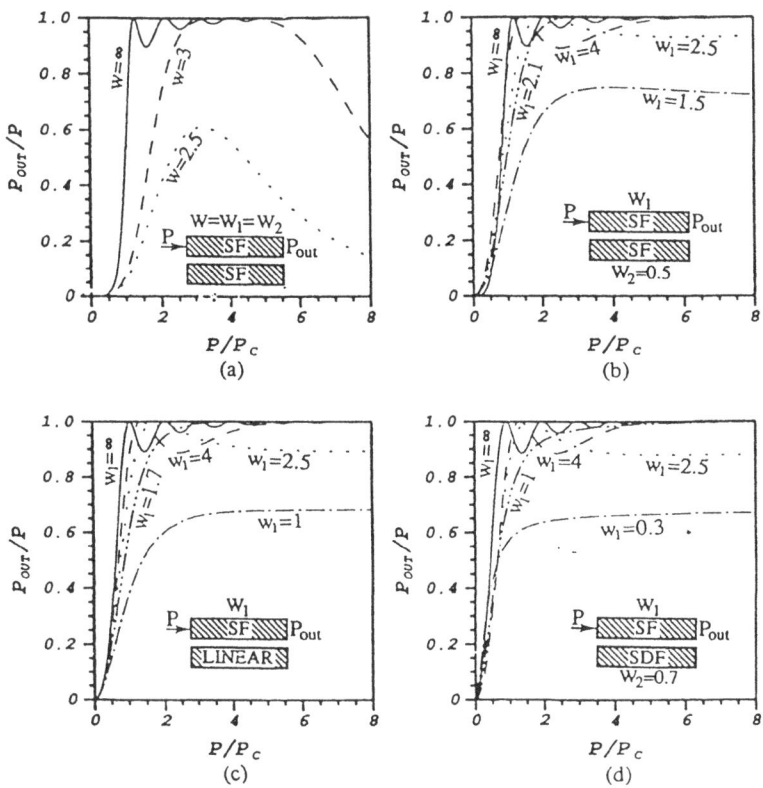

Fig. 22. Trasmissivity in channel 1 of exponentially saturable nonlinear couplers with length $L_c = \pi/2C$. SF denotes self-focusing core; SDF self-defocusing core, for different saturation values W. A Kerr-law nonlinearity has $w \to \infty$.

3.4. Spatial solitons

Spatial solitons are treated in detail elsewhere in this volume. We therefore give here only a brief sketch.

A spatial soliton can be defined as a beam which propagates without diffraction. To obtain this result it is necessary to compensate the spatial broadening of the beam due to diffraction during propagation with an equal amount of self-focusing which maintains a constant transverse dimension of the beam during propagation. A very simple calculation shows how this is possible.

At low intensities the nonlinear Kerr effect can be neglected. The beam spreads under diffraction[120] (see Fig. 23a) inducing a positive wavefront curvature. For an initial limited plane wave the wavefront sag after propagation through a distance z is

$$\Delta l_d = z[1 - \cos(\alpha/2)] \tag{47}$$

where the divergence angle α is $\alpha = \lambda/a\, n_o$, a being the initial beam width.

In contrast, an intense single mode beam at its waist, with a bell shaped intensity profile, generates a refractive index distribution looking like a graded index thick lens when the n_2 coefficient is positive. The beam focuses, leading to a negative wavefront curvature (see Fig. 23b). Assuming that diffraction is neglected, the wavefront sag Δl_{NL} results from the optical path increase by the nonlinear Kerr effect in the central part of the beam

$$\Delta l_{NL} = -n_2 EE^* z \qquad (48)$$

A balance can be expected between Kerr effect and diffraction beam spreading if

$$\Delta l_d = \Delta l_{NL}$$

The corresponding power density is

$$EE^* = \frac{\lambda^2}{8a^2 n_2 n_o^2} \qquad (49)$$

which corresponds to a total power carried by the beam P_c

$$P_c = a^2 EE^* = \frac{\lambda^2}{8 n_2 n_o^2} \qquad (50)$$

P_c does not depend upon geometrical parameters of the beam and therefore any fluctuation in the actual power P carried by the beam causes it to focus ($P > P_c$) or diverge ($P < P_c$). Three-dimensional propagation in nonlinear Kerr media is therefore unstable unless some special care, as a two-beam interference technique[121], is used to preclude the occurrence of filamentation.

In two-dimensional propagation, one transverse dimension of the beam is fixed. A beam in a waveguide with width a and a constant thickness b provides equilibrium between the two effects at a power density given by Eq. (50). The total power carried by the beam is however

$$P = abEE^* = \frac{\lambda^2 b}{8 a n_2 n_o^2} \qquad (51)$$

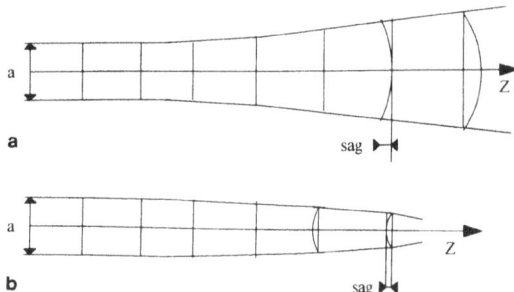

Fig. 23. a) Diffraction of limited plane wave induces a positive wavefront curvature; b) self-focusing of an intense beam induces a negative wavefront curvature.

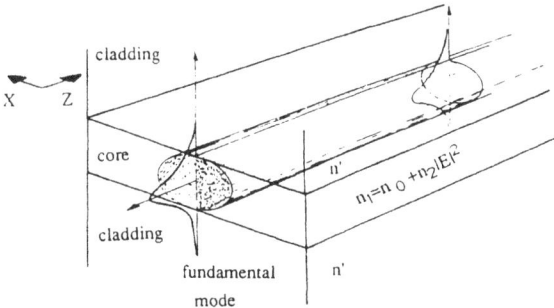

Fig. 24. A solition in a waveguide. Confinement is obtained along the x direction.

which depends on the beam width. If the power is too low, diffraction spreading increases the beam width until the product P_a equals $\lambda^2 b / 8 n_2 n_0^2$. Thus equilibrium is reached.

On the contrary, if the power is too high the width a decreases until a is such that P_a equals $\lambda^2 b / 8 n_2 n_0^2$, where equilibrium is also obtained.

A soliton of this kind, called the fundamental soliton, in a waveguide looks as shown in Fig. 24, and has been obtained experimentally[122].

A very interesting property of solitons is their ability to interact if they are close enough. If two fundamental solitons of respective amplitudes E_1 and E_2 propagate simultaneously, the nonlinear part of the refractive index can be written as

$$n_{NL} = n_2 \{ E_1 E_1^* + E_2 E_2^* + 2 \operatorname{Re}(E_1 E_2^*) \} \tag{52}$$

The mixed term with $E_1 E_2$ can be positive or negative according to the relative phase between the two solitons. The presence of this term gives rise to the possibility of interaction. Fig. 25 shows the refractive index profile induced when the relative phase is equal to 0 and π. If the optical fields E_1 and E_2 are in phase, the cross part of the nonlinear refractive index is positive. It delays the beam edges close to the symmetry axis and the two solitons converge. On the contrary, if the two solitons are opposite in phase they diverge[123]. If the relative phase between the two solitons takes any value in the range between 0 and π, the intensity of one of the solitons changes. Fig. 27 summarizes these results, giving the soliton output intensity versus the relative phase[123].

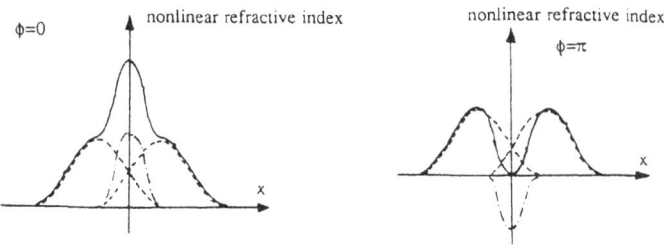

Fig. 25. Refractive index distributions induced by interacting soliton beams, a) $\phi = 0$; and, b) $\phi = \pi$.

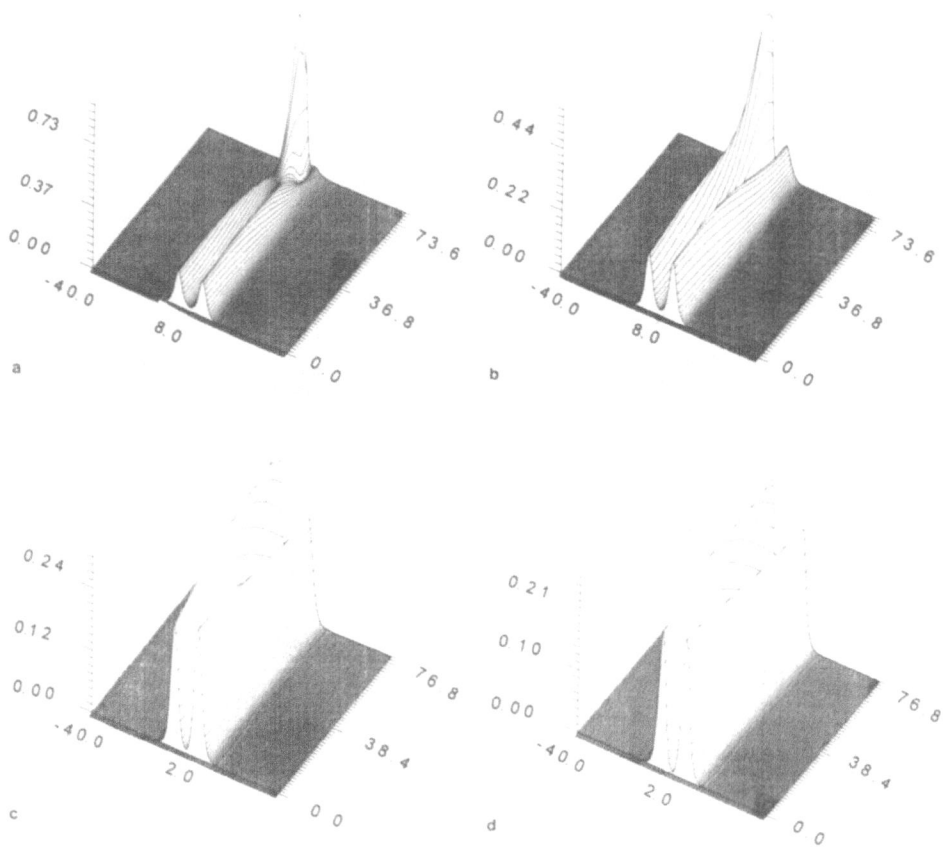

Fig. 26. Soliton propagation for different phase values: a) f = 0; b) f = 3/10π; c) f = 6/10π; d) f = 9/10π.

Solitons can also be emitted at the interface between a nonlinear and a linear medium. Perhaps the most interesting geometry is the three media case in which one or more of the media has a self-focusing nonlinearity. Spatial solitons can be emitted into a self-focusing bounding medium, in a way similar to the generation of spatial solitons at a linear-nonlinear interface. The simplest geometry of a nonlinear Kerr-type cladding and a linear substrate has been already discussed (see Fig. 8b). Up to the first maximum (P*) in the curve (branch I), the field distribution maintains a peak within the film and resembles closely that of the low-power (linear) normal mode. This is followed by an unstable region (dashed region) in which the β-curve has a negative slope, and then a further stable, positive-slope branch (branch II). In this latter stable branch, the field closely resembles that of a spatial soliton guided by a nonlinear-linear (cladding-film) interface, and the field maximum is situated inside the nonlinear self-focusing medium. For large values of guided wave flux the position of the field maximum becomes progressively more independent of the waveguide parameters and depends only on the film-cladding interface. Fig. 8b shows that, as a function of guided wave flux, there is a "jump" in the stable field configuration available to the system beyond P*[124]. It is precisely at this point that one might expect something critical to occur such as soliton emission.

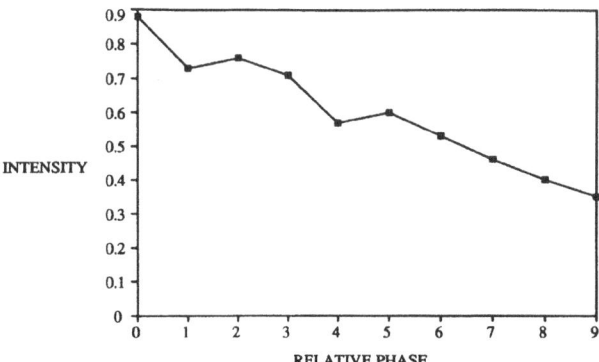

Fig. 27. Soliton intensity as a function of its relative phase with a plane wave background.

In Fig. 28 the flux trapped in the waveguide S_T versus the input flux S_{in} is shown and a sequence of thresholds is clearly seen[125]. At each discontinuous point a single self-focused channel, or spatial soliton, is emitted through the film-cladding interface into the nonlinear medium and propagates away from it.

This behaviour suggest that a second waveguide, near to the one emitting the soliton, can capture it. Such a device, which could be termed a soliton coupler, can potentially produce extremely sharp switching characteristics since soliton emission is a threshold effect. The exchange of solitons between adjacent waveguides[126] could lead to a whole new family of devices. When saturation is present one can no longer speak of solitons but rather of solitary waves. Since all media exhibit losses no true soliton can strictly exist. However, for the small distances involved in guided propagation, if losses are low enough one can consider the two waves indistinguishable. The possibility of producing "light bullets" in the form of spatially and temporarily limited pulses is the last case we wish to point out. A new type of spatial soliton, generated by the photorefractive effect of the medium, will be also described in this volume.

4. Cascading Effect

In nonlinear optics, it has generally been accepted that for purely quadratic nonlinear materials, one obtains various three-wave mixing effects, such as second harmonic generation. Intensity dependent effects are usually expected in cubic nonlinear materials, but not in purely quadratic nonlinear materials.

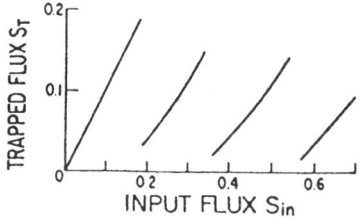

Fig. 28. The flux remaining in the waveguide of Fig. 8b as a function of input flux after a propagation distance of a thousand wavelengths.

The classic reason that intensity dependent effects are not expected in quadratic nonlinear materials is that the phase of the fundamental cannot be aligned with the phase of its harmonic. This notion of phase nonalignement is incorrect; a number of papers have considered phase modulation due to second-order nonlinear-optical processes, and intensity-dependent effects have been predicted in $\chi^{(2)}$ materials[126-132].

Large nonlinear phase shifts can be obtained from phase-mismatched second harmonic generation. A nonlinear phase shift can be experienced by a beam at the fundamental frequency during almost phase-matched second harmonic generation.

The evolution of this phase (and amplitude) of the fundamental with distance and power is quite different from that associated with an intensity-dependent refractive index n_2.

Starting from the wave equation driven by polarization sources including $\chi^{(1)}$ and $\chi^{(2)}$, and using the slowing varying phase and amplitude approximation, one obtains the usual coupled mode equations which describe second harmonic generation[131,133]

$$\begin{cases} \dfrac{da(2\omega,z)}{dz} = -ik(-2\omega;\omega,\omega)a^2(\omega,z)e^{i\Delta\beta z} - a(2\omega)\alpha(2\omega,z) \\[2mm] \dfrac{da(\omega,z)}{dz} = -ik(-\omega;2\omega,-\omega)a(2\omega,z)a^*(\omega,z)e^{-i\Delta\beta z} - a(\omega)\alpha(\omega,z) \\[2mm] k(-2\omega;\omega,\omega) = \dfrac{\omega d_{ijk}^{(2)}(-2\omega;\omega,\omega)e_i(2\omega)e_j(\omega)e_k(\omega)}{\sqrt{2n_i(2\omega)n_j(\omega)n_k(\omega)c^3\varepsilon_o}} \end{cases} \quad (53)$$

and a similar expression for $k(-\omega; 2\omega, -\omega)$. Here the wavevector mismatch is

$$\Delta\beta = 2k_o(\omega)[n(2\omega) - n(\omega)] \quad (54)$$

and the complex field amplitudes a(z) are normalized so that $|a(z)|^2$ is the intensity, the e_i's are the field unit vectors and α is the frequency-dependent linear loss. For frequency-independent loss, far from any material resonances, and in the absence of coupling to other fields, these equations can be solved in terms of Jacobi elliptic functions. Stegeman et al[134]

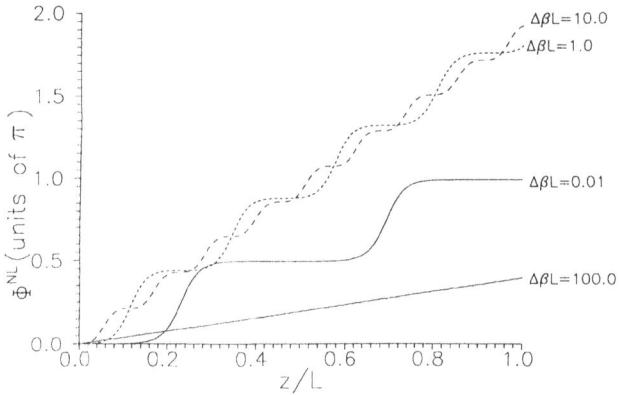

Fig. 29. The variation of phase shift with z for kL = 4 and $|a(\omega, 0)|^2 = 25$.

Nonlinear Guided Waves 51

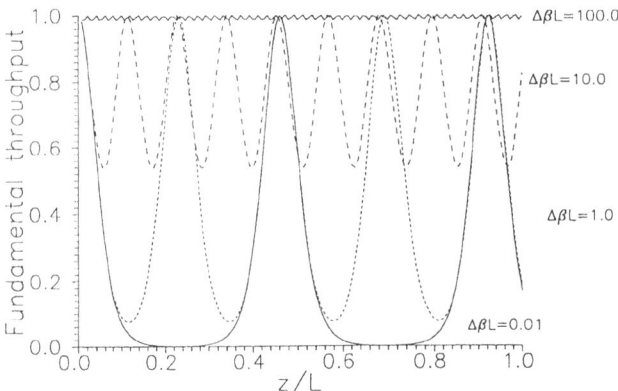

Fig. 30. The fractional fundamental intensity $|a(\omega, z)|^2 / |a(\omega, 0)|^2$ as a function of z for the same conditions as in Fig. 29.

have shown that also in this case a n_2 coefficient can be defined which can either be positive (self-focusing) or negative (self-defocusing).

The physical origin of this effective third order nonlinearity is straightforward. When the SHG process is mismatched, the second harmonic field propagates with the wavevector

$$2n(2\omega)k_o(\omega) \neq 2n(\omega)k_o(\omega) \tag{55}$$

Therefore the product $E(2\omega)E(\omega)^*$ in Eq. 53 produces a polarization source term with a component in quadrature with the fundamental, and hence slows it down ($\Delta\beta<0$) or speeds it up ($\Delta\beta>0$).

Shown in Fig. 29a[134] is the evolution of the nonlinear phase shift of the fundamental beam as a function of the normalized distance z/L for different values of wavevector detuning $\Delta\beta L$, where L is the sample length. Here Φ^{NL} is defined by

$$a(\omega, z) = |a(\omega, z)| \exp[i\Phi^{NL}(z)] \tag{56}$$

The corresponding spatial variation of the normalized fundamental power $|a(\omega,z)|^2 / |a(\omega,0)|^2$ is shown in Fig.30 (2a). If three beams of different frequency and appropriate polarization are almost phase-matched for sum or difference frequency generation, nonlinear phase shifts can occur for all interacting beams. These second order processes which mimic third order nonlinearities, seem able to produce large phase shifts without saturation and should be very interesting for making devices[135]. Large nonlinear phase modulation in quasi-phase-matched KTP waveguides as a result of this process has recently been demonstrated[136].

References

1. G.I. Stegeman, C. T. Seaton, W.M. Hetherington III, A.D. Boardman and P. Egan, in "Nonlinear Optics: Materials and Devices", C. Flytzanis and J.L. Oudar Eds., Springer-Verlag, Berlin 1986, p.31
2. D. Mihalache, M. Bertolotti and C. Sibilia, in "Progress in Optics", E. Wolf Ed., vol. XXVII, North-Holland 1989, p.227

3. A.D. Boardman, M. Bertolotti and T. Twardowski, "Nonlinear Waves in Solid State Physics", Plenum Press, London 1990
4. D. Marcuse, "Theory of Dielectric Optical Waveguides", Academic Press, New York 1974
5. P.K.Tien, R.Ulrich and R.J.Martin, Appl. Phys. Lett. 17, 447 (1970)
6. D.B.Anderson and T.J.Boyd, Appl. Phys. Lett. 19, 266 (1971)
7. Y.Suematsu, Y. Sasaki and K. Shibata, Appl. Phys. Lett. 23, 137 (1973)
8. Y. Suematsu, Y. Sasaki, K. Furuya, K. Shibata and S. Ibukuro, IEEE J. Quant. Electron. 10, 222 (1974)
9. W.K. Burns and A.B.Lee, Appl. Phys. Lett. 24, 222 (1974)
10. M.M.Hopkins and A. Miller, Appl. Phys. Lett. 25, 47 (1974)
11. H. Ito and H. Inaba, Opt. Commun. 15, 104 (1975)
12. B.U.Chen, C.C.Ghizoni and C.L.Tang, Appl. Phys. Lett. 28, 651 (1976)
13. J.P. van der Ziel, M. Ilegems, P.W. Fox, and R.M. Mikulyak, Appl. Phys. Lett. 29, 775 (1976)
14. N. Uesugi and T.Kimura, Appl. Phys. Lett. 29, 572 (1976)
15. H. Ito and H. Inaba, Opt. Lett. 2, 139 (1978)
16. N. Uesugi and T. Kimura, Appl. Phys. Lett. 33, 518 (1978)
17. W. Sohler and H. Suche, Appl. Phys. Lett. 33, 518 (1978)
18. W. Sohler and H. Suche, in "Integrated Optics III", L.D.Hutchenson and D.G.Hall, Eds., Proc. SPIE 408, 163 (1983) 83
19. M.M.Fejer, M.J.F.Digonnet and R.L.Byer, Opt. Lett. 11, 230 (1986)
20. R.Regener and W. Sohler, J. Opt. Soc. America B5, 267 (1988)
21. F.Laurell and G. Arvidsson, J. Opt. Soc. America B5, 292 (1988)
22. J.D.Bierlein and H.Vanherzeele, J. Opt. Soc. America B6, 622 (1989)
23. E.J. Lim, M.M. Fejer and R.L.Byer, Electron. Lett. 25, 174 (1989)
24. J. Webjorn, F. Laurell and G. Arvidsson, IEEE J. Lightwave Tech. (1989)
25. J.D.Bierlein, D.B.Laubacher, J.B.Brown and C.J.van der Poel, Appl. Phys. Lett. 56, 1725 (1990)
26. K. Hayata, K.Yanagawa and N.Koshiba, Appl. Phys. Lett. 56, 206 (1990)
27. W.P.Risk, Appl. Phys. Lett. 58, 19 (1991)
28. L.Babsail, G. Lifante and P.D. Townsend, Appl. Phys. Lett. 59, 384 (1991)
29. N.Asai, H.Tamada, I.Fujiwara and J.Seto, J. Appl. Phys. 72, 4521 (1992); L.Zhang, P.J.Chandler, P.D.Townsend, Z.T.Alwahabi and A.J. McCaffery, Electron. Lett. 28, 1478 (1992)
30. D. Fluck, B.Binder, M.Kupfer, H.Looser, Ch.Buchal and P. Gunter, Opt. Commun. 90, 304 (1992)
31. D. Fluck, J. Moll, P.Gunter, M.Leuster and Ch. Buchal, Electron. Lett. 28, 1092 (1992)
32. G.H. Hewig and K. Jain, Optics Commun. 47, 347 (1983)
33. K. Sasaki, T. Kinoshita, and N. Karasawa Appl. Phys. Lett. 45, 333 (1984)
34. H.Itoh, K.Hotta, H.Takara and K.Sasaki, Appl. Opt. 25, 1491 (1986)
35. H. Ito, K. Hotta, H. Takara and K. Sasaki, Opt. Commun. 59, 299 (1986)
36. M. Kull, J.L. Coutaz and R. Meyrueix, Opt. Lett. 16, 1930 (1991)
37. Z. Weissman, A. Hardy, E. Marom and S. R. J. Brueck, J. Appl. Phys. 69, 1201 (1991)
38. O. Sugihara, T. Toda, T. Ogura, T. Kinoshita and K. Sasaki, Opt. Lett. 16, 702 (1991)
39. O. Sugihara and K. Sasaki, J. Opt. Soc. Am. B9, 104 (1992)
40. D.Fluck, J.Moll, M.Fleuster, Ch.Buchal and P.Gunter, Electron. Lett. 28, 1092 (1992)
41. Y.Azumai and H.Sato, Jpn. J. Appl. Phys. 32, 800 (1993)
42. D.E.Thompson, J.D.McMullen and D.B.Anderson, Appl. Phys. Lett. 29, 113 (1976)
43. J.Webjorn, F.Laurell, and G.Arvidsson, IEEE Photonics Technol. Lett. 1, 316 (1989)
44. F.Armani, D. DeLaCourt, E.Lallier, M.Papuchon, Q. He, M. de Micheli and D.B.Ostrowski, Electron. Lett. 28, 139 (1992)
45. K.Yamamoto and K.Mizuuchi, IEEE Photonics Technol. Lett. 4, 435 (1992)
46. M.Yamada, N.Nada, M.Saitoh and K.Watanabe, Appl. Phys. Lett. 62, 435 (1993)

47. L. Zhang, P.J. Chandler, P.D.Towsend, Z.T.Alwahabi, S.L.Pityana and A.J. McCaffery, J.Appl. Phys. 73, 2695 (1993)
48. S.I. Bozhevolnyi and K.Pedersen, Appl. Opt. 31, 5813 (1992)
49. J.J.Kester, P.J.Wolf and W. Roc White, Opt. Lett. 17, 1779 (1992)
50. R.H.Stolen, in "Nonlinear Waves in Solid State Physics", A.D. Boardman, M.Bertolotti, and T. Twardowski, Eds., Plenum Press, New York 1990, p.297 ; B.Ehrlich-Hall, D.H.Krol, R.H.Stolen and H.W.K. Tom, Opt. Lett. 17, 396 (1992)
51. R. Normandin and G.I. Stegeman, Opt. Lett. 4, 58 (1979)
52. R. Normandin and G.I. Stegeman, Appl. Phys. Lett. 40, 759 (1982); R. Normandin and G.I. Stegeman, Appl. Phys. Lett. 36, 253 (1980)
53. D.Vakhshoori, M.C.Wu and S.Wang, Phys. Lett. 52, 422 (1988)
54. R. Normandin, S.Letourneau, F.Chatenoud and R.L.Williams, IEEE J. Quantum Electr. QE-27, 1520 (1991)
55. D. Vakhshoori, R.J.Fisher, M. Hong, D.L.Sivco, G.J. Zydzik, G.N.S.Chu and A.R.Cho, Appl. Phys. Lett. 59, 896 (1991)
56. H.Dai, S.Janz, R.Normandin, J. Nielsen, F.Chatenoud and R. Williams, IEEE Photonic Technol. Lett. 4, 820 (1992)
57. R. Normandin, H.Dai, S.Janz, A.Delage, J.Brown and F. Chatenoud, Appl. Phys. Lett. 62, 118 (1993)
58. N. Uesugi, K.Naikoku and M.Fukuma, J.Appl.Phys. 49, 4945 (1978)
59. N.Uesugi, Appl. Phys. Lett. 36, 178 (1980)
60. W.P.Risk, R.N.Payne, W.Lenth, C.Harder and H.Meier, Appl. Phys. Lett. 55, 1179 (1989)
61. J.C. Baumert, E.M.Schellenberg, W.Lenth, W.P.Risk, G.C. Bjorklund, Appl. Phys. Lett. 51, 2192 (1987)
62. M. Bertolotti, R. Horak, D. Mihalache, J. Perina and C. Sibilia, Inst. Phys. Conf. Ser. n.115, IOP Publ. 1991,p.215
63. A.G. Litvak and V.A. Mironov, Izv. VUZ Radiofiz. 11, 1911 (1968)
64. M. Miyagi and S. Nishida, Sci. Rep. Tohoku Univ. B24, 53 (1972)
65. A.E. Kaplan, Sov. Phys. JETP Lett. 24, 114 (1976)
66. A.E. Kaplan, Sov. Phys. JETP Lett. 45, 896 (1977)
67. W.J. Tomlinson, Opt. Lett. 5, 323 (1980)
68. A.A.Maradudin, Z.Phys. B41, 341 (1981)
69. A.A.Maradudin, in "Optical and Acoustic Waves in Solids: Modern Topics", M. Borissov, Ed. World Scientific Pu. Co., Singapore 1983, p.72
70. G.I.Stegeman, C.T. Seaton, J.Ariyasu, T.P. Shen and J.V. Moloney, Opt. Commun. 56, 365 (1986)
71. N.N.Akhmediev, V.I.Korneev and Y.V. Kuz'menko, Sov. Phys. JETP 61, 62 (1985)
72. G.I.Stegeman, E.M.Wright, C.T.Seaton, J.V.Moloney, T.P. Shen, A.A.Maradudin and R.F.Wallis, IEEE J. Quantum Electron. QE-22, 977 (1986)
73. A.A. Kolokolov, Lett. Nuovo Cimento 8, 197 (1973)
74. C.K.R.T. Jones and J.V. Moloney, Phys. Lett. A117, 175 (1986)
75. N.N. Akhmediev and N.V. Ostrovskaya, Sov. Phys. Tech. Phys. 33, 1333 (1988)
76. H. T. Tran, J. D. Mitchell, N.N. Akhmediev and A. Ankiewicz, Opt. Commun. 93, 227 (1992)
77. H.T. Tran and A. Ankiewicz, IEEE J. Quantum Electron. 28, 488 (1992)
78. N.V. Vysotina, N.N. Rozanov and V.A. Smirnov, Sov. Phys. Tech. Phys. 32, 104 (1987)
79. J. Ariyasu, C.T. Seaton, G. I. Stegeman and J.V. Moloney, IEEE J. Quantum Electron. QE-22, 984 (1986)
80. A.B. Aceves, J.V. Moloney and A.C. Newell, Phys. Rev. A39, 1809 (1989)
81. H.T. Tran, Opt. Commun. 93, 202 (1992)
82. J.D. Mitchell and A.W. Snyder, Opt. Lett. 13, 1132 (1988)

83. H.T. Tran, Opt. Lett. 17, 1767 (1992); D.Hart and E.M.Wright, Opt. Lett. 17, 121 (1992)
84. R.A.Sammut, Q.Y. Li and C. Pask, J. Opt. Soc. Am. B9, 884 (1992)
85. C.T.Seaton, X. Mai, G.I. Stegeman and H.G. Winful, Opt. Eng. 24, 593 (1985)
86. J.V.Moloney, J. Ariyasu, C.T. Seaton and G.I. Stegeman, Appl. Phys. Lett. 48, 826 (1986)
89. G.I.Stegeman, C.T.Seaton, C.N.Ironside, T.Cullen, and A. C. Walker, Appl. Phys. Lett. 50, 1035 (1987)
90. R.Ulrich, J. Opt. Soc. Am. 60, 1337 (1970)
91. J.D.Valera,C.T.Seaton, G.I.Stegeman, R.L.Shoemaker, Xu Mai and C. Liao, Appl. Phys. Lett. 45, 1013 (1984)
92. S.Patela, H.Jerominek, C.Delisle and R.Tremblay, J. Appl. Phys. 60, 1591 (1986)
93. R.M.Fortenberry, R. Moshrefzadeh, G.Assanto, Xu Mai, E.M. Wright, C.T.Seaton and G.I.Stegeman, Appl. Phys. Lett. 49, 687 (1986)
94. G.Assanto, B.Svenson, D.Kuchibhatla, U.J.Gibson, C.T. Seaton and G.I.Stegeman, Opt. Lett. 12, 419 (1986)
95. M.J.Goodwin, R.Glenn and I.Bennion, Electron. Lett. 22, 450 (1986)
96. G.M.Carter and Y.J.Chen, Appl. Phys. Lett. 42, 643 (1983) 83
97. G.M.Carter, Y.J.Chen and S.K.Tripathy, Appl. Phys. Lett. 43, 891 (1983)
98. Y.J.Chen, G.M.Carter, G.J.Sonek and J.M.Ballantyne, Appl. Phys. Lett. 48, 272 (19886)
99. F.Pardo, H.Chelli, A.Koster, N.Paraire and S.Laval, IEEE J. Quantum Electron. QE-23, 545 (1987)
100. V.J.Montemayor and R.T.Deck, J. Opt. Soc. Am. B2, 1010 (1985); ibidem 3, 1211 (1986)
101. R.Reinisch, P.Arlot, G.Vitrant and E.Pic, Appl. Phys. Lett. 47, 1248 (1985)
102. G.Vitrant, R.Reinich, J.Cl.Paumier, G.Assanto and G.I.Stegeman, Opt. Lett. 14, 898 (1989)
103. N.Saiga, Appl. Opt. 31, 3378 (1992)
104. C.Liao and G.I.Stegeman, Appl. Phys. Lett. 44, 164 (1984)
105. C. Liao, G.I.Stegeman, C.T.Seaton, R.L.Shoemaker, J.D. Valera and H.G.Winful, J. Opt. Soc. Am. A2, 590 (1985)
106. G.Assanto, R.M.Fortenberry, C.T.Seaton and G.I.Stegeman, J. Opt. Soc. Am. B5, 432 (1988)
107. G.Assanto, B.Svensson, D.Kuchibhatla, U.J.Gibson, C.T. Seaton and G.I.Stegeman, Opt. Lett. 11, 644 (1986)
108. R.Ulrich, J. Opt. Soc. Am. 63, 1419 (1973); D.G.Dalgoutte and C.D.W.Wilkinson, Appl. Opt. 14, 2983 (1975); T.Tamir and S.T.Peng, Appl. Phys. 14, 235 (1977)
109. G.Assanto, M.B.Marques and G.I.Stegeman, J. Opt. Soc. Am. B8, 553 (1991)
110. P.Vincent, N.Paraire, M.Neviere, A.Koster and R.Reinich, J. Opt. Soc. Am. B2, 1106 (1985); R.Burzynsui, B.P.Singh, P.N.Prasad, R.Zanoni and G.I. Stegeman, Appl. Phys. Lett. 53, 2011 (1988); G.Assanto, A.Gabel, C.T.Seaton, G.I.Stegeman, C.N.Ironside and T.J.Cullen, Electron. Lett. 23, 484 (1987); Y.J.Chen and G.M.Carter, Appl. Phys. Lett. 41, 307 (1982); R.M.Fortenberry, G.Assanto, R.Moshrefzaden, C.T.Seaton and G.I.Stegeman, J. Opt. Soc. Am. B5, 425 (1988)
111. S.M.Jensen, IEEE J. Quantum Electron. QE-18, 1580 (1982)
112. G.I.Stegeman, C.T.Seaton, A.C.Walker and C.N.Ironside, Opt. Commun. 61, 277 (1987)
113. G.I.Stegeman, E.Caglioti, S.Trillo and S.Wabnitz, Opt. Commun. 63, 281 (1987)
114. Y. Chen, J. Opt. Soc. Am. B8, 986 (1991)
115. H.C.Hsieh and P.N.Robson, J. Appl. Phys. 64, 1696 (1988); P.L.Kam Wa, J.E.Sitch, N.J.Mason, J.S.Roberts and P.N. Robson, Electron. Lett. 21, 26 (1985); M.Cada, B.P.Keyworth, J.M.Glinski, A.J.Springthorpe and P. Mandeville, J.Opt. Soc. Am. B5, 462 (1988); R.Jin, C.L.Chuang, H.M.Gibbs, S.W.Koch, J.N.Polky and G.A. Pubanz, Appl. Phys. Lett. 53, 1791 (1988); P.R.Berger, Y.Chen, P.Bhattacharya, J.Pamulapati and G.C. Vezzoli, Appl. Phys. Lett. 52, 1125 (1988)
116. A.Villeneuve, C.C.Yang, P.J.Wigley, G.I.Stegeman, J.S. Aitchinson and C.N.Ironside, Appl. Phys. Lett. 61, 147 (1992); K.Al-hemyari, J.S.Aitchinson, C.N.Ironside, G.T.Kennedy, R. S. Grant and W.Sibbett, Electron. Lett. 28, 1091 (1992)

117. N.Finlayson, W.C.Banyai, E.M.Wright, C.T.Seaton, G.I. Stegeman, T.J.Cullen and C.N.Ironside, Appl. Phys. Lett. 53, 1144 (1988)
118. J.Atai and Y.Chen, J. Appl. Phys. 72, 24 (1992)
119. Y.Silberberg and B.G.Slez, Opt. Lett. 13, 1132 (1988); Y.Silberberg, P.Perlmutter and J.E.Baran, Appl. Phys. Lett 51, 1230 (1987); J.P.Sabini, N.Finlayson, C.T.Seaton and G.I.Stegeman, Appl Phys. Lett. 55, 1176 (1986); P.Li Kam Wa, J.E.Stich, N.J.Mason, J.S.Roberts and P.N.Robson, Electron. Lett. 21, 26 (1985)
120. F. Reynaud and A. Barthelemy, in Guided Wave Nonlinear Optics, B. Ostrowsky and R. Reinish, Eds., Kluwer., 1992, pp.319
121. A.Barthelemy, S.Maneuf and C.Froehly, Opt. Commun. 55, 201 (1985); M. Shalaby and A.Barthelemy, Opt. Lett. 16, 1472 (1991)
122. J.S. Aitchinson, A.M.Weiner, Y.Silberberg, M.K.Oliver, J.L.Jackel, D.E.Leaird, E.M. Vogel and P.W.E.Smith, Opt.Lett. 15, 421 (1990); J.S. Aitchinson, Y.Silberberg, A.M. Weiner, D.E.Leaird, M.K.Oliver, J.L.Jackel, E.M. Vogel and P.W.E.Smith, J.Opt. Soc Am B8, 1290 (1991); S.Maneuf, R.Desailly and C.Frohly, Opt. Commun. 65, 193 (1988)
123. J.S. Aitchinson, A.M. Weiner,Y.Silberberg, D.E.Leaird, M.K.Oliver, J.L.Jackel and P.W.E. Smith, Opt.Lett. 16,15 (1991); M.Shalaby and A.Barthelemy, Opt.Lett. 16, 1472 (1991); M.Shalaby, F.Reynaud and A.Barthelemy, Opt.Lett. 17,778 (1992); M.Bertolotti, F.Garcia, C.Sibilia, R.Horak and H.Bayer, submitted for publication
124. E.M.Wright, D.R.Heatley and G.I.Stegeman, Phys. Report 194 309 (1990)
125. E.M.Wright, G.I.Stegeman, C.T.Seaton, J.V.Moloney and A.D. Boardman, Phys. Rev. A34, 4442 (1986)
126. M.A.Gubbels, E.M.Wright, G.I.Stegeman, C.T.Seaton and J.V. Moloney, J. Opt. Soc. Am. B4, 1837 (1987); D.R.Heatley, E.M.Wright and G.I.Stegeman, Appl.Phys. Lett. 53, 172 (1988); T.K.Gustafson, J.P.E.Taran, P.L.Kelley and R.Y.Chiao, Opt. Commun. 2, 17 (1970); M.Segev, B.Crosignani and A.Yariv, Phys Rev. Lett. 68, 923 (1992)
127. J.M.R.Thomas and J.P.E.Taran, Opt. Commun. 4, 329 (1972)
128. C.Flytzanis, in "Quantum Electronics", H.Rabin and C.L. Tang, Eds., Academic Press, NY, 1975 vol.1 A
129. R.C.Eckardt and J.Reintjes, IEEE J. Quantum Electron. QE- 20, 178 (1984)
130. J.T.Manassah, J. Opt. Soc. Am. B4, 1235 (1987)
131. N.R.Belashenkov, S.V.Gagarskii and M.V.Inochkin, Opt. Spectrosc. 66, 1383 (1989)
132. H.J.Baker, P.C.M.Planken, L.Kuipers and A.Lagendijk, Phys. Rev. A42, 4085 (1990)
133. R. DeSalvo, D.J.Hagan, M. Sheik-Bahae, G.I.Stegeman and E. W. VanStryland, Opt. Lett. 17, 28 (1992)
134. G.I.Stegeman, M.Sheik-Bahae, E.VanStryland and G.Assanto, Opt. Lett. 18, 13 (1993);
135. G.Assanto, G.Stegeman, M.Sheik-Bahae and E.Vanstryland, Appl. Phys. 62, 1323 (1993); D.C.Hutchings, J.S.Aitchinson and C.N. Ironside Opt.Lett.18, 793 (1993);
136. M.L.Sendheimer, Ch.Bosshard, E.W. Van Stryland, G.I.Stegeman and J.D.Bierlein Opt.Lett. 18, 1397 (1993).

Chapter 3

NONLINEAR OPTICAL MATERIALS FOR INTEGRATED OPTICS

C.FLYTZANIS

1. Introduction

Despite progress in nonlinear waveguided optics[1,2] over the last two decades the implementation of related devices in large scale technology is still not satisfactory. This can be traced to several causes, both structural and conceptual, but the most severe cause is the inadequate present performance[3] of nonlinear optical materials. Nature has not been generous with the optical nonlinearities of bulk optical materials; in the case of integrated nonlinear optics the situation is aggravated by additional requirements[1,2] on the materials such as processability, adaptability and interfacing with other materials. These additional requirements are intrinsic to the fabrication of nonlinear integrated devices, which besides efficiently performing the expected nonlinear operation, must be miniaturized, compact, reliable and with precisely reproducible characteristics in large scale production and long term operation. These problems acquire a particular twist because one of the goals of ongoing research and development is the replacement of integrated electronic devices by optical ones. However, progress in electronics has been remarkable over the last decade and seemingly does not show any signs of fatigue; standards and priorities there are well established and a consensus about future targets has been established. This competition with integrated electronics however has guided nonlinear integrated optics towards a development path that does not allow full exploitation of the intrinsic advantages of optics.

Notwithstanding these and other considerations, at the core of the future evolution of integrated nonlinear optics is the performance that the nonlinear optical materials can attain[4]. Below we summarize the present status of nonlinear optical materials[2,4] and the trends that are relevant for future development in integrated optics. In this chapter and in the Tables 1, 2 we succinctly describe the principal representatives of different classes of nonlinear optical materials that hold promise in this field, and discuss the relation between optical nonlinearities and the structural and dynamic characteristics of valence electron distribution.

2. Classification of Nonlinear Optical Effects and Materials

Nonlinear optical effects, irrespective of their order, can be divided into two major classes:

C. Flytzanis - Laboratoire d'Optique Quantique, Ecole Polytechnique, 91128 Palaiseau cédex, France

Table 1. Typical values of second order nonlinear coefficients. *

Cristal	Symmetry	n_0	d_{21}	d_{14}	d_{33}	d_{eff}	C	I_{ob} (GW/cm²) ***	Transm. (μ) ****
					$(10^{-12}$m/V) **				
$LiNbO_3$	3m	2.232	-2.1	0	-27	5.1	70	10	0.35 - 5
$BaBO_4$	3m	1.655	-2.3	0	0	1.9	16	14	0.2 - 2.6
$LiIO_3$	6m	1.857	0	0	4.5	1.8	13	2	0.34 - 4
KDP	$\bar{4}$2m	1.493	0	0.37	0	0.35	1	5	0.18 - 1.8
ADP	$\bar{4}$2m	1.509	0	0.47	0	0.39	1.2	6	0.18 - 1.5
$AgGaGe_2$	$\bar{4}$2m	2.594	0	33	0	28	81	0.3	0.78 - 18
$ZnGeP_2$	$\bar{4}$2m	3.073	0	69	0	70	292	0.05	0.74 - 12
KTP	mm2	1.737	0	0	8.3	3.2	47	15	0.35 - 4.5
$KNbO_3$	m2	2.119	0	0	-19.5	-11	312	7	0.4 - 5.5
				d_{11}					
Urea				12				5	0.2 - 1.4
MAP		1.508	16.8					3	0.5 - 2.5
POM		1.663		9.2				2	0.5 - 1.7
MNA		2.0			168		1000		0.48 - 2.0
NPP					85			0.05	0.48 ~ 2.0

*	Values selected from Ref.17
**	divide each value by $4.2 \cdot 10^{-4}$ to convert to esu
***	optical breakdown
****	transmission range

Table 2. Typical values of third order nonlinear coefficients.

Material	$\chi^{(3)}$ 10^{12} esu
glass	10^{-4}
NaCL	10^{-2}
CS2	1
Si	1
GaAs	10
Ge	10^2
PTS	10^2
SDG* (resonant)	$>10^4$

*   semiconductor doped glasses

1) Frequency preserving effects, such as the optical Kerr effect, that can be used for bistable operation, phase conjugation, soliton formation and other operations on the spatiotemporal profile of coherent light pulses; here one can also include hybrid effects such as electrooptic, magnetooptic and acoustooptic effects that can be used for parametric devices.

2) Frequency shifting effects, such as sum or difference frequency generation processes, which allow one to up or down convert the frequencies of existing light sources.

There is a major reason for this division: the efficiencies[5] of the latter effects are conditioned by the ability to match optical characteristics in the material, such as phase and group velocities, at widely different frequencies. This is a formidable task and drastically restricts the possible choices, and the situation is further complicated by the fact that material characteristics cannot be optimized over the wide frequency range often required in frequency up and down conversion processes; furthermore, constant progress in lasers with tunable frequency output will greatly reduce the need for frequency shifting nonlinear effects.

In contrast, the efficiency of the frequency preserving effects is not affected by such problems, since only frequencies within a very narrow spectral range are involved and the matching conditions can be automatically satisfied; as a matter of fact, to a large extent the interactions here can be assumed local and are in general much easier to analyze.

As a consequence, the nonlinear frequency preserving effects are the ones that are most seriously considered[1,2] for integrated optical devices. They can be either all optical or hybrid (parametric) effects. The all optical nonlinearities[6] essentially involve valence electron motion and are in general weaker than the hybrid ones, where the ionic motion, vibrational, orientational or translational, can set up very large nonlinearities; the situation however is reversed as regards the speed of establishing and erasing these nonlinearities, the electronic motion being much faster than ionic motion. The magnitude and speed of the nonlinearities are essential characteristics in any assessment of materials for nonlinear optical devices and must be properly introduced in any figure of merit.

The magnitude and speed of the nonlinearity, however, are strongly dependent on how close to a material resonance is the operating frequency or a multiple thereof, as this introduces resonant enhancement of the nonlinearity and absorption losses; furthermore, the speed is limited by the related relaxation processes close to a resonance. One may have linear (one photon) or nonlinear (multiphoton) absorption losses which can be related to the imaginary parts of odd order susceptibilities; we stress the fact that there are no absorption losses related to even order nonlinear processes[7]. Thus the amount of linear and/or nonlinear absorption losses is another important criterion for assessing nonlinear optical materials, along with the magnitude and speed of the nonlinearity at the frequency of interest.

Clearly the above criteria are relevant to the extent that the material is not irreversibly modified by the optical radiation. Such modifications can result from photochemical or thermal bond breaking, state selective photochemistry, optical breakdown and many other processes directly related to the interaction of light and matter. In addition, material growth, doping and aging considerations, processability, interfacing and packaging and many other aspects will heavily weight on the final choice, since for a wide class of materials the nonlinear optical figures of merit are of comparable magnitude.

On the basis of the types of cohesive forces that bind the charges and polarizable units together, nonlinear optical materials can be assigned to one of the following classes:
- ionic crystals, essentially oxygen-polyedra based solids;
- covalent crystals, essentially semiconductors;
- molecular crystals, in particular organic and polymeric crystals;
- disordered and amorphous solids, in particular glasses and polymers; and,

- composites and inhomogeneous artificial solids.

We shall succinctly review the main features of some materials representative of these classes. At this stage however it is appropriate to make certain remarks concerning the nonlinear mechanisms and the evaluation criteria for the corresponding nonlinearities.

We recall that for the applications we have in mind the relation between the electric field E and its induced polarization in a material is written in a power series

$$\underline{P} = \underline{\underline{\chi}}^{(1)}\underline{E} + D_2 \underline{\underline{\chi}}^{(2)}\underline{EE} + D_3 \underline{\underline{\chi}}^{(3)}\underline{EEE} + \ldots \qquad (1)$$

where $\chi^{(1)}$, $\chi^{(2)}$, $\chi^{(3)}$... are the linear, second, third ... order susceptibilities respectively, whose expressions can be derived [6] using quantum mechanical perturbation techniques; D_2, D_3 ... are factorials that keep track of the number of combinations of electric field modes that produce a given polarization mode. A rough order of magnitude estimate of $\chi^{(n)}$ can be obtained through

$$\chi^{(n+1)} \approx 1/E_c^{-n} \qquad (2a)$$

where E_c is the effective cohesive field that keeps the polarizable charges or units attached to one another. In the final analysis, all arguments concerning the improvement of nonlinear coefficients amount to appropriately modifying this field.

The mechanisms that underlie the nonlinear polarization terms in Eq. 1 can be qualitatively discussed without making appeal to the detailed quantum mechanical expressions of $\chi^{(2)}$, $\chi^{(3)}$... A classification can be obtained by setting up the expression of the linear susceptibility

$$\chi_{ij}^{(1)}(\omega) = N \sum_{g,e} \Delta\rho_{ge} \left\{ \frac{\langle\mu_i\rangle_{ge}\langle\mu_j\rangle_{eg}}{\hbar(\omega_{eg} - \omega - iT_2^{-1})} + \frac{\langle\mu_j\rangle_{ge}\langle\mu_i\rangle_{eg}}{\hbar(\omega_{eg} + \omega - iT_2^{-1})} \right\} \qquad (2b)$$

where $\Delta\rho_{ge}$ is a population difference for states g and e, and $\hbar\omega_{eg} = E_e - E_g$, $<\mu>_{eg}$ and T_2 are the transition energies, dipole moments and coherence times respectively. In the presence of an intense electric field E all these quantities are perturbed and become E-field dependent and the higher order terms in Eq. 1 can be thought to arise from the modification of $\chi_{ij}^{(1)}(\omega)$ by the electric field.

Thus the population difference $\Delta\rho_{ge}$ can be perturbed by real transitions with energy (photons) supplied by the field, and hence it depends on powers of the beam intensity $I \sim EE^*$ and coherence here is irrelevant; hence the lowest order effect related to this mechanism is cubic in the electric field intensity and is sometimes termed photoinduced. In contrast, the quantities $<\mu>_{eg}$ and E_{eg} in general depend on both even and odd order powers of the field E and coherence here is generally relevant; these mechanisms give rise to both quadratic and cubic terms in the electric field intensity in Eq. 1. This distinction is very important to keep in mind because the decay time of the photoinduced effects that exploit population changes following real transitions is T_1, the longitudinal or population relaxation time, which is much longer than T_2, the transverse or coherence relaxation time, and the exploitation of incoherent processes strongly reduces the operation al speed of the material.

To the extent that in device applications one essentially exploits[1,2,8] the phase shifts induced by the nonlinear terms, the above approach also clearly shows that the relevant parameters are not the bare nonlinear susceptibilities but appropriately renormalized expressions thereof that measure the induced relative phase shift; these are called figures of merit and are also directly related to the efficiency of the nonlinear interaction. Their precise definition is actually closely related[9] to the operation or the device one has in mind.

Thus if we concentrate[1,2] our attention on all optical guided wave switching devices such as the nonlinear directional coupler, nonlinear Bragg reflector, nonlinear Mach-Zehnder interferometer, nonlinear mode sorter or nonlinear X-junction, the relevant figure of merit can be derived by the requirement that the nonlinear phase shift per unit absorption length be larger than 2π. The total phase shift over a length L of the nonlinear material is given by

$$\Delta\phi = (n_0 + n_2 I)k_0 L = \Delta\phi_0 + \Delta\phi_{NL} \tag{3}$$

where n_0 and k_0 are the linear refractive index and wave vector respectively, and n_2 is the optical Kerr effect coefficient to be defined in Section 4. The absorption coefficient is

$$\alpha = \alpha_0 + \alpha_2 I \tag{4}$$

where α_0 is the one photon (linear) loss and α_2 the two photon (nonlinear) loss. The criterion of Eq. 3 implies that

$$\Delta\phi_{NL}/2\pi = n_2 I / \lambda_0 \alpha_0 \geq 1 \tag{5}$$

if one photon absorption is dominant and

$$n_2 / \lambda_0 \alpha_2 > 1 \tag{6}$$

if two photon absorption is dominant. Thus possible figures of merit are $F_0 = n_2/\alpha_0$ and $F_2 = n_2/\alpha_2$ respectively; these considerations pertain to steady-state operation. In actual devices using high repetition rate pulse, one instead uses[10] the figure of merit

$$F_d = \frac{n_2 I}{\alpha_0 c \tau} \tag{7}$$

where τ is the recovery time of the nonlinearity, which in practice is the population relaxation time T_1. One can similarly introduce figures of merit for quadratic nonlinearities. One such figure of merit is

$$C = d_{eff}^2 / n^3$$

where $2 d_{eff} = \chi^{(2)}$ and n is the refractive index. Arguments based on figures of merit should be made with some caution[9] as they may lead to controversies.

We discuss now the different material classes that can have been considered for applications in integrated nonlinear optical devices.

3. Second Order Nonlinearities

There is only one second order frequency preserving effect, namely the electrooptic or Pockels effect [11], which is related to the nonlinear second order polarization:

$$\underline{P}_\omega^{(2)} = 2\underset{=EO}{\chi^{(2)}}(\omega,0)\underline{E}_\omega \underline{E}_0 \tag{8}$$

where ω is an optical frequency within the transparency region of the crystal and "0" is a frequency well below any vibrational or orientational frequency. The second order susceptibility $\chi_{EO}^{(2)}(\omega,0)$ contains[6,4] two contributions: one purely electronic, the same as for optical second order harmonic generation $\chi_E^{(2)}(0,0)$, and another which arises from a rearrangement of the valence electron distribution consecutive to nuclear displacements induced by the static field or

$$\chi_{EO}^{(2)}(\omega,0) = \chi_E^2(0,0) + \chi_Q^{(2)}(0,0) \tag{9}$$

The first term is directly related to the characteristics of the valence electronic charge distribution, while the second is related to the modulation of the latter by the phonons and can be evaluated from infrared and Raman spectroscopic data. Except for the ferroelectrics[11], where the ionic displacements related to the driven soft mode can be substantial, the latter contribution[6] is in general smaller although comparable to the purely electronic one and of either relative sign; the overall signs of $\chi_{EO}^{(2)}$ and $\chi_E^{(2)}$ (or $\chi_Q^{(2)}$) are of certain relevance and must be explicitly stated together with the axis orientation adopted. In the following we will concern ourselves only with the electronic contribution $\chi_E^{(2)}$ (0,0) to the electrooptic coefficient, which is also relevant for second harmonic generation.

In the early days of nonlinear optics it was suggested[12] that an estimate of $\chi_{ijk}^{(2)}$ could be obtained through the conjecture that the coefficient

$$\delta_{ijk}(\omega_1 + \omega_2) = \chi_{ijk}^{(2)}(\omega_1,\omega_2) / \chi_{ii}^{(1)}(\omega_1+\omega_2)\chi_{jj}^{(1)}(\omega_1)\chi_{kk}^{(1)}(\omega_2) \tag{10}$$

is a constant for all crystals. However careful examination[13] of the nonlinear polarization mechanisms and values of $\chi^{(2)}$ for different crystal classes showed that this conjecture does not apply; δ may differ by orders of magnitude among the different crystal classes and does not constitute an useful figure of merit for identifying crystals with large nonlinearity. However, the fact that the spread of δ_{ijk} values is narrower than that of $\chi_{ijk}^{(2)}$ implied[6,13] that among the noncentrosymmetric crystals those with large $\chi^{(1)}$, which implies crystals with covalent bonding, would also have large $\chi^{(2)}$ in general. Actually, the lack of inversion symmetry is more or less connected with the existence of heteropolar covalent bonds, since these naturally introduce a preferred direction and asymmetry in the structure while purely ionic forces favor centrosymmetric crystalline configurations. The question is then to what extent the bond configuration can be optimized for a crystal to produce a large second order nonlinearity; this is the key problem in noncentrosymmetric material research.

Concerning the electrooptic effect, for evident reasons it is convenient to consider the static field E_0 as an external parameter and introduce[8,11] the r-coefficients in the expansion of the optical indicatrix as

$$\Delta\left(\frac{1}{n^2}\right)_i = \sum_j r_{ijk} E_{0k} + \sum_{kl} s_{ijkl} E_{0k} E_{ol} \tag{11}$$

with the relation

$$\chi^{(2)}_{EO_{ijk}}(\omega,0) = n_i^2 n_j^2 r_{ijk} \tag{12}$$

the s-coefficients in [11] are related to the static (DC) Kerr effect or quadratic Pockels effect.

The EO-effect being second order in the electric field amplitudes, the relevant materials must be crystals lacking inversion symmetry or partially ordered structures of asymmetric polarizable units. As previously stated, all such media contain heteropolar covalent bonds or related molecular complexes, and to a good approximation one may assign all their second order nonlinearity to these covalently bonded complexes. Concomitantly one may also assume[6] that the second order polarizability coefficients of these complexes, after certain provisions regarding local field corrections are made, satisfy the additivity and transferability property as was found to be the case for the linear polarizabilities. Accordingly one may write

$$\chi^{(2)}_{ijk} = \frac{1}{v}\beta_{ijk} \tag{13}$$

where v is the volume available to the repeat unit cell and β_{ijk} is the second order polarizability tensor of the polarizable unit, which eventually can be expressed in terms of second order bond polarizabilities and their direction coefficients. This implies that the second order susceptibility is entirely determined by the asymmetric charge distribution within the constituent repeat polarizable unit, and not by charge delocalization across such units. This greatly simplifies the search for materials with large second order nonlinearities on one hand and, on the other, points the road for fabricating new materials starting with molecules that possess large second order polarizabilities. These ideas have been essential in improving the second order nonlinear coefficients of existing materials by judicious modification of the constituent unit, and in conceiving new materials by using appropriate molecular engineering techniques.

3.1. Covalent Crystals. Heteropolar Semiconductors

This is the simplest class of noncentrosymmetric crystals[14]. The building block unit in this case is one heteropolar bond that serves to form the whole crystal through tetrahedral connection. The bonding is sp^3-type and the structure is cubic zinc-blende (e.g., GaAs) or hexagonal wurzite (e.g., CdS); the former is optically isotropic, while the latter is uniaxial and differs from the former by compression or expansion of the bonds along the 111-direction which then becomes the c-axis. The valence electronic structure of these crystals has been extensively studied by various approaches. Using the bond additivity conjecture and making provisions for local field corrections one may express[6,13] their linear and second order optical coefficients in terms of those of the heteropolar bonds namely the linear and second order polarizabilities α and β respectively; one actually may take the bonds to be unidimensional. The values of $\chi^{(2)}$ can be then predicted to high accuracy and related to the charge asymmetry along the heteropolar bond. One can show[15] that for an isoelectronic series of IV-IV, III-V, II-VI and I-VII compounds the III-V compound has the optimal second order nonlinearity; furthermore, d-hybridization of the sp^3-orbitals, (as in CuCl), leads to a reversal of sign of $\chi^{(2)}_E$.

As a general rule $\chi^{(2)}$ is large in these materials but so also is $\varepsilon_\alpha = n^2 = 1 + 4\pi\chi^{(1)}$, so that their figures of merit are not impressively larger than those of other materials we will discuss below. Their most severe drawbacks however are their optical isotropy, which prevents phase matching, and their narrow transparency region that hardly extends beyond 2 eV for any of them. Although the first of these drawbacks can now be compensated by fabricating periodic stacked structures with the polar axis direction periodically reversed from layer to layer, the second one remains a major drawback because real transitions and photocarrier generation can easily take place in these materials, greatly reducing their electrooptic modulation efficiency.

With new fabrication techniques[16,17] that have been introduced in semiconductor technology these materials have surfaced again in the competition for the development of nonlinear integrated optical devices; thus compounds such as InP, which possesses a large electrooptic coefficient and a wide energy gap, are now seriously considered.

3.2. Ionic Crystals. Crystals with oxygen polyhedra

This is the class of materials that presently sets the standard in designing prototype integrated nonlinear devices. These materials can be viewed as ionic crystals[11,14] where one of the ions is replaced by an oxygen polyedron AO_n, in a ionized state, and stabilized in a noncentrosymmetric configuration by the surrounding ions (usually metal ions). The second order optical nonlinearity in this case results from the unbalanced system of heteropolar AO-bonds which has a rather complex configuration around the element A; in contrast to the heteropolar semiconductors, the bonds here cannot be assumed to be unidimensional. Indeed in all models, in order to account for the magnitude and sign of $\chi^{(2)}$, one must introduce[11] for each bond a substantial transverse contribution $\beta_\perp$ for the second order polarizability; this is because the AO bonds are usually counterdirected in pairs and their longitudinal components $\beta_\parallel$ cancel to a large extent. In fact, the additivity of bond polarizabilities has never been convincingly established for these compounds. Preferably, one should treat the entire AO_n polyedron as a single unit; however, this makes the calculation and prediction of the nonlinearities quite cumbersome and subtle, since the modeling of the electronic distribution must be compatible with that expected from the phase transitions that almost all these materials undergo, e.g. para- to ferroelectric.

As a general rule, the values of $\chi_E^{(2)}$ for these materials are lower than those of the heteropolar semiconductors, however most of their other bulk properties are superior to those of the semiconductors. In particular: they are optically anisotropic, allowing for phase matching of nonlinear interactions; their transparency region can extend up to the near ultraviolet; they are robust against photochemical degradation or optical breakdown; and they can be produced in high optical quality. However, because of their complex chemical structure, a superposition of ionic and covalent, they cannot be processed by the same sophisticated fabrication techniques as the semiconductors; furthermore their surfaces, because of their high polarity and the formation of charged layers, are easily attacked by moisture and other degradation and must be properly treated, encapsuled or packaged[2].

After weighing all these factors, at present only $LiNbO_3$ seems to gain the favour of potential manufacturers and actually constitutes the reference[1,2] for comparing and evaluating all other potentially useful materials. However, the progress in growing other oxygen polyhedron based crystals with larger second harmonic efficiency than $LiNbO_3$ has been spectacular[17] during recent years; this progress has led to many applications[18] in ultra short laser pulse technology and may well lead to similar improvements in nonlinear integrated optical devices that exploit the electrooptic effect. The introduction of new crystal growth and fabrication techniques will certainly be needed here in order to reach the standards required for integrated optical devices.

3.3. Molecular Crystals. Organics

Many expectations were placed[19÷22] on noncentrosymmetric molecular crystals, in particular the organic ones, for second order nonlinear effects and a considerable effort has been concentrated the last two decades on identifying such crystals or growing new ones with second order nonlinearities higher than those of $LiNbO_3$. The results have been mixed; some molecular crystals with large values of $\chi^{(2)}$ formed by organic molecules have now been grown, but their other characteristics fall short of all expectations and their potential implementation in nonlinear integrated optical devices seems problematic and still open to debate. The major limitation is their photochemical instability and aging against long term exposure to intense light pulses at high repetition rate, as is encountered in most applications.

All the advantages, but also all the disadvantages, of molecular organic crystals compared with covalent and ionic crystals for applications in nonlinear integrated optics can be traced to the difference between inter- and intramolecular forces that prevail in these systems. As a consequence, the constituent molecules preserve their main characteristics even when assembled to form the molecular crystal. Indeed, in contrast to covalent and ionic crystals, where inter-and intra-cell forces are similar and of equal strength, in molecular crystals the intramolecular forces are covalent directive and already saturate with the formation of the molecule while the intermolecular forces are essentially unsaturated van der Waals or dipole-dipole interactions which are more than one order of magnitude weaker than the intramolecular ones and do not significantly alter the intramolecular electron distribution. In several cases other forces such as hydrogen bonding[19], intermediate in strength between the previous two, are operative.

Thus the process of search for and growth of noncentrosymmetric molecular crystals with large second order nonlinearities is at first sight greatly facilitated by these facts, since the search is reduced to that of the identification and synthesis of stable asymmetric molecules with large second order polarizabilities β and wide transparency range between the highest vibrational and lowest electronic transitions. The latter requirement is very restrictive for organic molecules and excludes the majority of them for further consideration in nonlinear optics. Among the remaining organic candidates a large proportion is also excluded because they cannot form stable noncentrosymmetric crystals. This is mainly because asymmetric molecules most frequently carry a dipole moment in their ground electronic state, and in order to reduce the dipole-dipole interaction, which is dominant over the van der Waals in the lattice, a head- to-tail antiparallel configuration will be most frequently favored when forming the crystal, resulting in centrosymmetric crystalline structures and consequently vanishing $\chi_E^{(2)}$ unless certain precautions[19] are taken to prevent this to happen. There are several features that can be exploited in this respect:
- poling in an externally applied field;
- synthesis of chiral molecules;
- synthesis of asymmetric molecules with vanishing dipole moment in ground electronic state;
- hydrogen bonding.

The first of these is not very favorable for crystalline structures but has been successfully applied and exploited in polymers[23,25] as will be discussed below. The other three, on the other hand, have been successfully used for growth of good optical quality asymmetrie crystals with large second order susceptibilities. The most representative crystal cases are MAP (methyl-(2,4-dinitrophenyl)-aminopropanoate), POM (3- methyl-4-nitropyridine-1-oxide) NPP (N-(4-nitrophenyl)-(L)-prolinol) and urea $(CO(NH_2)_2)$.

Because of the wide difference between inter- and intramolecular forces in molecular crystals one may safely assume that the molecular polarizability coefficients α and β satisfy[6,19] the additivity property after provisions are made to take into account local field corrections. For most organic molecules the dominant contribution to β comes from an asymmetrie part of the electronic distribution which can be modelled with a quantum two-level system. The expression of β for a quantum system consisting of two levels g and e of energies E_g and E_e and permanent electronic dipole moment μ_{gg} and μ_{ee} respectively is

$$\beta = \frac{3\Delta\mu\, \mu_{eg}^2}{\Delta E} \qquad (14)$$

where $\Delta\mu = \mu_{gg} - \mu_{ee}$, $\Delta E = E_g - E_e$, and μ_{eg} is the transition dipole moment between states g and e. The coefficient β is a third rank tensor and clearly its maximum value is obtained for directions where $\Delta\mu$ and μ_{eg} are maximal but not necessarily collinear. For most organic molecules the most favorable case occurs when a charge transfer complex is formed[19] and this can be modelled with a two level system involving a nonbonding π-donor orbital and a vacant π-antibonding acceptor.

The important point to notice in Eq. 14 is that β depends only on the difference $\Delta\mu$ and not on the individual dipole moments μ_{gg} and μ_{ee}; in particular, μ_{gg} can be zero and yet have $\beta \neq 0$ as long as $\mu_{ee} \neq 0$. Accordingly, one can have molecules with large β but vanishing μ_{gg} in which case dipole-dipole interactions in the lattice are weak and cannot prevent the noncentrosymmetric lattice configuration from being formed. Such is the case of POM [26], and good quality large crystals for this compound have been grown and used in second harmonic generation.

This situation however is not common, and one must also deal with molecules having large β- values but also appreciable μ_{gg} -values; the remedy then is either to render the molecule chiral by attaching an appropriate molecular unit, either left- or right- handed, in which case the lattice structure can never be centrosymmetric (this is the case for MAP [27] and for MNA [28]), or one can judiciously introduce hydrogen bonding between the molecules to overcome the dipoledipole forces and combine with chirality to favor noncentrosymmetric structures as in the case of NPP [29]. Presumably this situation prevails also in urea [30]. In all three cases very good quality crystals with large $\chi_E^{(2)}$ have been grown and used in several applications; urea seems to be the most stable of them. Clearly these approaches can be used for a wide range of organic materials.

Despite a formidable multidisciplinary effort [19+22], the bulk organic and molecular crystals, with the possible exception of urea, have not reached yet the required standards for applications in nonlinear optics and much less so in integrated optics. The reasons are several and can be traced back, as was previously pointed out, to the weakness of the intermolecular forces with respect to the intramolecular ones and to the sensitivity of organic molecules to light. As a consequence, the organic crystals have much larger defect densities than the covalent or ionic crystals; moreover, their mechanical and optical resistance is weaker than in the inorganic crystals. A second issue is even more serious, as all organic molecules easily undergo bond breaking, bond arrangement or other photochemical reactions once the photon energy or a multiple thereof is close to a photosensitive transition. In addition, organic crystals are subject to aging even without exposure to light. Other serious drawbacks regarding the use of organic crystals in nonlinear integrated optical devices are the difficulties encountered in doping, processing, polishing or interfacing them with techniques currently used for semiconductors, ionic and other inorganic crystals.

3.4. Disordered Oriented Media

Crystallinity is not an imperative requirement [20,23,24] for second order effects to occur in a material. It is sufficient for the constituent asymmetric molecules to be oriented on the average so that the medium macroscopically lacks inversion symmetry. This implies that the requirement of translational periodicity is irrelevant, so long as the molecules are uniformly distributed but with their axis pointing on the average in the same direction. This can also occur in an otherwise centrosymmetric medium, if one uniformly induces oriented noncentrosymmetric molecular complexes. These two approaches have been successfully put into practice in the case of poled polymers [20,23,24,25] and in certain glasses [31,32]. While the first class [23,24,25] was the outcome of a well thought strategy that led to materials and prototype devices that compete or even surpass those based on $LiNbO_3$, the second class was an unexpected [31,32] finding and the origin of the second harmonic generation there is still not well understood and accounted for by the different models that have been proposed. We recall that solid polymers [2] and glasses can be prepared with excellent optical quality and all show a phase transition around a temperature T_g to a plastic phase where collective flexibility and molecular motion at large distances can take place; below this temperature, the medium is rigid and only slight local motion is allowed as in a crystal. T_g can be very high, well above room temperature.

3.4.1. Poled Polymers.
Here organic molecules with large charge transfer type polarizability β are introduced and uniformly dispersed [25] into an excellent optical quality amorphous organic polymer. The solution of the polymer and guest molecules is then evaporated by spin coating, film casting or other techniques; at this stage the guest molecules are randomly oriented. They are oriented by poling, namely by applying a strong static electric field (electrode poling) or corona discharge (corona poling) at a temperature T close to T_g; by proper choice of the polymer T_g can be very high, much higher than the extreme temperatures that a transparent nonlinear material will be expected to operate in practical applications. Several other precautions allow one to obtain excellent optical quality poled polymers, photoresistant and long aging. Concerning the guest molecule relation to the polymer several situations are possible [25,33]:
- guest-host system: here the nonlinear molecule is not attached to the polymer chain and thus can freely rotate close to T_g;
- side-chain (co) polymers: here one end of the nonlinear molecule is chemically reacting and is covalently attached or grafted onto the polymer chain, but its other end is free;
- cross-linked polymers: here both ends of the nonlinear molecule chemically react and are covalently attached to the polymer chains; and,
- main-chain systems: here the guest molecule is inserted in the polymer chain.

The poling efficiency depends on several factors, but the most important one is the ratio $\mu_{gg}E/kT$ where E is the poling electric field; corona and electrode poling each have their own advantages and disadvantages. The techniques are well understood and controlled, and allow one to obtain poled polymers with large orientational order for the guest molecules, namely large $<\cos^3\theta>$ where θ is the angle between the guest molecule axis and the poling field E direction.

Poled polymers exhibit [25,33] substantial advantages over all crystalline materials regarding most requirements, nonlinearity, optical quality, processability, interfacing, phase matching etc..., except for one: their stability against aging is not yet satisfactorily solved. Clearly this is connected to the degree to which T_g exceeds the highest temperature that the material reaches under practical operation. The best results have been obtained with cross-linked polymers, in which aging can be retarded to any desirable length of time, but more recently equally good results were also obtained with side-chain polymers. The aging manifests itself as a reduction in $<\cos^3\theta>$, that is, the guest molecules lose their preferential orientation and $\chi^{(2)} \sim N\beta<\cos^3\theta>$ decreases. Such a decrease of $\chi^{(2)}$ always occurs immediately after poling is interrupted and the material is cooled to the room or the operating temperature; however, after a few hours or days, the average orientation parameter $<\cos^3\theta>$ and $\chi^{(2)}$ both stabilize to values that are well above those of most crystalline noncentrosymmetric materials inclusive of $LiNbO_3$, and can remain so for years with a judicious choice of the guest molecule and the polymer.

Another problem that recently surfaced in connection with the implementation of poled polymers in electrooptic waveguided devices is the occurrence [34] of a photorefractive effect which is not well understood and clearly sets limitations for long term operation since it implies the existence of charged molecular defects.

Notwithstanding these and some other drawbacks, poled polymers presently constitute [25] the most serious and advantaged contenders in the competition and development of nonlinear integrated devices.

3.4.2. Glasses. Glasses and in particular fused silica have now reached outstanding optical quality, more than any other optical material, as their introduction in modern technology becomes increasingly sophisticated and irreversible. These materials were always expected to be centrosymmetric macroscopically and the observation [31] of second harmonic generation in Si-Ge glass fibers, as well as in other glassy media, came as a surprise and aroused immense interest and expectations [35]. Despite the substantial and understandable effort in delineating the mechanisms that provoke this effect, its understanding is still not satisfactory and falls short of our capability to control and exploit it. The proposed models, such as electric-field induced nonlinearities (the static electric field being provided by a third order rectification process), or a photovoltaic effect based on the interference between the fundamental and harmonic field, or other mechanisms, do not simultaneously account for all observations related to this effect.

Often a glass fiber of a particular composition, for instance Si-Ge, will initially not show second order generation; however, after irradiation with a laser beam, within a certain time lapse and over a certain distance that never exceeds a few tens of cms, evolves to a fiber that subsequently allows instantaneous and highly efficient second harmonic generation at the same wavelength as was used in its preparation. This occurs because of a built in photoinduced spatial modulation, along with noncentrosymmetricity, that allows phase matching at a given wavelength. The SH conversion efficiency can be very large (more than 20 %). It has not been proved yet that the same fiber can also be used for efficient electrooptic modulation or other applications related to the electrooptic effect.

More recently it has been shown [32] that a large second order nonlinearity, less than but still comparable to that of $LiNbO_3$, can be induced within a layer near the surface region of commercial fused silica optical plates at a temperature close to the Orbach temperature by poling in the presence of a static field; the layer can be of the order of a few µm, namely of the order of or larger than typical optical wavelengths.

4. Third Order Nonlinearities

The main frequency preserving third order effect, is the optical Kerr effect which is related to the third order polarization [5,6]

$$\underline{P}_\omega^{(3)} = 3\underline{\underline{\chi}}^{(3)}(\omega, -\omega', \omega') |\underline{E}_{\omega'}|^2 \underline{E}_\omega \qquad (15)$$

where ω and ω' can both be optical frequencies; however, ω' can also be zero, in which case the effect reduces to the static or second order Pockels effect also defined in Eq. 11. To the extent that the term of Eq. 15 can be compounded with the linear polarization term induced at frequency ω

$$\underline{P}_\omega^{(1)} = \underline{\underline{\chi}}^{(1)}(\omega) \underline{E}_\omega \qquad (16)$$

one can also describe the optical Kerr effect as a photoinduced change of the refraction or absorption indices of the medium. By writing

$$\underline{P}_\omega = \underline{P}_\omega^{(1)} + \underline{P}_\omega^{(3)} = \tilde{\chi}(\omega, I')\underline{E}_\omega \qquad (17)$$

where I' is the light intensity at frequency ω', one obtains the intensity dependent indices

$$\tilde{n}(I) = n_0 + n_2 I \qquad (18)$$

$$\tilde{\alpha}(I) = \alpha_0 + \alpha_2 I \qquad (19)$$

for refraction and absorption respectively, where

$$n_2(\omega) = 3\chi^{(3)}(\omega, -\omega', \omega')/n_0^2 \qquad (20)$$

and similarly for α_2. There have been attempts to relate α_2 and n_2 by a Kramers-Kronig type relation analogous to the one that relates α_0 to n_0. In contrast to the α_0-n_0 relation, which is a rigorous one, the α_2-n_2 is not and in fact is incorrect in a strict sense; however, in some simple models a rough agreement has been obtained.

There are several mechanisms [6] that can lead to photoinduced changes of the refraction or absorption: electronic, vibrational, orientational and translational. With the exception of the first one, all these involve ionic or molecular motion and can be substantial in magnitude; however, when all factors are taken into consideration, these nonlinear effects are not exploitable in applications. Hence the following discussion will only concern electronic cubic nonlinearities.

In line with the introductory discussion of Section 2, electron delocalization over several identical units has been identified [19,36] as the principal material feature that markedly influences the magnitude of the cubic nonlinearities; this is in contrast to the case of quadratic nonlinearities, where charge asymmetry within a single unit is the material feature that most matters. The main implications are that the structure must be periodic at least over the electron delocalization length; polarizability additivity, which is a real space property, is irrelevant for materials with large cubic nonlinearities such as the inorganic semiconductors and the linear conjugated polymers. Instead, in such materials where the electron states are delocalized band states, one can establish [37] an additivity in k-space which involves the contributions to $\chi^{(3)}$ from a small number of critical points in the joint density of states. This allows one to derive scaling laws that strongly depend on the electron charge dimensionality. In such materials with very delocalized electrons, the cubic nonlinearities [38] can be sensitive to many body effects such as the Fermi exclusion principle or charge screening.

It has been suggested [6,41,42] that third order nonlinear processes can also take place through two cascading second order processes if the material is noncentrosymmetric. Here the electric field generated by a second order polarization, induced by any pair of incident beams inside the material, interacts with one of the incident beams to create a second order polarization which however is effectively equivalent to a third order one and can be cast in the form [15] and combined with this term; the corresponding effective third order nonlinearity, which is now nonlocal and depends on the wavelengths, can be comparable or larger than the intrinsic cubic nonlinearity if phase matching is achieved in the intermediate second order process.

In general the values of $\chi^{(3)}$ $(\omega,-\omega',\omega')$ due to electrons, when ω and ω' are in the transparency region of the material, are too weak for applications involving interactions that occur over a short distance, as is the case in miniaturized single pass integrated devices. The only alternative is to resonantly enhance the nonlinearities. The use of resonances however introduces two drawbacks which in general can be very severe: linear and/or nonlinear absorption losses on one hand, and a long nonlinearity recovery time on the other. One can resonantly enhance $\chi^{(3)}$ $(\omega,-\omega',\omega')$ by having ω or ω' (or both) close to dipole allowed resonances of the material; here the absorption losses are linear in the light intensity and nonlinear at high intensities (saturation effect). One can also enhance $\chi^{(3)}$ $(\omega,-\omega',\omega')$ by having $\omega + \omega'$ (or $\omega - \omega'$) close [6] to a two-photon (or Raman) allowed resonance without any of the frequencies ω or ω' being close to a resonance; here the losses are nonlinear in the light intensity.

These drawbacks related to the resonant enhancement of the cubic nonlinearities can be brought under control to some extent by exploiting [10,39,43] morphological resonances whose position, width, and oscillator strength can be externally controlled and tailored to meet certain requirements related to the desired performances of the device. Such morphological resonances can be introduced through confinement of the delocalized electrons, for instance by quantum and dielectric confinements. The new engineered material fabrication techniques allow [16,17] one to control the interfacing and confinement to better than an atomic layer. Several classes of such artificial semiconductor based microstructures are now being studied for this purpose and in the case of quantum wells these studies have reached a high degree of sophistication. For applications, however, the artificial materials that emerge as the most promising [10,43] are composites obtained by uniformly dispersing semiconductor or even metal nanocrystals in a glass or in a transparent polymer matrix of high optical quality.

Based on the above general remarks, the materials that presently show promise in devices that exploit cubic nonlinearities are:
- covalent semiconductors,
- composite materials,
- linear conjugated polymers,
- noncentrosymmetric media with large $\chi^{(2)}$

We shall briefly discuss the principal nonlinear mechanisms that are responsible for large cubic nonlinearities and how they relate to electron distribution and dynamics.

4.1. Semiconductors

Covalent semiconductors such as Ge or GaAs possess [44] the largest nonresonant cubic nonlinearities among all known crystals but these are unexploitable for several reasons, the most severe being the low absorption threshold in these compounds. On the other hand, large and exploitable cubic optical nonlinearities can be produced [38,45,46] by generating a finite concentration of electron-hole pairs by resonant excitation above the band gap.

These large resonant nonlinearities have their origin in the band filling mechanism whereby the photocreated electrons and holes quickly thermalize and fill all states in the bottom of the conduction and the top of the valence bands respectively up to levels that depend on the light intensity and pulse or recombination times, thus excluding these states from further occupation because of the Fermi principle. This blocking mechanism [38,46] appears as a repulsion of the states on either side of the forbidden energy gap, or equivalently as a blue shift in the absorption threshold; at very high intensities this mechanism leads to a saturation of the nonlinearity.

This band filling mechanism is not the sole cause of nonlinearities; it is accompanied [38,39] by additional many-electron effects, such as exciton formation and band renormalization, that have an additional effect on the global nonlinearity. These additional effects have received a lot of attention, but their impact on device applications is overestimated and to some extent unreliable; in particular their relative contribution to the total nonlinearity is greatly reduced at room temperature, compared with that of band filling.

As previously stated, the major problems with resonant nonlinearities, and in particular those in semiconductors, are the concomitant absorption losses and the long times required for the photoinduced nonlinearities to decay. These decay times are related to the evolution of the electron and hole populations. These times can be modified to some extent by appropriate doping or through quantum confinement by using the sophisticated but well established fabrication techniques of engineered materials.

4.2. Composites. Semiconductor and Metal Nanocrystals in Glasses

One way to enhance the cubic nonlinearities of materials with very delocalized electrons, such as metals, semiconductors, or conjugated polymers, is to artificially confine the valence electrons in regions much shorter than their natural delocalization length in the bulk, which extends over many unit cells or even to infinity. One conspicuous feature of this artificial confinement is the appearance[10,39,43] of broad but discrete optical resonances whose position, oscillator strength, and dynamics depend on the extent of the artificial confinement and hence can be modified to meet certain requirements. These morphology[43] related resonances result from two types of confinement: quantum and dielectric. The first type prevails in semiconductor nanocrystals, while the second one dominates in metal nanocrystals.

The dielectric confinement depends[47,43,10] on the difference in dielectric constants between the crystallites and their surrounding transparent medium. Because of this dielectric inhomogeneity, the electric field $E_{\omega\ell}$ that effectively polarizes the charges in these crystallites can be substantially different from the macroscopic Maxwell field E_ω in the composite. Under certain simplifying conditions one finds that the relation between these two fields is

$$E_{\omega\ell} = \frac{3\varepsilon_0}{\varepsilon_m(\omega) + 2\varepsilon_0} E_\omega \equiv f_\ell(\omega) E_\omega \tag{21}$$

and the effective cubic susceptibility of the composite is[47]

$$\tilde{\chi}^{(3)}(\omega) = p|f_\ell(\omega)|^2 f_\ell(\omega)^2 \chi^{(3)}(\omega) \tag{22}$$

for spherical particles of volume concentration p<<1, dielectric constant $\varepsilon_m(\omega)$ and cubic susceptibility $\chi^{(3)}$ embedded in a transparent dielectric of dielectric constant ε_0. To the extent that only $\varepsilon_m(\omega)$ is complex and frequency dependent in the optical range, the field E_ℓ is enhanced close to the resonance ω_s, such that

$$\text{Re}\,\varepsilon_m(\omega_s) + 2\varepsilon_0 = 0 \tag{23}$$

which is the condition for the surface plasmon frequency; the cubic nonlinearity is enhanced by the fourth power of the same resonance. The position of the resonance is controlled by modifying ε_0, while its width depends on the size of the crystallites.

Careful studies[48,10] of the nonlinearities of composites formed by uniform dispersion of gold particles of different average sizes in glass matrices revealed that quantum confinement is irrelevant in these compounds and that the nonlinearity $\chi^{(3)}$ of the crystallites results essentially from saturation of the interband transition and the rearrangement of the hot conduction electron population.

The quantum confinement occurs[49,43] when the electron and hole envelopes are restricted within a region of extension L equal to or smaller than the electron and hole Bohr radii, a_e and a_h respectively; the latter are defined in the bulk by the condition that the average value of the electron or hole kinetic energy roughly equals that of the potential. The confinement perturbs this balance, since these energies now vary as $(a_c/L)^2$ and a_c/L respectively, where $a_c = a_e, a_h$. The characteristic energy of the confinement is the kinetic energy

$$E_c = \frac{1}{2}\frac{e^2}{a_c}\left(\frac{a_c}{L}\right)^2 \tag{24}$$

and as L decreases E_c increases and gradually suppresses the effect of the other interactions.

Otherwise stated, the charges behave as free within the confinement region, which to a good approximation can be considered as a spherical quantum well of infinite potential height. The main consequence is that the initially continuous energy spectrum is replaced[49] by a discrete one with a spacing that varies as in Eq. 24; the widths of these resonances as well as their oscillator strengths can dependent on L. Because of the selection rules that prevail for the transitions between these quantum confined states, each crystallite essentially behaves as a quantum two level system and the principal nonlinearity results[50] from the saturation of these transitions. The interface and impurity states, however, can drastically modify this behavior. In this respect, the technique employed to fabricate these composites plays a crucial role since it allows one to control these states.

These materials are formed[10] by uniformly dispersing semiconductor nanocrystals such as CdS_xSe_{1-x} in a glass where x varies from 1 to 0, but other II-VI and I-VIII compounds as well can also be used. The technique is a more or less thermally or chemically controlled precipitation in transparent solid matrices such as glasses or polymers. It evolves in two steps; the nucleation and striking stages. In the nucleation stage one first forms widely supersaturated glass melts with uniformly dispersed semiconductor clusters; at this stage, the resulting batch material is as transparent as the initial undoped material, the glass on the polymer. In the ensuing striking stage, which occurs close to the Orbach temperature, these clusters grow to crystallites by coalescence; that is, larger clusters grow further at the expense of smaller ones, and reach sizes where volume properties overtake surface properties and solid-state-like features are acquired. Several analytical techniques have revealed that these crystallites beyond the nucleation stage have the same structural features as bulk crystals. The size distribution can to some extent be controlled by the temperature and duration of the striking process, and this also fixes the color of the doped glass. It has been found that the optical properties of these materials are strongly affected by a photodarkening process that takes place when they are exposed to high fluences[51]. More sophisticated growth techniques such as growth in zeolites, on polymers or on colloids, together with controlled doping, allow one to produce composites with narrower size distribution and improved characterization of the interface. Similar techniques are used to obtain metal nanocrystals uniformly dispersed in glasses or polymers.

Although the nonlinearities of these artificial composites and their response times can be tailored to arbitrary values by varying several parameters during the fabrication process, their figures of merit are not significantly improved over those that prevail in bulk semiconductors. However, many of their other properties such as robustness, interfacing and processability are superior to those of bulk semiconductors, and this has led to intensive studies to use composites in device applications in integrated nonlinear optics; with narrower size distribution of the crystallites, improvement in the doping process and interface characterization, these materials will become the leading candidates for applications.

4.3. Linear Conjugated Polymers

There is a large class of conjugated polymers that can be obtained by several stereochemical processes. The most studied type is the polydiacetylene class[52,20,21]. Here one initially forms good quality crystals with diacetylene monomers $R_1 - C \equiv C - C \equiv C - R_2$, where R_1 and R_2 are appropriately chosen radicals; then one polymerizes them by heat, mechanical shear force or UV radiation and obtains conjugated polymer crystals, which however contain defects of several types irrespective of the fabrication technique employed. The obtained crystals are photochemically rather stable and of good optical quality but occur only in small sizes and still have poor mechanical properties.

The nonresonant cubic nonlinearity along the chain direction in these compounds is among the largest known[53], and comparable to that of bulk semiconductors such as Ge or GaAs; in the direction transverse to the chain direction $\chi^{(3)}$ is two or three orders of magnitude lower. The magnitude and anisotropy of $\chi^{(3)}$ is due to the extensive delocalization that the valence π electrons undergo along the chain direction. Visualizing these chains as unidimensional semiconductors and using a simple tight-binding model to describe[36] their electron states, one can show that

$$\chi^{(3)} \sim \left(\frac{E_F}{E_g}\right)^6 \tag{25}$$

while

$$\chi^{(1)} \sim \left(\frac{E_F}{E_g}\right)^2 \tag{26}$$

where E_F is the Fermi level and E_g the energy gap; the parameter E_F/E_g plays a very important role, as it measures the optical delocalization length and also the minimum chain length beyond which the nonlinearities are insensitive to the chain length. This is certainly important, since these chains are never infinite even in the best quality crystals. The power behavior of the nonlinearity [25] is intrinsicly related[37] to the unidimensional character of the electron distribution, and is also reproduced in more sophisticated descriptions of the electron states. On the other hand conjugation defects, such as Pople-Wamsley or solitons, seem irrelevant to the magnitude of the nonresonant cubic nonlinearities, contrary to certain claims which were never substantiated. As in the case of bulk semiconductors, the nonresonant cubic nonlinearities of the conjugated polymers are not exploitable, mainly because of the inconvenient transparency range of these compounds.

The origin of the resonant cubic nonlinearity in these one dimensional organic semiconductors is still a question of debate[40,54] as is also the origin and width of the main absorption peak. This controversy is related to the uncertainties that still persist, regarding the assignment of the levels in these conjugated chains and the inclusion of the delocalized π-electron correlation effects; another uncertainty is the strong electron-phonon or vibronic coupling in these chains, which introduces a polaron-like character in the electron states. Both these features lead to complicated state configurations and nonlinear mechanisms that are only partially verified by experimental findings. Thus state-state interaction mediated by phonons, similar to inverse Raman scattering, has been claimed[54] to be the origin of the resonant cubic nonlinearity. Whatever the mechanism may be, these π-electron conjugated polymers are photochemically instable[40] and their use in device applications becomes rather questionable.

4.4. Cascading of Second-order Nonlinearities

In noncentrosymmetric media besides the intrinsic contribution the cubic nonlinearity also contains a contribution that results from cascading of second-order nonlinearities. This is actually a retardation effect[6] that leads to a nonlocal or wave vector dependent nonlinearity. Its origin was briefly described in the beginning of this section. Thus in the case of the optical Kerr effect in a noncentrosymmetric material, the third order polarization at frequency ω besides the intrinsic contribution of Eq. 15 also contains two additional contributions: 1) a contribution that results from a second order polarization induced by a field of frequency -ω and another at 2ω, the latter being the field generated by a second order polarization at frequency 2ω induced by the field at frequency ω; and 2) a contribution that results from a Pockels effect induced by a static electric field generated by an optical rectification process. One can easily show that all these contributions lumped together lead to [6] an effective third order polarization of the same form as in Eq. 15 but with a nonlinear susceptibility that now depends on the wave vector mismatch at the intermediate second order process.

These cubic nonlinearities can be equal to or larger [6,42] than the intrinsic ones in the nonresonant regime, and are related to coherent processes. All noncentrosymmetric materials discussed in relation to the second order nonlinearities can be used here, and the advantage with respect to the centrosymmetric materials is evident: they benefit from the more developed material technology related to noncentrosymmetric materials with large second order nonlinearities, and in addition the same class of materials can be used for quadratic and cubic nonlinearity based devices.

5. Hybrid Nonlinearites

Besides applications that exploit the photoinduced or all optical modifications of the optical characteristics of a material, a whole class of other or similar applications in integrated nonlinear optics can be envisaged that exploit optical parametric effects, where the modifications of the optical characteristics of the material are provoked by an external agent. Here we have in mind modifications and modulations of the characteristics mediated[5] through electrooptic, acoustooptic or magnetooptic coupling. The first two cases and their related devices and materials are well documented in the literature but much less so the magnetooptic case; we briefly discuss here certain aspects of interest for integrated devices, and the related materials.

Nonlinear magnetooptical effects can be treated within the framework of nonlinear optics in general, but in line with the criteria stated in this section we shall only consider the principal frequency preserving effects, namely photoinduced Faraday rotation [55,56] and gyrotropy. These effects allow a very efficient photoinduced control of the polarization state of an optical field and the development of nonreciprocal optical devices such as optical valves and others.

The photoinduced Faraday effect originates from the combined effect of Faraday rotation and the optical Kerr effect. In an isotropic medium in the presence of a static magnetic field H the two eigenmodes of frequency ω in the direction of H are the left and right circularly polarized waves with indices n_- and n_+ respectively; through the optical Kerr effect the latter become intensity dependent for high light intensities. Accordingly, the polarization direction of a linearly polarized input wave E of frequency ω, after propagation through a length L in such a medium collinear with H, rotates through an angle

$$\tilde{\theta}_F = \theta_F + \Delta\theta_{NL}(I) \tag{27}$$

where θ_F is usual linear Faraday rotation angle $\theta_F = \omega L(n_- - n_+)/2c$; and $\Delta\theta_{NL}(I)$ is the photoinduced change of the latter, and is proportional to the difference of the optical Kerr coefficients for left and right circular polarizations.

These photoinduced changes are in general very weak, even in materials with large optical Kerr effect coefficients, such as the semiconductors, unless certain provisions are made so that the difference of the optical Kerr coefficients for left and right polarizations is large. This can be done in II-VI semiconductors like CdSe or CdTe doped with magnetic impurities, the so-called [57] semimagnetic semiconductors. Through the spin exchange interaction of the band and impurity electrons, the Lande factor of the band electrons is enhanced by almost two orders of magnitude, as is the magnetooptical coupling and the Zeeman splitting. Otherwise stated, the magnetic impurities act as "local amplifiers" of the static magnetic field. Without optimizing the interaction configuration, giant photoinduced Faraday rotations exceeding 90° degrees have been measured in these compounds under resonant conditions near liquid nitrogen temperature and in magnetic fields of the order of one half Tesla; this performance can be substantially improved to meet device requirements.

Similar effects are expected without the presence of a magnetic field, in media with rotatory (gyratory) power; here one has a wide choice of organic materials but also their severe problems such as photochemical stability. At present, photoinduced natural gyrotropy has been studied in a couple of isolated cases [58] but the measured rotations were not sufficiently large for application.

6. General Remarks

It would be an understatement to say that there is yet no consensus regarding the class of nonlinear optical materials that will be used in integrated nonlinear optics. It is a fact that the choice of such materials cannot rest only on the figures of merit for nonlinear operation, but must also take into account many other criteria crucial for the design, production and maintenance of the nonlinear devices. Actually, in view of the rather limited nonlinear optical performance of available materials, these other criteria will be the most crucial ones in making a choice. And this choice must be made with less ambitious goals than one would wish or has wished as our understanding of nonlinear integrated optics has been progressing.

There are now certain materials that can be used not only to conceive prototype nonlinear optical devices, but also to proceed to their massive development and production and their insertion into large scale technology. These materials and devices exploit quadratic (second order) optical processes. $LiNbO_3$ and poled polymers seem to provide a good starting point for such development. It is doubtful whether organic crystals can be used at this time for such practical application. There are still severe problems associated with their stability, long term optical quality, etc.

The situation is less bright for materials and devices that exploit cubic (third order) optical processes. Either these nonlinearities are weak, or they can be made large but with a high price paid in absorption losses and reduced operation speed. At present, only the composites offer some limited possibilities, but their fabrication technique must be substantially improved. The advantage of these materials is that one can artificially modify several of their properties by external control. Another possibility is to exploit cascading processes, but here we need a better control of the different factors that affect these processes. The advantage with these processes is that one can use the same class of materials in devices that exploit quadratic and cubic nonlinear optical effects.

However, one can still ask whether there is room for improvement in the figures merit, by an amount that will allow one to use much lower beam intensities and thus avoid some of the problems related to material stability under high fluences. This seems a difficult task, because nonlinearities are associated with very loosely bound electrons and hence the material is apt to undergo irreversible photochemical changes. The relation between nonlinearity and photochemical stability is still not understood, and much work remains to be done there.

References

1. See, for instance, "Guided Wave Nonlinear Optics." D.B. Ostrowsky and R. Reinisch, Eds, NATO ASI Series, Kluwer Publ., Dordrecht 1992
2. See, for instance, "Nonlinear Optical Materials and Devices and their Applications in Information Technology", A. Miller, B. Daino and K. Welford, Eds, NATO ASI Series, Kluwer Publ., Dordrecht 1993 (to appear)
3. See, for instance, "Nonlinear Optics; Materials and Devices", C. Flytzanis and J.L. Oudar, Eds, Springer Verlag, Berlin 1986
4. See, for instance, "Nonlinear Optical Materials Principles and Applications", V. Degiorgio and C. Flytzanis, Eds, North Holland Co. Amsterdam 1994
5. See, for instance, Y.R. Shen, "Principles of Nonlinear Optics", J. Wiley, New York 1984
6. C. Flytzanis in "Quantum Electronics: a Treatise", Vol. Ia, H. Rabin and C.L. Tang, Eds, Academic Press, New York 1975
7. See, for instance, C. Flytzanis in Ref. 2
8. See, for instance, A. Yariv, "Introduction to Optical Electronics", Holt Rinehart and Winston, New York 1971
9. See, for instance, G. Stegeman in Ref. 2 or G. Assanto in Ref. 4
10. C. Flytzanis, F. Hache, M.C. Klein, D. Ricard and Ph. Roussignol, in "Progress in Optics", Vol. XXIX, E. Wolf, Ed, Elsevier, Amsterdam 1991, p. 321-411 See also D. Ricard in Ref. 4

11. See, for instance, "Principles and Applications of Ferroelectrics and Related Materials", M.E. Lines and A.M. Glass, Clarendon Press, Oxford 1979
12. D. Miller, Appl. Phys. Lett. 5, 17 (1964)
13. C. Flytzanis and J. Ducuing, Phys. Rev. 178, 1218 (1969)
14. See, for instance, C. Kittel, "Introduction to Solid State Physics", J. Wiley, New York 1984
15. C. Flytzanis and C.L. Tang, Phys. Rev. B4, 2520 (1971)
16. See, for instance, "The Physics and Fabrication of Microstructures and Microdevices", M.J. Kelly and C. Weisbuch, Eds, Springer Verlag, Berlin 1988; "Quantum Semiconductor Structures: Fundamentals and Applications", Academic Press New York 1991
17. See J. Bierlein in Ref. 1; F.C Zumsteg, J.D. Bierlein and T.E. Dier, J. App. Phys. 47, 4986 (1976); for a recent review see, for instance, F. Bordui and M.M. Fejer, Ann. Rev. Mat. Sci. 23, 321 (1993); G.I. Chen and G.Z Liu, Ann. Rev. Mat. Sci. 16, 203 (1986)
18. See C.L. Tang in Ref. 4
19. See, for instance, "Nonlinear Optical Properties of Organic Molecules and Crystals", Vols. 1 and 2, D.S. Chemla and J. Zyss, Eds, Academic Press, New York 1987
20. See, for instance, "Organic Molecules for Nonlinear Optics and Photonics" J. Messier, F. Kajzar and P. Prasad, Eds, NATO ASI Series Kluwer, Publ. 1991; P.N. Prasad and D.J. Williams, "Introduction to Nonlinear Optical Effects in Molecules and Polymers", J. Wiley Interscience, New York 1990
21. See, for instance, "Principles and Applications of Nonlinear Optical Materials", R.W. Munn and C.N. Ironside, Eds, Blackie Academic, London 1993
22. B. Davydov, L. Derkacheva, L.D. Duna, V.V Zhabotinskii and M.E. Zolin, Sov. Phys. JETP. Lett. 12, 16 (1970)
23. G.R. Meredith, J.G. Van Dusen and D.J. Williams, Macromolecules 15, 1385 (1982)
24. K.D. Singer, J. E. Sohn and J. Lalama, Appl. Phys. Lett. 49, 248 (1986)
25. See Mohlmann in Ref. 4
26. J. Zyss, J.F. Nicoud and D.S. Chemla, J. Chem. Phys. 74, 4800 (1981)
27. J.L. Oudar and R. Hierle, J. Appl. Phys. 48, 2699 (1977)
28. B.F. Levine, C.G. Bethea, C.D. Thurmond, R.T. Lynch and J.L. Bernstein, J. Appl. Phys. 50, 2523 (1979)
29. J. Zyss, J.F. Nicoud, and M. Coquillay, J. Chem. Phys. 81, 4160 (1984); R. Mase and J. Zyss, Mol. Eug. 1, 141 (1991)
30. J.M. Halbout, S. Blit, W. Donaldson and C.L. Tang, EEE J. Quant. Electr. 15, 1176 (1979)
31. V. Osterberg and W. Margulis, Opt. Lett. 11, 516 (1986)
32. R.A. Myers, N. Mukherjee and S.R.J. Brueck, Opt. Lett. 16, 1734 (1991)
33. See also J. Kajzar in Ref. 1
34. M.C.J. Donckers, S.M. Silence, C.A. Walsh, F. Hache, D.M. Burlaud, W.E. Moerner and R.J. Twieg, Opt. Lett. 18, 1()44 (1993)
35. R.H. Stolen and H. Tom, Opt. Lett. 12, 587 (1987); D.Z. Audersson, V. Mizcahi and J.E. Sipe, ibid. 16, 796 (1991)
36. G.P. Agrawal and C. Flytzanis, Chem. Phys. lett. 44, 366 (1976), G.P. Agrawal, Cojan and C. Flytzanis, Phys. Rev. B17, 776 (1978)
37. See, C. Flytzanis in Ref. 19, also M. Cardona and F.H. Pollack in Optoelectronic Materials, G.A. Albers, Ed, Plenum, New York 1971
38. See, for instance, "Optical Nonlinearities and Instabilities in Semiconductors" H. Haug, Ed, Academic Press, New York 1988
39. S. Schmitt-Rink, D.S. Chemla and D.A.B. Miller, Adv. Phys. 38, 89 (1989)
40. B.I Greene, J. Orenstein and S. Schmitt-Rink, Science 247, 679 (1990)
41. E. Yablonovitch, C. Flytzanis and N. Bloembergen, Phys. Rev. Lett. 29, 865 (1972)
42. G.I. Stegeman, M. Sheik-Bahae, E. Van Stryland and G. Assanto, Opt. Lett. 18, 13 (1993)

43. See, for instance, C. Flytzanis and J. Hutter in "Contemporary Nonlinear Optics", G.P. Agrawal and R.W. Boyd, Eds, Academic Press, New York 1992
44. J.J. Wynne, Phys. Rev. 178, 1295 (1969)
45. D. Weaire, B.S. Wherett, D.A.B. Miller and S.D. Smith, Opt. Lett. 4, 331 (1979); A. Miller, D.A.B. Seaton and S.D. Smith, Phys. Rev. Lett. 47, 197 (1981)
46. B.S. Wherett, Proc. Roy. Soc. (London) A390, 373 (1983)
47. D. Ricard, Ph. Roussignol and C. Flytzanis, Opt. Lett. 10, 511 (1985), K.C. Rustagi and C. Flytzanis, Appl. Phys. Lett. 2, 344 (1984)
48. F. Hache, D. Ricard, C. Flytzanis and U. Kreibig, Appl. Phys. Lett. A47, 347 (1988)
49. Al.L. Efros and A.L Efros, Sov. Phys. Semic. 16, 772 (1982)
50. D.A.B. Miller, D.S. Chemla and S. Schmitt-Rink, Phys. Rev. B35, 8113 (1987)
51. Ph. Roussignol, D. Ricard, J. Lukasik and C. Flytzanis J. Opt. Soc. Am. B4, 5 (1987)
52. See, for instance, G. Wegner, Makromol. Chem. 154, 35 (1971); see also Refs 20 and 21
53. C. Sauteret, J.P. Hermann, R. Frey, F. Pradere, J. Ducuing, R.R. Chance and R.H. Baughman, Phys. Rev. Lett. 36, 956 (1976)
54. B.I. Greene et al, Phys. Rev. Lett. 61, 325 (1988), G.J. Blanchard et al, ibid 63, 887(1989)
55. See J. Frey, R. Frey and C. Flytzanis in Ref. 1
56. J. Frey, R. Frey and C. Flytzanis, Phys. Rev. B45, 4056 (1992)
57. See, for instance, J.K. Furdyna, Appl. Phys. 64, R29 (1988)
58. H. Ashitaka, Y. Yokoh, R. Shimizu, T. Yohozawa, K. Morita, T. Snehiro and Y. Matsumoto, Nonlinear Optics 4, 281 (1993)

Chapter 4

INTEGRATED OPTICS ON LITHIUM NIOBATE

D. DELACOURT

Introduction

Since the demonstration of optical waveguiding using metallic indiffusion in $LiNbO_3$[1], this material has become the favorite substrate for integrated optics primarily because of its electrooptic, acoustooptic and nonlinear properties. Thanks to a simple processing technology, many waveguide geometries such as the Mach-Zehnder interferometer and the directional coupler have been studied and the major basic functions of integrated optics have been worked out. Today, some devices are commercially available, for example, it is possible to find on the market amplitude modulators with bandwidth up to 18 GHz. Nevertheless, research in the field is still active and is oriented to higher bandwidth components and to switching matrices (8x8, 16x16,...). Moreover, some new paths for investigation seem very promising, as is the particular case for optical frequency converters based on Quasi-Phase Matching techniques and for rare-earth doped $LiNbO_3$ waveguide lasers. In this Chapter, we shall recall some basics concerning $LiNbO_3$, then in the second part discuss modulation and switching. The third part will be focussed on optical frequency conversion, in particular the possibility of efficient Quasi-Phase Matched blue Second Harmonic Generation in $LiNbO_3$ waveguides from a near-infrared laser diode. Finally, we shall describe some of the numerous advantages that are offered by the combination of electrooptical effect and laser properties in waveguides fabricated in rare-earth doped $LiNbO_3$.

1. Basics of Integrated Optics on $LiNbO_3$

1.1. General Data

$LiNbO_3$ belongs to the (3m) symmetry group. The third order symmetry axis corresponds to the ferroelectric axis (Curie point [2] around 1165°C) and to the optical axis. $LiNbO_3$ is transparent in the range 0.4-4.5 µm and has negative birefringence (for example, [3] $n_e = 2.1741$ and $n_o = 2.2598$ at $\lambda = 0,8$ µm).

D. Delacourt -Thomson-CSF, Laboratoire Central de Recherches, Domaine de Corbeville, 91404 Orsay Cédex, France

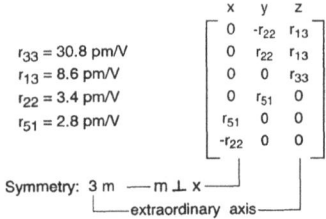

Fig. 1. Electrooptic tensor of $LiNbO_3$.

The electrooptic tensor of $LiNbO_3$ is given in Fig. 1 from Ref. 2. As is well known, the largest coefficient is r_{33}, and in general electrooptic devices on $LiNbO_3$ are designed to exploit this coefficient as we shall see below. Concerning Second Harmonic Generation coefficients, d_{31} can be used with birefringence assisted phase matching in a limited range of the spectrum ($d_{31} \approx$ 5-6 pm/V). This classical technique of phase matching cannot be implemented using d_{33} which is about 6 times higher ($d_{33} \approx$ 30-40 pm/V). Nevertheless, Quasi-Phase Matching (QPM) methods that will be detailed below, lead to very high conversion efficiencies, for example from near-infrared to blue, by using d_{33} in waveguides.

1.2. Waveguide Realization

Two techniques are mainly used to realize waveguides in $LiNbO_3$. The first approach, introduced by Schmidt and Kaminow in 1974 [1], consists of localized Titanium indiffusion. The second, proposed by Jackel in 1982 [4], is based on proton exchange between Li+ ions of the crystal and H+ obtained from an acid.

1.2.1. Titanium Indiffusion. This waveguide fabrication technique leads to an increase in both extraordinary and ordinary indices. The index change typically lies between 10^{-2} and 10^{-3}, depending on the thickness of the Titanium before diffusion and on the operating conditions. This diffusion is generally performed at a temperature between 800 and 1100°C during several hours (typically between 8 and 12 hours). The diffusion of Titanium stripes leads to graded index channel waveguides as schematized in Fig. 2. The process that is currently used to obtain these stripes involves classical photolithographic etching or lift-off.

1.2.2. Proton Exchange. This fabrication technique consists of a substitution of Lithium ions of $LiNbO_3$ by H+ ions. This exchange leads to an increase of the extraordinary index and to a slight decrease of the ordinary index; as a result, the waveguides obtained by this method are single polarization (TE or TM). To simplify, we can distinguish two types of proton exchanged waveguides. The first type corresponds to high H+ concentrations and high change in refractive index ($\approx 10^{-1}$). This high concentration leads in general to strains and dislocations and high propagation losses ($\geq$ 1 dB/cm). Moreover, this high proton concentration is associated with a drastic decrease of the nonlinear coefficients (in particular r_{33} and d_{33}) which is probably related to the fact that $HNbO_3$ is centro-symmetric. Note also that the waveguide of this first type can be considered as a step-index waveguide. In the second type of proton exchanged waveguide, the waveguides have a lower proton concentration and a lower index change ($\leq 10^{-2}$). The crystal structure is less perturbed than in the first case and the decrease of nonlinear coefficients can be small. The waveguides of this second type are graded index waveguides.

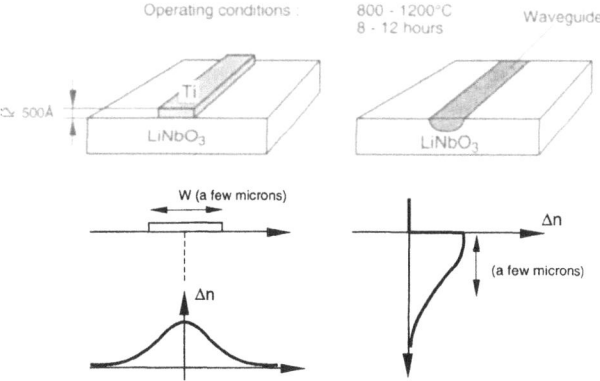

Fig. 2. Titanium indiffused waveguides.

In practice, proton exchange is achieved from an acid acting through a mask (Ta, SiO_2 for example...). The waveguides of the first type are generally obtained with a strong exchange. To decrease the proton concentration in order to produce waveguides of the second type, one possibility consists of annealing the substrate after this exchange. Under certain conditions, this annealing is able to mostly restore the nonlinear coefficients, as studied for example by Bortz et al.[5] who have carried out an exchange in a benzoic acid bath at 173°C during 66 min followed by an annealing at 333°C for varying times (several hours). This technique, which leads to waveguides of the second type, is commonly called "Annealed Proton Exchange" (APE). To realize these second type waveguides, it is also possible to slow down the kinetics of the exchange. For example, De Micheli[6] proposes introducing a small concentration of lithium in the acid bath, for example via lithium benzoate in a benzoic acid bath. Whatever solution is chosen to realize waveguides in $LiNbO_3$, it is necessary to polish the end-facets of these waveguides to permit in- and out-coupling of light. Then the waveguides can be characterized as regards propagation losses and mode profiles by the usual methods.

1.3. Electrode Realization

To use the electrooptic properties of $LiNbO_3$, for example in order to construct a phase modulator, it is necessary to provide electrodes. In general, coplanar electrodes are used, to minimize the gap between these electrodes and thereby reduce drive voltage needed to create a given electric field. The two typical configurations which are commonly implemented are shown in Fig. 3. The first one uses a horizontal electric field and the second one uses a vertical

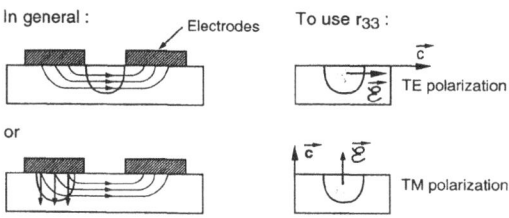

Fig. 3. Typical electrode configurations.

field. The choice between these two geometries depends on the orientation of the substrate. As a matter of fact, in order to use r_{33}, it is convenient to implement the first configuration to modulate the TE modes of the waveguide when the substrate is y or x-cut with z in the plane of this substrate and perpendicular to the direction of propagation. In the case of z-cut substrates, the waveguide must be covered by one of the electrodes and the modulator works in the TM polarization (see again Fig. 3). To minimize the drive voltage of the devices (for example the drive voltage V_π to reach a phase-shift of π), the overlap between the optical mode to modulate, and the applied electric field needs to be optimized. In practice, when the configuration and the position of electrodes are defined, their realization generally involves photolithographic processes and metal deposition techniques as evaporation, sputtering, electroplating, and so on.

2. Amplitude Modulation and Switching

2.1. Basic Geometries

Amplitude modulation and switching can be obtained with the electrooptic effect using convenient configuration of waveguides. Fig. 4 indicates some typical geometries, where the waveguides are assumed to be single-mode.

Concerning the Mach-Zehnder interferometer (a), optical power at the output depends on the interference state between two independent arms that can be phase-shifted relative to each other. The "push-pull" electrode configuration that is shown in Fig. 4(a) gives the possibility of inducing opposite phaseshifts in the two arms with the same drive voltage. The directional coupler (b) is based on the interaction between two single mode waveguides. This interaction causes a periodic exchange of light between the two guides, and by modifying the conditions of this exchange via electrooptical effect it is possible under certain conditions to switch the light from one guide to the another at the output. Note that the interaction region, consisting of two single mode waveguides, can be considered as a global two mode guiding region and the output of the device can be interpreted as the result of interferences between these two modes in the interaction region.

The operating principle is the same in the case of the two mode interference switch (c), where the interaction occurs in a two mode waveguide and the output depends on the interference state between these two modes. Nevertheless, the modulation of the optical power at one output of the switch is closer to a sine shaped function than in the case of the directional coupler. The Y fed directional coupler (d) is optically self-biased. As a matter of fact, in-coupling using a Y junction permits exciting the coupler in a symmetric way. Without any drive voltage the optical power is the same in the two output waveguides. Moreover, the response of the device is linear with drive voltage around this operating point. The X switch (e) is also based on two mode interference but within a short interaction length. This leads to a very compact element, which is convenient for integrating numerous switches on the same substrate, for example in the case of switch matrices. Nevertheless the drawback of this compactness is a much higher drive voltage compared with the previous examples. The Y switch (f) can be understood as an adiabatic mode converter. As a matter of fact, the electrooptical effect is used to create an asymmetry of the Y junction to favor the output of the light from one or by the other waveguide.

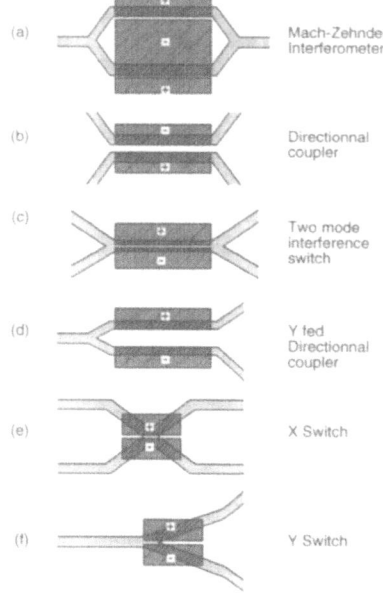

Fig. 4. Amplitude modulator and switch geometries.

2.2. Speed Limitation of Integrated Optical Devices

Let us consider the example of a phase modulator to illustrate what can limit the bandwidth of electrooptical components.

2.2.1. Lumped Electrodes. This is the simplest configuration of electrodes. Assuming that these electrodes are equipotential, the optical wave is submitted to a varying drive voltage during its transit time under them. If this transit time is equal to the modulation period, the average index variation induced by the electrooptical effect will be equal to zero. On lithium niobate, the order of magnitude of the corresponding modulation frequency is between 10 and 15 GHz for a 1 cm long interaction. Another limitation is related to the capacitance of electrodes. This capacitance is in general in the range of 1pF for the same interaction length. This capacitance leads to cut off frequencies of several GHz. Nevertheless, these values are optimistic because at these frequencies, the lumped electrodes can no longer be considered as equipotentials. To overcome these limitations, it is convenient to implement travelling wave electrodes.

2.2.2. Travelling Wave Electrodes. In this case, the electrodes are designed to guide the electrical wave used to drive the modulator. The speed limitation is then related to the phase mismatch between the optical and electrical waves. In fact, the phaseshift $\Delta\Phi$ obtained at the output of the modulator is proportional to $\sin(x)/x$, where x is given by:

$$x = \frac{\pi FL}{c}\left(n_M - n_o\right) \quad (1)$$

for a copropagating interaction between the two waves. The parameters appearing in Eq. 1 are: F = modulation frequency; L = interaction length; c = velocity of light in vacuum; n_M = effective index of electrical wave; and, n_O = effective index of the optical mode.

For typical travelling wave electrode configurations on lithium niobate, we have: $n_M - n_0 \approx 2$. The order of magnitude of the lower modulation frequency for which $\Delta\Phi = 0$ is then around 15 GHz for a 1 cm long interaction.

Fig. 5 illustrates an example of the travelling wave electrode geometry that is usually associated with a directional coupler. In this configuration, the "hot" electrode corresponds to the microstrip line and the planar electrode is grounded. These two electrodes form a transmission line, the impedance of which has to be matched with that of the electric power supply (generally 50 Ω) by adjustment of the geometrical parameters. Moreover, the propagation losses of the transmission line must be reduced to limit the absorption of the electrical wave; this is why thick electrodes (several microns), which can be realized by electroplating, are generally used for these high bandwidth devices. Note also that a dielectric buffer layer can be implemented to reduce the optical propagation losses due to the metal electrodes.

In all cases, the shorter the interaction length the higher the bandwidth but the higher the drive voltage. Thus, in practice it is important to find the trade-off that best fits the specific application. Nevertheless high working frequencies can be reached without high bandwidths, using resonant devices or periodic electrodes [7].

2.3. State of the Art

Today several components based on integrated optics technology are commercially available. As an example, consider the case of high bandwidth amplitude modulators. Most of these devices are based on a Mach-Zehnder interferometer with travelling wave electrodes. Bandwidths up to 18 GHz at -3 dB electrical response are mentioned on the data sheets of different companies. Generally these components are designed to operate either at 1.3 µm or at 1.55 µm. The drive voltage V_π, which depends on both operating wavelength and bandwidth, typically varies from about 3 V for low bandwidth modulators (3 GHz) to about 13 V for 18 GHz bandwidth devices at $\lambda = 1.3$ µm. These components are pigtailed and their order-of-magnitude insertion loss is around 5 dB.

Some companies offer devices designed for a specific application such as CATV modulation, for which the linearity and the flatness of the response across the electrical bandwidth are critical parameters. Devices integrating several functions, such as the Fiber Optic Gyroscope circuit, can also be found on the market. Among the other applications of integrated optics components we should also mention antenna remoting and instrumentation.

Concerning research, work is principally oriented towards higher bandwidth devices and switching matrices. For example, the authors of Ref. 8 propose covering the travelling wave electrodes used to drive a Mach-Zehnder interferometer with a metallic shielding plane, in order to speed up the electrical wave. Upon doing so, the difference $n_M - n_0$ mentioned in Section 1 is lower and the bandwidth reaches 40 GHz at -3dB optical response. It is noteworthy that the drive voltage V_π of this modulator is very low (3.6 V) because of improved overlap between the applied electric field and the optical waves. This improvement has been obtained by etching the $LiNbO_3$ on both sides of the waveguides. Among switching matrices let us mention the 16x16 single chip switch described in Ref. 9.

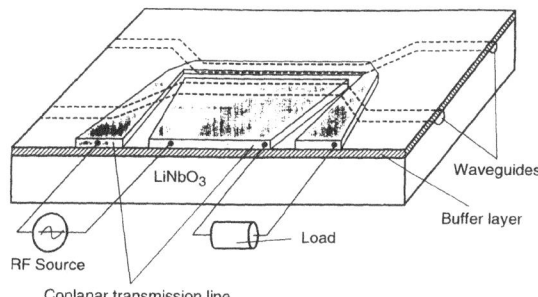

Fig. 5. Travelling wave electrodes implemented on a directional coupler.

Let us also take note of the tree-structured polarization independent 8x8 LiNbO$_3$ switch matrix based on Y switches with digital response, as presented at ECIO 93 [10]. The total chip size is 80x15 mm², the maximum total insertion losses are lower than 15 dB and the drive voltage is about 100 V. Also presented at ECIO 93 [11] was a strictly non blocking Extended Generalized Shuffle Network comprising 23 LiNbO$_3$ modules, incorporating 448 directional couplers (mean drive voltage 12.5 V) and 512 fiber connections. This network is designed for TM polarization at 1.5 µm.

These examples give an idea of the trends of research in the field of modulation and switching using LiNbO$_3$. Concerning other fields of investigation such as Quasi-Phase Matched optical frequency conversion and waveguide lasers based on rare-earth doped LiNbO$_3$, some recent results are very promising, as will be described in the next paragraphs.

3. Quasi-Phase Matched Nonlinear Guided Wave Interaction

Today it would be very important to achieve a compact coherent blue source. In particular, this would give the possibility of significantly increasing the capacity of optical data storage systems based at the present time on near infrared GaAs laser diodes. As a matter of fact, such a decrease in the operating wavelength would permit us to divide by four the area of the focused spot, which limits the information density on a disc for example. To obtain this blue source, several approaches are being considered. Among them, direct laser emission in II-VI materials, which has been demonstrated in the last few years [12], is very attractive even though several technological problems still need to be solved. Up-conversion in rare-earth doped fibers or crystals is also of interest (see for example reference [13]). Nevertheless, the solution that seems to be able to meet the need in the shortest time is frequency doubling of near-infrared laser diodes which have been already developed and are very low cost devices. To be viable, such a solution has to show a very high conversion efficiency. This is why the Quasi-Phase Matching (QPM) technique should be considered: it permits one to profit both from the highest nonlinearities of materials and from a geometry of integrated optics which is very favorable to nonlinear processes, since a strong confinement of light can be maintained over long interaction lengths. As previously seen, integrated optics technology is well known in LiNbO$_3$. Moreover, in this material, the QPM technique can be implemented to use d_{33}.

3.1. Quasi-Phase Matching

As is well known, phase matching between the nonlinear polarization and the harmonic wave is necessary to obtain high conversion efficiencies in frequency doubling. This means that the following expression must be fulfilled:

$$\Delta\beta = \beta_{2\omega} - 2\beta_\omega = 0 \qquad (2)$$

where $\beta_{2\omega}$ and β_ω are the propagation constants of the harmonic and the fundamental modes respectively. $\Delta\beta$ can also be written

$$\Delta\beta = \frac{4\pi}{\lambda^\omega}\left(\tilde{\beta}_{2\omega} - \tilde{\beta}_\omega\right) \qquad (3)$$

where λ^ω is the fundamental wavelength (the harmonic wavelength $\lambda^{2\omega}$ is equal to $\lambda^\omega/2$); and $\tilde{\beta}_{2\omega}$ and $\tilde{\beta}_\omega$ are the effective indices of the harmonic and fundamental modes respectively.

Because of refractive and effective index dispersion, $\Delta\beta$ is different from zero. After the optical waves propagate through the so-called coherence length L_c, the phase-shift between the nonlinear polarization and the harmonic waves reaches π ($\Delta\beta L_C = \pi$) and the interaction becomes destructive. This behavior is illustrated by curve (a) of figure 6, which shows the evolution of the harmonic power in the case of phase mismatch, versus the interaction length graduated in coherence length. The harmonic power in this case cannot be higher than the value proportional to L_c^2 which is obtained after one coherence lenght. In the case of blue generation from a near-infrared source via d_{33} in $LiNbO_3$, this coherence length is about 1.5 µm. This is why a phase matching technique has to be implemented. QPM consists of periodically perturbing the nonlinear interaction by modulation of a parameter of this interaction. Among the different possibilities (see for example Ref. 14), one of the most efficient is the periodic modulation of the nonlinear coefficient d involved in the interaction. In that case d can be written as follows:

$$d(x) = \sum_{n=-\infty}^{n=+\infty} d_n e^{-inKx} \qquad (4)$$

with: x = coordinate along the propagation axis; n = integer; d_n = Fourier coefficient corresponding to the nth harmonic of the modulation ; and, $K=2\pi/\Lambda$ where Λ is the period of the modulation of d. To achieve QPM, Λ has to be chosen such that a spatial harmonic (n=m) of the Fourier decomposition of d satisfies

$$\Delta\beta = mK \qquad (5)$$

If this is the case, QPM is obtained in the mth order and the spatial harmonic mK is the only one which is used. The conversion efficiency will be directly related to the value of the d_m coefficient and in general, the higher the order m, the lower the coefficient d_m and the lower the conversion efficiency. Another consequence is the impossibility of achieving QPM in the mth order with a modulation which does not contain the corresponding spatial harmonic mK ($d_m=0$), as will be the case with a sine shaped modulation using an order higher than one. When $d_m \neq 0$ the period Λ_m of the modulation has to satisfy:

$$\Lambda_m = 2m\, L_c \tag{6a}$$

in particular, to achieve QPM in the first order, the period of the perturbation has to satisfy:

$$\Lambda_1 = 2L_c. \tag{6b}$$

Curve (b) of Fig. 6 gives the evolution of the harmonic power when the first order QPM is reached by a periodic change of the nonlinear coefficient sign (this is the most efficient way to achieve QPM). As a comparison, the curve (c) illustrates the case of real phase matching.

2.2. The Case of LiNbO₃

In this material, a change in the sign of d_{33} can be obtained by Titanium indiffusion on the +c face of the crystal via a ferroelectric polarization reversal [15]. In our case, a 50 Å thick Titanium grating (the period of which is in the range 3-4 µm to achieve QPM in the first order) is diffused at 1100°C during 5 min. Unfortunately, this process does not lead to deep domains with abrupt boundaries perpendicular to the surface of the substrate (+c). The typical shape of ferroelectric polarization obtained after such a process is schematically shown in Fig. 7. In fact, probably because of Titanium lateral diffusion and Lithium outdiffusion, we obtain two homogeneous regions with opposite polarization (that means opposite d_{33}), separated by a triangle shaped periodic boundary. Nevertheless, the modulation shape that will directly influence the conversion efficiency can be completely different from this triangle shape. As a matter of fact, the only modulation shape that has to be taken into account is the modulation shape resulting from the global nonlinear overlap integral which can be written in a simplified form as follows:

$$I(x) = \int_{-\infty}^{+\infty} d(x,z)\, E_{2\omega}^{*}(z)\, E_{\omega}^{2}(z)\, dz \tag{7}$$

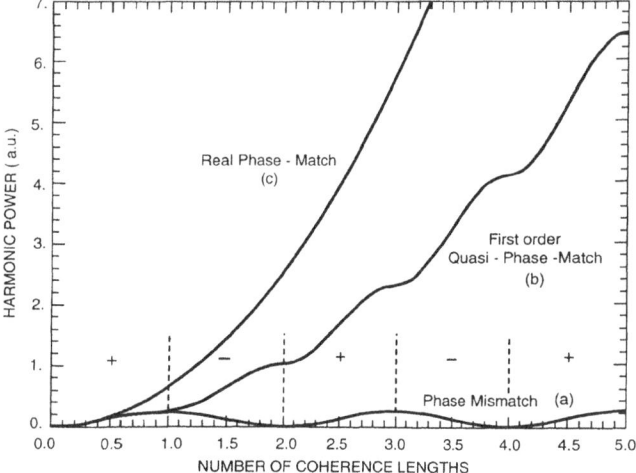

Fig. 6. Harmonic power evolution versus interaction length in the case of phase mismatch (a), first order Quasi-Phase-Match (b) and real phase match (c).

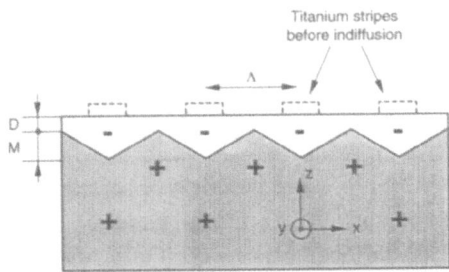

Fig. 7. Ferroelectric pattern obtained after periodic titanium indiffusion.

where: $d(x,z)$ is the nonlinear coefficient depending on x and z as it can be seen in Fig. 7; $E_\omega(z)$ and $E_{2\omega}(z)$ are respectively the fundamental and harmonic guided TM mode profiles in the z direction.

In fact, this overlap integral makes evident an effective nonlinear coefficient $d_{eff}(x)$, the modulation shape of which will govern the conversion efficiency of the interaction [16]. This point is illustrated by Fig. 8, where mode profiles (MP) of varying extent are considered.

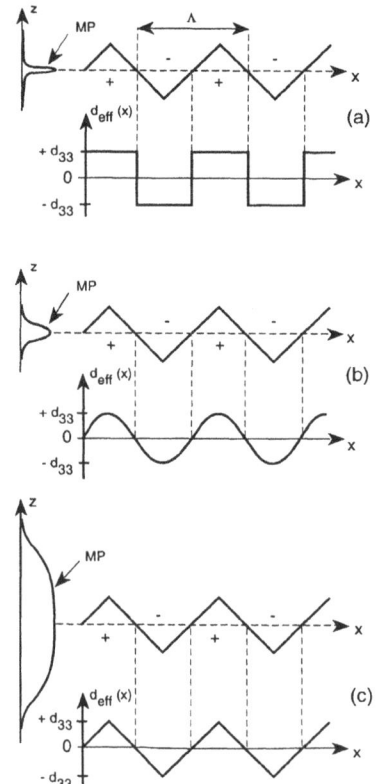

Fig. 8. Influence of the mode profiles on the effective nonlinear coefficient used in Quasi-Phase-Matching.

To simplify, these profiles are centered on the triangle shaped boundary and are supposed to represent the product $E^*_{2\omega}(z) \, E^2_{\omega}(z)$.

In the case of very confined modes (a), it can be readily understood that the effective modulation of d that will be used to reach QPM is square shaped. In the case of very extended modes (b) we obtain a triangle shaped d_{eff}. In intermediate cases, in particular where the extent of the modes is about that of the triangle shaped boundary, the result is an effective nonlinear coefficient not far from a sine shape function, which cannot be used for QPM in an order higher than one.

This simplified approach shows how important it is to operate in the first order. This has been done, for example, in obtaining 0.56 mW at $\lambda = 0.434$ µm from 67 mW in the near-infrared after a 1 cm long interaction. This experimental result was obtained in a proton exchanged waveguide with a period of 3.5 µm for the nonlinear grating [17] (see also Ref. 18 and 19 for other results in that field).

Some other techniques can be used to periodically reverse the ferroelectric polarization in $LiNbO_3$. For example, the electron bombardment has been successfully applied [20].

Very interesting results concerning ferroelectric polarization reversal at room temperature have also been demonstrated by the application of a pulsed periodic electric field [21]. Using this technique, 20.7 mW were measured in the blue at the output of a 3 mm long proton exchanged waveguide, from a power of 195.9 mW in the near-infrared.

3.3. Quasi-Phase Matching of Other Second Order Nonlinear Interactions

In the case of frequency mixing or optical parametric oscillation, it is also very attractive to try to combine the advantages of integrated optics and the use of d_{33} in $LiNbO_3$. A nonlinear grating is then used to fulfill the condition

$$k_p - (k_s + k_i) = m K \tag{8}$$

where k_p, k_s and k_i are the propagation constants of the three waves (pump, signal and idler), the optical frequencies of which are respectively ν_p, ν_s, ν_i and have to satisfy

$$\nu_p = \nu_s + \nu_i \tag{9}$$

As an example, let us mention the results we have obtained in difference frequency generation in 5 µm wide waveguides realized by proton exchange. Again, the nonlinear grating was induced by periodic Titanium indiffusion.

Periods Λ of 19 and 20 µm were used to phasematch the difference frequency generation between a pump mode the wavelength of which was tunable around $\lambda_p \approx 0.8$ µm (Ti: Al_2O_3 laser) and a signal mode at a wavelength $\lambda_s = 1.5515$ µm (GaInAsP DFB laser). Using the 19 µm period, QPM was obtained with $\lambda_p = 0.7833$ µm ($\lambda_i = 1.5819$ µm). With the 20 µm period, the pump wavelength was tuned to $\lambda_p = 0.8056$ µm ($\lambda_i = 1.6759$ µm) to obtain QPM. In the first case ($\Lambda = 19$ µm) a power of 0.48 µW was measured at the idler wavelength λ_i from a pump power of 6.4 mW and a signal power of 1.5 mW.

Today, the possibilities of working out a laser diode pumped Quasi-Phase Matched Optical Parametric Oscillator are under investigation.

4. Waveguide Lasers in Rare-Earth Doped $LiNbO_3$

Rare-earth doped $LiNbO_3$ combines the laser gain of the dopant ions (Nd^{+3}, Er^{3+} for example) with the well known electrooptic, acoustooptic and nonlinear properties of $LiNbO_3$. This can yield waveguide lasers integrated with modulators, lasers with internal frequency conversion, loss compensated integrated optical circuits, etc. The results that we have obtained on both modelocked and Q-switched waveguide lasers on Nd: $LiNbO_3$ are good illustrations of these numerous possibilities.

4.1. CW Nd: $LiNbO_3$ Waveguide Laser

The 1.08 μm transition in Nd: $LiNbO_3$ corresponds to a very efficient four level laser system.[22] To maintain a good optical quality of $LiNbO_3$, the Nd doping is limited to 0.3 % (atomic). Under this condition, more than 90 % of a pump at 0.814 μm polarized along the c axis can be absorbed in a 1 cm long propagation path. Moreover, a MgO codoping (3-5 % mol.) of $LiNbO_3$ prevents optical damage which is also limited by the use of annealed proton exchange to realize the waveguides. These waveguides must present very low propagation losses to ensure a low threshold for the lasers to be fabricated. The next fabrication step consists of polishing the input and output facets of the device. The dielectric mirrors forming the cavity are then coated directly on these facets.

With 99 % and 50 % reflectivity respectively for the input and output mirrors, thresholds of about 1.5 mW and slope efficiencies around 50 % have been achieved with diode laser pumping in waveguides exhibiting propagation losses around 0.2 dB/cm at the lasing wavelength.

4.2. Mode-locked Nd: $LiNbO_3$ Waveguide Laser

By intracavity modulation of the laser at the cavity round-trip frequency, it is possible to phase-lock together, the different longitudinal modes of a laser.[23] If this is done, the output of the laser is then pulsed and the repetition rate is precisely the modulation frequency. In the case of a 11 mm long waveguide laser on Nd:$LiNbO_3$, this frequency is about 6 GHz. To obtain the mode-locking, a 6 mm long travelling wave electrooptic phase modulator has been implemented as schematized in Fig. 9. Because of the higher efficiency of this modulator for co-propagating optical and electric waves than for counter-propagating interactions, the phase shift induced on the laser field during a round-trip in the cavity is not equal to zero. This approach to mode locked operation is easy to implement and does not add losses in the laser.

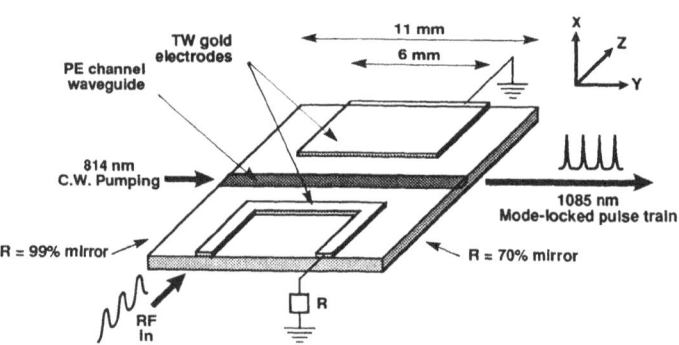

Fig. 9. Mode-locked Nd:$LiNbO_3$ waveguide laser.

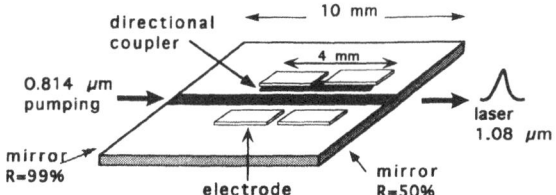

Fig. 10. Q-switched Nd:LiNbO$_3$ waveguide laser.

Using this configuration and a 70 % reflection output mirror, the following characteristics were measured: coupled pump power (at λ = 0,814 μm) = 50 mW; pulse width = 7 ps; peak power = 250 mW; and, repetition rate = 6 GHz.

4.3. Q-switched Nd: LiNbO$_3$ Waveguide Laser

As described above, mode-locking of the laser offers the possibility of high repetition rate pulsed operation. For coupled pump power of the same order of magnitude, Q-switching of the cavity can also lead to pulsed operation but with much higher peak power. The Q-switch technique consists of creating high losses in the laser cavity to avoid the laser action. This is a way to store the pump energy, taking advantage of the long lifetime of these rare-earth ions (between 50 and 100 μsec for Nd in LiNbO$_3$). When the losses are suddenly removed, a giant pulse is emitted and the energy storage can then begin again. To realize this modulation technique we implemented a "step Δβ" directional coupler as schematized in Fig. 10. This electrode configuration permits to reaching both the "cross" and "bar" states using the electrooptic effect, even if the coupler is not exactly in the "cross" state when the drive voltage is equal to zero. One waveguide of the coupler does not extend to the facets of the sample that serve as the mirrors of the laser. Thus, when the coupler is in the "cross" state the losses of the cavity are very high and energy storage occurs. When the coupler is switched into the "bar" state, theses losses are canceled and a large laser pulse is emitted. Again, annealed proton exchange was used to realize the waveguides and gold electrodes were applied. For 15 mW of coupled pump power, using a 50 % reflection output mirror, peak powers of 350 W associated with pulse widths of 300 ps were measured, at a repetition rate of 1 kHz (the switching voltage of the coupler was about 35 V). In this case the coupler was not optimized and higher peak powers could be reached by improving the loss modulation ratio. Nevertheless, these results illustrate the possibility of high peak power pulse emission in these rare-earth doped waveguide lasers. Such characteristics are interesting for time-multiplexed fiber sensor systems and should permit a drastic enhancement of the conversion efficiencies related to nonlinear intracavity interactions. For example it could be very efficient to integrate a nonlinear grating with such a Q-switched laser to achieve QPM intracavity Second Harmonic Generation, in order to generate green from the 1.08 μm transition or blue from the 0.93 μm transition. Other ions are of interest, as in the case of Er^{3+}. The authors of Ref. 25 have demonstrated Er: LiNbO$_3$ Q-switched laser operation at 1.53 μm. Broadband amplification in the range 1.53-1.61 μm has also been observed, showing the potential for tunable sources [26].

5. Conclusions

LiNbO$_3$ is integral to the history of integrated optics. This technology itself took shape thanks to LiNbO$_3$, in which it is possible to realize low loss waveguide with simple processes, making accessible the very good electrooptic, acoustooptic and nonlinear properties of this material. Today some integrated optic components are commercially available as is the case for example with high bandwidth pigtailed amplitude modulators presenting low drive voltage and low insertion losses. These devices which are offered for sale by more and more companies worldwide, find application in domains where external modulation rather than direct modulation of the source is more suitable. Several companies also offer devices especially designed for given applications as CATV or fiber optic gyroscopes. This illustrates how Integrated Optics in LiNbO$_3$ already seem to be able to meet many specific needs. Besides these commercially available components, research is going on not only in the field of high bandwidth modulators but also in the area of n x n switching matrices.

In parallel, important work is being done to develop a compact blue source that would permit increasing the capacity of optical storage systems and would also find applications in laser printers or in biology. For such a source, Second Harmonic Generation via Integrated Optics on LiNbO$_3$ seems to be an attractive technical approach because of the efficiencies that can be attained, thanks to the periodic reversal of ferroelectric polarization that permits profiting from Quasi-Phase-Matching involving the highest nonlinear coefficient of LiNbO$_3$. This artificial phase-matching technique could also be used to demonstrate diode pumped integrated optical parametric oscillators. Also in the domain of new compact sources, the recent results obtained with rare-earth doped LiNbO$_3$ waveguide lasers are very promising. In particular, the combination of the electrooptic or nonlinear effect of LiNbO$_3$ with the laser properties of rare-earth ions offers numerous opportunities. Among them, the pulsed emission of these lasers using either integrated mode-locking or Q-switching techniques is very attractive for many applications. As an example, a Q-switched Er:LiNbO$_3$ waveguide laser could be a suitable compact source for eye-safe range finders. Thanks to the possibilities of optical amplification in rare-earth doped LiNbO$_3$, the insertion of zero-loss components in telecommunication networks could be also of interest.

In summary, the association between integrated optics and LiNbO$_3$ is today the source both of commercially available products and of important research programs, mainly because of the very wide range of applications that can be impacted, from the simplest modulation function to the Q-switched source. Nevertheless, the technology remains relatively simple, and in many domains LiNbO$_3$ based integrated optics is a credible alternative to other competitive solutions.

ACKNOWLEDGEMENTS. The author is indebted to M. Papuchon, E. Lallier, M. de Micheli and M. Doisy for the help in the preparation of this chapter and for providing some of the results presented here.

References

1. R.V. Schmidt and I.P. Kaminow, "Metal-diffused optical waveguides in LiNbO$_3$", Appl. Phys. Lett. vol.25 n°8 pp 458-460 (1974)
2. A.M. Prokhorov and Y.S. Kuz'minov, In chapter 6 of "Physics and chemistry of crystalline lithium niobate" Adam Hilger Ed, IOP publisher, New-York, Bristol 1990.

3. G.D. Boyd, W.L. Bond and H.L. Carter, J. Appl. Phys. 38 (4) pp 1941-1943 (1967).
4. J.L. Jackel, C.E. Rice and J.J. Veselka, Appl. Phys. Lett. 41 (7) pp 606-607 (1982).
5. M.L. Bortz, L.A. Eyres and M.M. Fejer, Appl. Phys. Lett. 62 (17) pp 2012-2014 (1993).
6. M. de Micheli, J. Botineau, S. Neveu, P. Sibillot and D.B. Ostrowsky, Opt. Lett. 8 (2) pp 114-115 (1983).
7. R.C. Alferness, S.K. Korotky and E.A.J. Marcatili, IEEE J. Qant. Electron. QE 20 (3) pp 301-309 (1984).
8. K. Noguchi, O. Mitomi, K. Kawano and M. Yanagibashi, IEEE Phot. Tech. Lett. 5 (1) pp 52-54 (1993).
9. P.J. Duthie and M.J. Wale, Elect. Lett. 27 (14) pp 1265-1266 (1991).
10. P. Ganestrand, B. Langerstrom, P. Swensson, H. Olofsson, J.E. Falk and B. Stolz, "Tree-structured 8 x 8 $LiNbO_3$ switch matrix with digital optical switches", Proceedings of the European Conference on Integrated Optics, Neuchatel Switzerland, September 1993.
11. E.J. Murphy and T.O. Murphy, "Characteristics of twenty-three Ti:$LiNbO_3$ switch arrays for a guided wave phtotonic switching system", Proceedings of the European Conference on Integrated Optics, Neuchatel Switzerland, September 1993.
12. M.A. Haase, J. Qiu, J.M. Depuydt and H. Cheng, Appl. Phys. Lett. 59 (11) pp 1272-1274 (1991).
13. S.G. Grubb, K.W. Bennett, R.S. Cannon and W.F. Humer, Electron. Lett. 28 (13) pp 1243-1244 (1992).
14. J. Khurgin, S. Colak, R. Stolzenberger and R.N. Bhargava, Appl. Phys. Lett. 57 (24) pp 2540-2542 (1990).
15. S. Miyazawa, J. Appl. Phys. 50 (7) pp 4599-4603 (1979).
16. D. Delacourt, F. Armani, M. Papuchon, To be published in IEEE J. Quantum Electron.
17. F. Armani, D. Delacourt, E. Lallier, M. Papuchon, Q. He, M. de Micheli and D.B. Ostrowsky, Electron. Lett. 28 (2) pp 139-140 (1992).
18. E.J. Lim, M.M. Fejer, R.L. Byer and W.J. Kozlovsky, Electron. Lett 25 (11) pp 731-732 (1989).
19. X. Cao, B. Rose, R.V. Ramaswamy and R. Srisvastava, Opt. Lett. 17 (11) pp 795-797 (1992).
20. H. Ito, C. Takyu and H. Inaba, Electron. Lett. 27 (14) pp 1221-1222 (1991).
21. M. Yamada, N. Nada, M. Saitoh and K. Watnabe", Appl. Phys. Lett. 62 (5), pp 435-436 (1993).
22. E. Lallier, J.P. Pocholle, M. Papuchon, M. de Micheli, M. J. Li, Q. He, D.B. Ostrowsky, C. Grezes-Besset and E. Pelletier, IEEE J. Quant. Electron. 27 (3) pp 618-625 (1991).
23. E. Lallier, J.P. Pocholle, M. Papuchon, M. de Micheli, Q. He, D.B. Ostrowsky, C. Grezes-Besset and E. Pelletier, Electron. Lett. 27 (11) pp 936-937 (1991).
24. E. Lallier, D. Papillon, J.P. Pocholle, M. Papuchon, M. de Micheli and D.B. Ostrowsky, Electron. Lett. 29 (2) pp 175-176 (1992).
25. R. Brinkmann, W. Sohler and H. Suche, Electron. Lett. 27 (5) pp 415-417 (1991).
26. P. Becker, R. Brinkmann, W. Sohler and H. Suche, "Erbium-doped integrated optical amplifiers and lasers in lithium niobate", in Optical Amplifiers And Their Applications, vol 17 of 1992 OSA Digest Series, invited paper ThB4, pp 109-112.

Chapter 5

PROPAGATION OF SELF-TRAPPED OPTICAL BEAMS IN NONLINEAR KERR MEDIA AND PHOTOREFRACTIVE CRYSTALS

B. CROSIGNANI

1. Introduction

It has been well-known for many years that, under specific conditions, "......an electromagnetic beam can produce its own dielectric waveguide and propagate without spreading"[1] in a suitable nonlinear material.

The original result, obtained in a Kerr medium in which the refractive index increases with field intensity, was mainly concerned with a monochromatic one-dimensional case (i.e., field propagating along the z-direction and exhibiting a transverse dependence on y) and gave rise to an important area of research in nonlinear optics.

Several years later, it was shown[2] that propagation of self-trapped beams found its natural description in the frame of the theory of spatial solitons which describes solutions of the so-called nonlinear Schroedinger equation. These solitons possess the property of evolving without change in their intensity profile (or with, at most, a periodic change) and of passing through each other without changes in their amplitudes. One of the important issues was, naturally, whether this finely-tuned balance between electromagnetic diffraction and self-focusing produced by the Kerr nonlinearity was stable or not, that is whether a small perturbation of the self-trapped beam would lead to collapse or divergence of its waist. The answer to this question is rather straightforward in the case in which the refractive index is proportional to the instantaneous optical intensity I (stability in the one dimensional case, instability in the two dimensional case, that is transverse dependence on x and y).[3] However, the solution can become very involved if the self-focused intensity becomes so high as to give rise to more complicated higher-order nonlinear and diffractive contributions.

The subject of distortionless propagation acquired great technological relevance for optical telecommunications since it was recognized[4] that the same mechanism capable of holding together the beam in the space-domain could work out as well in the time-domain. More precisely, the temporal broadening of a pulse in a dispersive material due to chromatic dispersion can be compensated by the narrowing associated with self-phase modulation , so that a narrow pulse can propagate without temporal spreading. Of course, the equation describing this process is, *mutatis mutandis*, the same one-dimensional nonlinear Schroedinger equation valid in the space-domain (with time replacing the z-component), and it is precisely this equation that has received a great deal of attention in the last twenty years.[5]

B. Crosignani - Dipartimento di Fisica, Universita' dell'Aquila, 67010 L'Aquila, Italy

More recently, the subject of self-trapped optical beams and spatial solitons has undergone a renewal of interest (mainly associated with experimental demonstrations of feasibility [5-11]), both in the one-dimensional and two-dimensional case [12-17] (and even in the three-dimensional case, light bullets [18], that is when both two-dimensional diffraction in space and chromatic dispersion in time are taken into account).

Very recently, the existence of a new kind of spatial soliton, associated with the nonlinearity present in photorefractive crystals in connection with the photorefractive effect, has been predicted [19,20] and experimentally observed [21], its mathematical description relying on a nonlinear equation completely different from the usual Schroedinger equation.

The purpose of this Lecture is to provide an analytic approach to the problem of self-trapped propagation of optical beams, starting from the derivation of the nonlinear equations which describe the field evolution and pointing out the difficulties associated with their extension beyond the paraxial approximation. The case of photorefractive materials proves particularly interesting since turns out to be nonlocal in space, that is spatially dispersive.

Different methods for solving the relevant equations will be discussed, with particular enphasis on the finite dimensional Lagrangian method, which provides a particularly intuitive approach to the problem.

2. Beam Evolution in a Homogeneous Medium : Helmholtz Equation and Coupled-mode Theory

We consider here the evolution of a monochromatic electromagnetic field in a homogeneous isotropic bulk medium characterized by a linear refractive index n_1; the medium becomes inhomogeneous due to the presence of a (usually small) contribution $\delta n(x,y,z)$ to the refractive index, i.e., $n = n_1 + \delta n$, which, in the particular case of a Kerr medium, is proportional to the instantaneous optical intensity I ($\delta n = n_2 I$, with $n_2 > 0$ through this paper).

The problem of describing the propagation in such a medium is usually dealt with by factorizing the electromagnetic field $E(x,y,z,t)$ into the product of a fast-varying function and a slowly-varying one, in the form

$$E(x,y,z,t) = e^{i\omega t - ikz} E(x,y,z) \tag{1}$$

where $k = (\omega/c)n_1 = k_0 n_1$, and looking for an equation for the evolution of $E(x,y,z)$ (the factorization appearing in Eq.(1) is, of course, particularly useful whenever the beam propagates along the z-direction).

In order to derive an equation for $E(x,y,z)$, one can resort either to the Helmholtz equation or to the coupled-mode theory.

In the first case, let us recall that, from Maxwell's equation, it is possible to derive in full generality the vectorial Helmholtz equation[22]

$$\nabla^2 E + 2\nabla(E \nabla \ln n) + k_0^2 n^2 E = 0 \tag{2}$$

which, after neglecting the term containing $\nabla(\ln n)$ and writing $n^2 \cong n_1^2 + 2n_1 \delta n$, takes the scalar form

$$\nabla^2 E + k^2 E + 2k^2 \frac{\delta n}{n_1} E = 0 \tag{3}$$

Eq. (3) does not contain any paraxial approximation and, in principle, its solution can describe situations in which the "slowly varying" part of the field E(x,y,z) exhibits fast variations[23,24] In practice, in order to make it analytically tractable, it is customary to introduce Eq.(1) into Eq.(3) and take advantage of the slowly-varying approximatiom hypothesis (SVA), that is to assume that E(x,y,z) actually varies, as a function of z, on a scale much slower than 1/k . Under this condition, Eq.(3) reduces to the parabolic equation (Fock-Leontovich equation)

$$\frac{\partial}{\partial z} E + \frac{i}{2k}\left(\frac{\partial^2}{\partial x^2} + \frac{\partial^2}{\partial x^2}\right) E = -i \frac{k}{n_1} \delta n E \qquad (4)$$

valid in the paraxial approximation.

In the frame of coupled-mode theory,[25] the field is written as a superposition of a continuum of radiation modes E(ξ, σ; r), that is

$$E(x,y,z) = \sum_{\sigma=1}^{2} \int d\xi E(\xi,\sigma,r) c(\xi,\sigma,z) e^{-i(\beta_\xi - k)z} \qquad (5)$$

where r =(x,y) , $\beta_\xi = (k^2 - \xi^2)^{1/2}$ and

$$E(\xi,1,r) = N_1 e^{-i\xi r}\left(\hat{x} - \frac{\xi_x}{\beta_\xi}\hat{y}\right) \qquad (6)$$

$$E(\xi,2,r) = N_2 e^{-i\xi r}\left[\frac{\xi_x \xi_y}{\beta_\xi}\hat{x} - \left(\beta_\xi + \frac{\xi_x^2}{\beta_\xi}\right)\hat{y} + \xi_y \hat{z}\right]$$

N_1 and N_2 being two suitable normalization coefficients.

The evolution of the expansion coefficients c(ξ,σ;z) is governed by the set of coupled-mode equations

$$\frac{d}{dz} c(\xi,\sigma,z) = \sum_{\sigma'=1}^{2} \int d\xi' \, K(\xi,\sigma,\xi',\sigma',z) e^{i(\beta_\xi - \beta_{\xi'})z} c(\xi',\sigma',z) \qquad (\sigma = 1,2) \qquad (7)$$

where the coupling coefficients K(ξ,σ;ξ',σ';z) are given by

$$K(\xi,\sigma,\xi',\sigma',z) = \frac{\omega\varepsilon_0}{2i} \int_{-\infty}^{+\infty} dr \left[n^2(r,z) - n_i^2\right] E^*(\xi,\sigma,r) E(\xi',\sigma',r) \qquad (\sigma,\sigma' = 1,2) \qquad (8)$$

It is worthwhile to note that the validity of the set of Eqs.(7) does not require neglecting any term of the type $\nabla(\ln n)$ (as in the scalar Helmholtz equation) and that the equations are inherently first order in the z-derivative without the necessity for any slow-variation hypothesis.[26] Starting from Eqs.(7), multiplying both sides by the mode configuration provided by Eq.(6) times the factor exp(-iβ_ξz) and integrating over ξ, it is possible to go back to the space domain and obtain an equation for E(x,y,z). More precisely, if the field is polarized in the x-z plane this procedure yields, in the unidimensional case E = E(x,z) , the following equation for the component E_T transverse to z

$$\left(\frac{\partial}{\partial z}+\frac{i}{2k}\frac{\partial^2}{\partial x^2}-\frac{i}{8k^3}\frac{\partial^4}{\partial x^4}+...\right)E_T = -i\frac{k}{n_1}\delta n E_T + \frac{i}{2kn_1}\left(\delta n\frac{\partial^2}{\partial x^2}E_T - E_T\frac{\partial^2}{\partial x^2}\delta n\right)+... \quad (9)$$

Note that this equation goes beyond the usual paraxial approximation, both in the diffractive part on the left-hand side and in the interaction part on the right-hand side.

3. The Nonlinear Schroedinger Equation

If we insert in Eq.(4) the expression for δn pertinent to the optical Kerr effect, one obtains, for a linearly polarized field E for which $\delta n = n_2|E|^2$, the so-called nonlinear Schroedinger equation (NLSE), that is

$$\frac{\partial}{\partial z}E + \frac{i}{2k}\left(\frac{\partial^2}{\partial x^2}+\frac{\partial^2}{\partial y^2}\right)E = -ik\frac{n_2}{n_1}|E|^2 E \quad (10)$$

By introducing the dimensionless quantities $(\xi,\eta,\zeta)=(kx,ky,kz)$ and $u=(n_2/n_1)^{1/2}E$, this equation takes the normalized form

$$\frac{\partial}{\partial \zeta}u + \frac{i}{2}\left(\frac{\partial^2}{\partial \xi^2}+\frac{\partial^2}{\partial \eta^2}\right)u = -i|u|^2 u \quad (11)$$

while Eq.(9) becomes

$$\left(\frac{\partial}{\partial \zeta}+\frac{i\partial^2}{2\partial \xi^2}-\frac{i\partial^4}{8\partial \xi^4}\right)u = -i|u|^2 u - \frac{i}{2}\frac{\partial}{\partial \xi}\left(u^2\frac{\partial}{\partial \xi}u^*\right) \quad (12)$$

It is possible to generalize Eq.(11) to include a time dependence accounting for the influence of chromatic dispersion on a narrow-band polychromatic field, thus obtaining[18]

$$\frac{\partial}{\partial \zeta}u + \frac{i}{2}\left(\frac{\partial^2}{\partial \xi^2}+\frac{\partial^2}{\partial \eta^2}+\frac{\partial^2}{\partial \tau^2}\right)u = -i|u|^2 u \quad (13)$$

where the dimensionless variable $\tau = (t-z/V)(-kA)^{1/2}$ has been introduced (we assume $A<0$) and we have defined $1/A = d^2k/d\omega^2$ and $1/V = dk/d\omega$.

In the one dimensional case (1D), that is $\partial/\partial \eta = \partial/\partial \tau = 0$, the NLSE is exactly solvable by means of the so-called inverse scattering method.[2] Particular solutions (spatial solitons) can be obtained by imposing the boundary condition (with N integer)

$$u(\zeta = 0,\xi) = N \operatorname{sech}(\xi) \quad (14)$$

In particular, for N=1 one obtains the fundamental bright soliton which propagates along z without changing its shape,

$$u(\zeta,\xi) = e^{-i\zeta/2} \operatorname{sech}(\xi) \quad (15)$$

Self-trapped Optical Beams

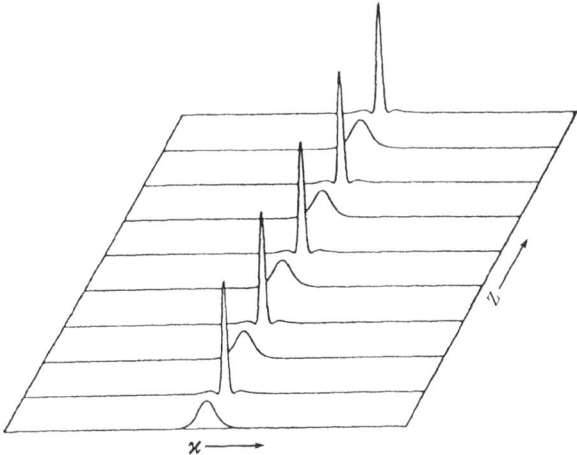

Fig. 1. Evolution of the N = 2 soliton over half soliton-period.

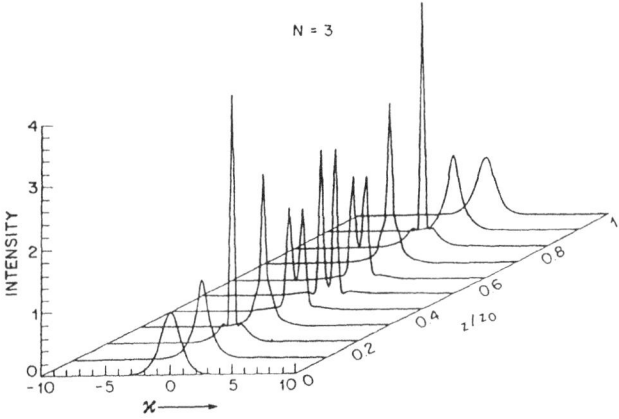

Fig. 2. Evolution of the N = 3 soliton over one soliton-period.

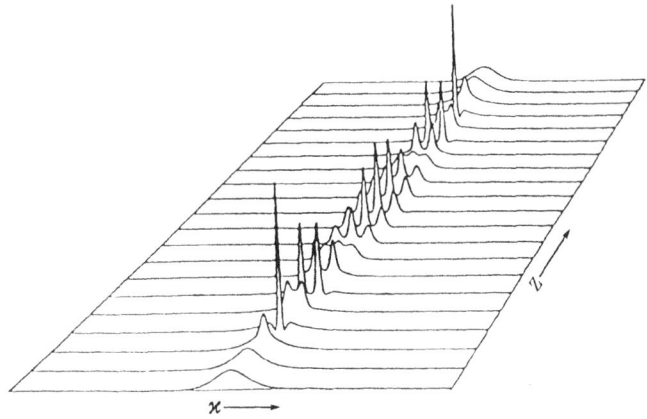

Fig. 3. Evolution of the N = 4 soliton over one soliton-period.

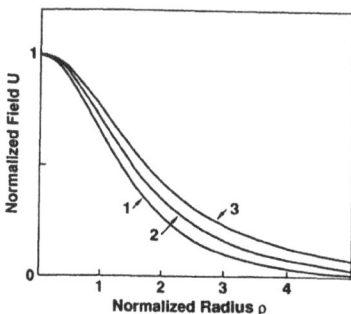

Fig. 4. Radial dependence of U(ρ) for the 1D, 2D and 3D case. The 1D case corresponds to U(ρ) = sech(ρ) (after Ref. 18).

while for N>1 one obtains a class of higher-order solitons whose intensities $|u(\zeta,\xi)|^2$ are periodic in ζ with a common period $\zeta_0 = \pi/2$ (see Figs. 1,2, and 3).

Note that, in general, the NLSE is invariant under the simultaneous transformations $\zeta \to \varepsilon^2\zeta$, $(\xi,\eta,\tau) \to (\varepsilon\xi,\varepsilon\eta,\varepsilon\tau)$ and $u \to \varepsilon u$, a circumstance which allows us, for example, to choose the scale of variation of x in such a way that (with ε small) higher-order contributions (such as those appearing in Eq.(12)) are indeed negligible and the paraxial approximation is valid. However, if ε becomes comparable with unity, that is the transverse variation of the field takes place on the scale of a wavelength, the paraxial approximation fails. Thus, if we consider the fundamental soliton in ordinary units, one has

$$E(z,x) = E_0 e^{-iz/2kd^2} \operatorname{sech}(x/d) \tag{16}$$

with $E_0 = (n_1/n_2)^{1/2} 1/kd$, where d is a measure of the beam radius (d>>λ), while the period of higher-order solitons is $z_0 = \pi k d^2/2$.

Although true soliton solutions exist only in the 1D case, it is however possible, in the 2D and 3D cases, to investigate the existence of self-supporting solutions of the type $u(\zeta,\xi,\eta) = \exp(i\beta\zeta)U(\xi,\eta)$ or $u(\zeta,\xi,\eta,\tau) = \exp(i\beta\zeta)U(\xi,\eta,\tau)$ where, due to the radial symmetry of the diffraction process, one can look for radially symmetric self-supporting solutions of the kind $U(\xi^2+\eta^2)$ or $U(\xi^2+\eta^2+\tau^2)$ (light bullets). These solutions have been studied numerically in Ref.(18) and the dependence of U as a function of the radial coordinate $\rho = (\xi^2+\eta^2+\tau^2)^{1/2}$ is reported in Fig. 4. A simple argument, based on dimensionality, shows that while the 1D case is stable, the 2D and 3D cases are unstable, that is any small deviation from the self-trapped solutions leads to either collapse or divergence.[3]

A very general, if approximate, method for studying self-trapped solutions is based on the finite dimensional Lagrangian approach, which shall be outlined in next section.

4. The FiniteDimensional Lagrangian Approach

This approch, which has proved very useful in the study of a number of nonlinear propagation problems,[14,15,17,27,28] requires the field equations for u and u* to be expressible as a Euler-Lagrange equation of a suitable Lagrangian density $L(u,u^*,u_\zeta,u_\zeta^*,u_\xi,u_\xi^*,u_\eta,u_\eta^*,u_\tau,u_\tau^*)$ (where $u_\zeta=\partial u/\partial \zeta$, and analogously for the other coordinates), that is

$$\frac{\partial}{\partial \zeta}\frac{\partial L}{\partial u_\zeta^*} + \frac{\partial}{\partial \xi}\frac{\partial L}{\partial u_\xi^*} + \frac{\partial}{\partial \eta}\frac{\partial L}{\partial u_\eta^*} + \frac{\partial}{\partial \tau}\frac{\partial L}{\partial u_\tau^*} - \frac{\partial L}{\partial u^*} = 0 \qquad (17)$$

$$\frac{\partial}{\partial \zeta}\frac{\partial L}{\partial u_\zeta} + \frac{\partial}{\partial \xi}\frac{\partial L}{\partial u_\xi} + \frac{\partial}{\partial \eta}\frac{\partial L}{\partial u_\eta} + \frac{\partial}{\partial \tau}\frac{\partial L}{\partial u_\tau} - \frac{\partial L}{\partial u} = 0 \qquad (18)$$

corresponding to a vanishing variation of L

$$\delta \int_{-\infty}^{+\infty} d\zeta \int_{-\infty}^{+\infty} d\xi \int_{-\infty}^{+\infty} d\eta \int_{-\infty}^{+\infty} d\tau L\!\left(u, u^*, u_\zeta, u_\zeta^*, u_\xi, u_\xi^*, u_\eta, u_\eta^*, u_\tau, u_\tau^*\right) = 0 \qquad (19)$$

Considering for example Eq.(13), the associated L reads

$$L = \frac{1}{2i}\left(u * u_\zeta - u u_\zeta^*\right) + \frac{1}{2}|u_\xi|^2 + \frac{1}{2}|u_\eta|^2 + \frac{1}{2}|u_\tau|^2 + \frac{1}{2}|u|^4 \qquad (20)$$

We look now for an approximate analytical solution of Eq.(11) or (13) within a set of suitably chosen trial functions, respectively of the form

$$u(\xi,\eta,\zeta) = \frac{M_0^{1/2}(\zeta)}{\sigma^{1/2}(\zeta)\mu^{1/2}(\zeta)} F^{1/2}\!\left(\frac{\xi^2}{\sigma} + \frac{\eta^2}{\mu^2}\right) e^{ia_0(\zeta)+ia_1(\zeta)\xi^2+ia_2(\zeta)\eta^2} \qquad (21)'$$

$$u(\rho,\tau,\zeta) = \frac{M_0^{1/2}(\zeta)}{\sigma^{1/2}(\zeta)\mu^{1/2}(\zeta)} F^{1/2}\!\left(\frac{\rho^2}{\sigma} + \frac{\tau^2}{\mu^2}\right) e^{ia_0(\zeta)+ia_1(\zeta)\rho^2+ia_2(\zeta)\tau^2} \qquad (21)''$$

where F is a well-behaved prescribed function and M_0, σ, μ, a_0, a_1, a_2 are *a priori* unknown ζ-dependent parameters. Referring for simplicity to the 2D case, we can now substitute Eq. (21)' into Eq. (20) (where we put $u_\tau = 0$) and insert the resulting equation into Eq.(19) (where the τ integration is omitted) thus getting the reduced variational principle

$$\delta \int_{-\infty}^{+\infty} d\zeta L_r = 0 \qquad (22)$$

where

$$L_r = \int_{-\infty}^{+\infty} d\xi \int_{-\infty}^{+\infty} d\eta L \qquad (23)$$

Accordingly, the exact field Lagrangian L is substituted by the reduced finite-dimensional Lagrangian L_r, whose Euler-Lagrange equations for the q_i's $\equiv$ (M_0, σ, μ, a_0, a_1, a_2)

$$\frac{d}{dt}\frac{\partial L_r}{\partial q_i} - \frac{\partial L_r}{\partial q_i} = 0$$

describe their evolution.

If, for example, we assume $|u(\zeta=0,\xi,\eta)|$ to possess a Gaussian shape, that is

$$F^{1/2}\left(\frac{\xi^2}{\sigma^2}+\frac{\eta^2}{\mu^2}\right)=e^{-\left[\xi^2/2\sigma^2(\zeta)+\xi^2/2\mu^2(\zeta)\right]} \tag{24}$$

the reduced Lagrangian takes the form

$$L_r = \dot{a}_0 M_0 + M_0\left(\dot{a}_1\sigma^2 + \dot{a}_2\mu^2\right) - \frac{1}{4\pi}\frac{M_0^2}{\sigma\mu} + M_0\left(a_1^2\sigma^2 + a_2^2\mu^2\right) + \frac{M_0}{4}\left(\frac{1}{\sigma^2}+\frac{1}{\mu^2}\right) \tag{25}$$

where the dot stands for the derivative with respect to ζ.

The corresponding Euler-Lagrange equations read (omitting, for the sake of conciseness, the equation describing the evolution of a_0)

$$\dot{M}_0 = 0 \tag{26}$$

$$\dot{\sigma} - 2\sigma a_1 = 0 \tag{27}$$

$$\dot{\mu} - 2\sigma a_2 = 0 \tag{28}$$

$$\dot{a}_1 + \frac{1}{4\pi}\frac{M_0}{\sigma^3\mu} + 2a_1^2 - \frac{1}{2\sigma^4} = 0 \tag{29}$$

$$\dot{a}_2 + \frac{1}{4\pi}\frac{M_0}{\sigma\mu^3} + 2a_2^2 - \frac{1}{2\mu^4} = 0 \tag{30}$$

where Eq. (26) expresses energy conservation. By differentiating Eqs.(27) and (28) with respect to ζ and taking advantage of Eqs.(26), (29) and (30), we can obtain the two nonlinear coupled equations describing the ζ-evolution of the widths σ and μ of an asymmetric pulse, that is of a pulse possessing different widths along the transverse x and y-axes, as follows:[29]

$$\ddot{\sigma} = -\frac{1}{2\pi}\frac{M_0}{\sigma^2\mu} + \frac{1}{\sigma^3} \tag{31}$$

$$\ddot{\mu} = -\frac{1}{2\pi}\frac{M_0}{\sigma\mu^2} + \frac{1}{\mu^3} \tag{32}$$

whose solution requires the knowledge of the boundary conditions $\sigma(\zeta=0)=\sigma_0$ and $\mu(\zeta=0)=\mu_0$.

It is evident that the form of the finite dimensional equations that approximate the exact partial differential equation for $u(\zeta,\xi,\eta)$ depends on the choice of the trial function F. (We could have as well chosen a hyperbolic secant shape instead of the Gaussian one provided by Eq.(24) and we would have obtained the same equations but with different numerical coefficients.) However, for sensible choices of F, the variational approach provides analytical results in good agreement with the exact numerical ones, as has been for example shown [15] in connection with self-trapping of symmetric optical pulses (see Fig. 5).

Self-trapped Optical Beams

The standard 1D case is recovered by assuming that both nonlinearity and diffraction are ineffective in one of the two transverse dimensions (e.g., y), and corresponds to letting μ go to infinity in Eq.(32) and to setting $\mu=\mu_0$ in Eq.(31), thus getting

$$\ddot{\sigma} = -\frac{1}{2\pi}\frac{M_0}{\sigma^2\mu_0} + \frac{1}{\sigma^3} \tag{33}$$

where the first term on the right-hand side accounts for the Kerr nonlinearity and the second for diffraction. The solution of Eq.(33) can be investigated by taking advantage of the identity $2d^2\sigma/d\zeta^2 = d/d\sigma[(d\sigma/d\zeta)^2]$ and by assuming the beam waist to be located in the input plane $\zeta=0$, that is $(d\sigma/d\zeta)_{\zeta=0} = 0$. Under this hypothesis, Eq.(33) can be rewritten as

$$\sigma\frac{d\sigma}{d\zeta} = \frac{1}{\sqrt{2\sigma_0}}[(\sigma-\sigma_0)(\sigma_0-\alpha\sigma)]^{1/2} \tag{34}$$

where $\alpha = (M_0\sigma_0/\pi\mu_0) - 1$. Depending on the values assumed by this parameter, we obtain different propagation regimes. If $\alpha = 1$, that is for the critical value of the power $M_0 = <M_0> = 2\pi\mu_0/\sigma_0$, one has the self-trapped solution $\sigma(\zeta) = \sigma_0$. For $-1 < \alpha \leq 0$, one has a diffraction-dominated regime and $\sigma(\zeta) \to \infty$ as $\zeta \to \infty$. For α slightly different from unity, $\sigma(\zeta)$ is a periodic function undergoing small oscillations around the equilibrium value σ_0, which expresses the well-known stability properties of 1D solitons.

The set of Eqs. (31)-(32) can also be used to describe the evolution of the width μ along the "large" dimension once the width along the "small" dimension is assumed to remain constant in ζ (and equal to σ_0). Taking into account that $\sigma_0 \ll \mu_0$, Eq.(32) can be rewritten as

$$\ddot{\mu} = -\frac{1}{2\pi}\frac{M_0}{\sigma_0\mu^2} \tag{35}$$

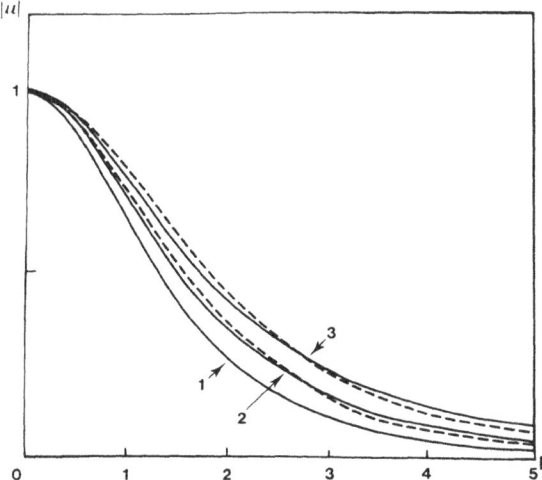

Fig. 5. Comparison of the radial profiles of the self-trapped analytical solutions provided by the variational approach (dashed curves) and numerical solutions for the 1D, 2D and 3D cases. (after Ref. 15).

where diffraction effects have been neglected. This equation can be easily rewritten in the form

$$\frac{d\mu}{d\zeta} = \left[\gamma \left(\frac{1}{\mu} - \frac{1}{\mu_0} \right) + q_0^2 \right]^{1/2} \tag{36}$$

where $q_0 = d\mu/d\zeta)_{\zeta=0}$ and $\gamma = M_0/\pi\sigma_0$, which can be easily integrated by quadrature. In particular, the quantity $d\mu/d\zeta$, which can be interpreted as self-induced angular divergence $\theta(\zeta)$ of the beam, can be approximately written as[29]

$$\theta(\zeta) \cong \theta(0) - \frac{n_2 Z_0 I}{2\pi n_1} (z/w_y) \tag{37}$$

where Z_0 is the vacuum impedance, w_x and w_y the transverse widths of the beam and $I=P/w_x w_y$ its intensity. The last expression can be used to interpret recent experimental observations of beam self-deflection.[30]

5. The Photorefractive Soliton

The preceding considerations were mainly concerned with the optical Kerr effect, a nonlinear effect based on modification, local in space and time, of the linear refractive index. We wish now to deal with a new kind of nonlinearity, associated with the photorefractive effect present in some crystals (like $BaTiO_3$, barium titanate), which turns out to be nonlocal in space (nonlocality in time is not relevant, since we consider a stationary monochromatic case).

The photorefractive mechanism[31] is schematically illustrated in Fig.6 for the case of two intersecting laser beams, forming an interference pattern inside the crystal. The associated periodic intensity distribution excites charge carriers into the conduction zone, where they are preferentially trapped in regions of low optical intensity; this leads to charge separation and to a periodic space charge which, in turn, modulates the refractive index via the electrooptic effect. In this simple case, the nonlinear contribution $\delta n(r,z)$ to the refractive index reads (note the different choice of the sign of the wavevector compared with the preceding paragraphs)

$$\delta n(r,z) = \frac{1}{I_0} [a_1(z) e^{i(\xi_1 r + \beta_{\xi_1} z)} a_2^*(z) e^{-i(\xi_2 r + \beta_{\xi_2} z)} \delta\hat{n}(\xi_1, \xi_2) + c.c.) \tag{38}$$

where the $a_i(z)$'s (i=1,2) are the amplitudes of the two interfering waves, $I_0 = |a_1|^2 + |a_2|^2$ is the light intensity ($I_0 \gg |a_1 a_2|$) and $\delta\hat{u}(\xi_1, \xi_2)$ is a complex factor depending on the material properties, the static externally applied electric field E_0 and the polarization of the waves.[20] In the more complicated situation in which a structured beam is incident on the crystal and a continuum of plane waves is present, Eq.(38) involves a summation over all possible pairs of interfering waves. More precisely, if one writes

$$E(r,z,t) = \frac{1}{2} \left[e^{i(kz - \omega t)} \int d\xi E(\xi, r) e^{i(\beta_\xi - k)} c(\xi, z) + c.c. \right] \equiv \frac{1}{2} \left[A(r,z) e^{i(kz - \omega t)} + c.c. \right] \tag{39}$$

Eq.(38) takes the form

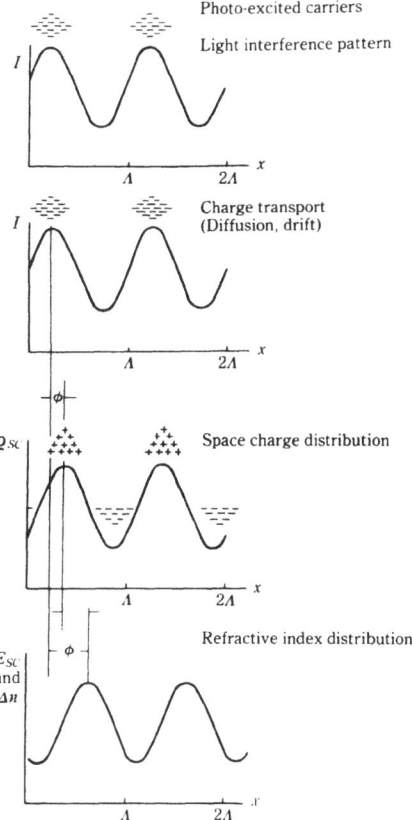

Fig. 6. The photorefractive effect.

$$\delta n(r,z) = \frac{1}{|A(r,z)|^2} \int d\xi_1 \int d\xi_2 E(\xi_1,r)E^*(\xi_2,r)e^{i(\beta_{\xi_1}-\beta_{\xi_2})z}c(\xi_1,z)c^*(\xi_2,z)\delta\hat{n}(\xi_1,\xi_2) \quad (40)$$

After introducing the Fourier transform of $\delta\hat{n}(\xi_1,\xi_2)$, that is

$$\delta\hat{n}(\xi_1,\xi_2) = \int d\rho \int d\rho'\, e^{-i(\xi_1\rho+\xi_2\rho')}g(\rho,\rho') \quad (41)$$

$\delta n(r,z)$ can be rewritten in the form

$$\delta n(r,z) = \frac{1}{|A(r,z)|^2} \int d\rho \int d\rho'\, A(r-\rho,z)A^*(r+\rho',z)g(\rho,\rho') \quad (42)$$

which makes evident the spatial nonlocal nature of the photorefractive mechanism. By inserting Eq.(42) into Eq.(4) we obtain, after recalling Eq.(39), the nonlinear integro-differential equation

$$\left[\frac{\partial}{\partial z}-\frac{i}{2k}\left(\frac{\partial^2}{\partial x^2}+\frac{\partial^2}{\partial y^2}\right)\right]A(r,z)=\frac{ik}{n_1}\frac{1}{A^*(r,z)}\int d\rho\int d\rho'\, A(r-\rho,z)A^*(r+\rho',z)g(\rho,\rho') \quad (43)$$

We now look for self-trapped solutions of the kind

$$A(r,z)=e^{i\gamma z}U(r) \quad (44)$$

Referring to the one dimensional geometry depicted in Fig.7, and assuming the beam diameter to be large compared with the typical scale d of nonlocality of the crystal, we expand $A(r-\rho,z)$ and $A^*(r+\rho',z)$ around $\rho,\rho' = 0$. Truncating the Taylor expansion after the second term, it is possible to obtain an ordinary nonlinear differential equation which reads[20]

$$\left(\frac{\partial}{\partial z}-\frac{i}{2k}\frac{\partial^2}{\partial x^2}\right)A(x,z)=-i\frac{k}{n_1}I_{11}\frac{1}{A^*}\left|\frac{\partial A}{\partial x}\right|^2+\frac{ik}{2n_1}I_{20}\frac{\partial^2 A}{\partial x^2}+\frac{ik}{2n_1}I_{20}\frac{A}{A^*}\frac{\partial^2 A^*}{\partial x^2} \quad (45)$$

where

$$I_{nm}=\int d\rho\int d\rho'\, g(\rho,\rho')\rho^m\rho'^n \quad (46)$$

Looking now for a self-trapped solution of the form given by Eq.(44), Eq.(45) can be cast, after some simple manipulations, in the compact form

$$\gamma-a\frac{U''}{U}+b\left(\frac{U'}{U}\right)^2=0 \quad (47)$$

where the prime stands for the derivative with respect to x and

$$a=\frac{1}{2k}+\frac{k}{n_1}\text{Re}(I_{20})\, ,\, b=\frac{k}{n_1}I_{11} \quad (48)$$

Equation (47) admits of the solution

$$U(x)=U_0[\text{sec h}(\alpha x)]^D \quad (49)$$

where U_0 and α can be chosen arbitrarily and

$$D=\frac{a}{b-a}\, ,\, \gamma=\frac{a^2}{a-b}\alpha^2 \quad (50)$$

Let us consider, as an example, $Sr_{1-x}Ba_xNb_2O_6$ (strontium barium niobate, SBN). If the optical field is linearly polarized along the x-axis, which we assume to coincide with the c-axis of the crystal (see Fig. 7), one has

$$\delta\hat{n}(q_1,q_2)=\frac{B}{1+d^2(q_1-q_2)} \quad (51)$$

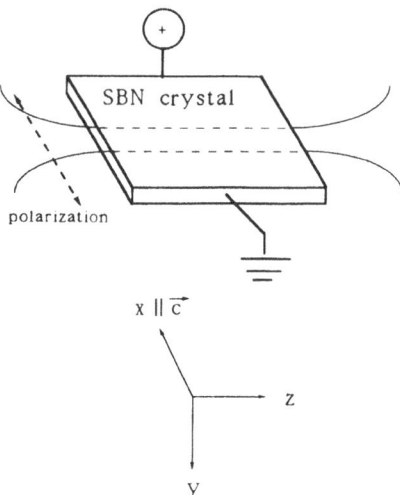

Fig. 7. The geometry of propagation.

where d is the scale of nonlocality and B a coefficient whose value depends on various crystal parameters and on the external field E_0. Equation (51) allows us to evaluate the I_{nm}'s ; in particular, one has $I_{11} = -\mathrm{Re}(I_{20}) = -2Bd^2$, which, inserted in Eq.(48), permits the evaluation of the coefficients a and b and , through Eq.(50), of the exponent D appearing in Eq.(49). In order that $U \to 0$ when $x \to \pm \infty$, D has to be positive, a condition which implies

$$-\frac{n_1}{4k^2d^2} < B < -\frac{n_1}{8k^2d^2} \qquad (52)$$

which in turn can be shown to provide a well defined interval of values of E_0 consistent with the possibility of propagating a self-trapped beam. Thus, any experimental verification of the existence of the photorefractive soliton has to satisfy these three conditions: independence of the beam amplitude; and transverse scale of variation $1/\alpha$ (as long as $1/\alpha \gg d$); and the existence of a limited range of admissible values of E_0.

Very recently, photorefractive solitons have been experimentally observed in SBN crystals.[21] These solitons preserve their profile, independently of the input power which can be less than 100 μW, for an external field E_0 in the range 200 - 400 V/cm.

References

1. R.Y. Chiao, E. Garmire, and C.H. Townes, Phys.Rev.Lett. 13, 479 (1964)
2. V.E. Zacharov and A.B.Shabat, Sov.Phys. JETP 34, 62 (1972)
3. V.E. Zacharov and V.S. Synakh, Sov.Phys. JETP 41, 465 (1976)
4. A. Hasegawa and F. Tappert, Appl.Phys.Lett. 23, 142 (1973)
5. See, e.g., C.P. Agrawal, "Nonlinear Fiber Optics", (Academic Press, San Diego, 1989)
6. A. Barthelemy, S. Maneuf, and C. Froehly, Opt.Commun. 55, 201 (1985)
7. S. Maneuf, R. Desailly, and C. Froehly, Opt.Commun. 65, 193 (1988)

8. S. Maneuf and F. Reynaud, Opt.Commun. 66, 325 (1988)
9. F. Reynaud and A. Barthelemy, Europhys.Lett. 12, 401 (1990)
10. J.S. Aitchison, A.M. Weiner, Y. Silberberg, M.K. Oliver, J.L. Jackel, D.E. Laeird, E.M. Vogel and P.E.W. Smith, Opt.Lett. 15, 471 (1990)
11. J.S. Aitchison, A.M. Weiner, Y. Silberberg, D.E. Laeird, M.K. Oliver, J.L. Jackel and P.E.W. Smith, Opt.Lett. 16, 15 (1991)
12. A.W. Snyder, D.J. Mitchell, L. Poladian and L. Ladouceur, Opt.Lett. 16, 21 (1991)
13. Q.Y. Li, C. Pask and R.A. Sammut, Opt.Lett. 16, 1083 (19..)
14. M. Karlsson, D. Anderson, M. Desaix and M. Lisak, Opt.Lett. 16, 1373 (1991)
15. M. Desaix, D. Anderson and M. Lisak, J.Opt.Soc.Am. B 8, 2082 (1991)
16. J.T. Manassah, Opt.Lett. 17, 1259 (1992)
17. R.A. Sammut, C. Pask and Q.Y. Li, J.Opt.Soc.Am. B 10, 485 (1993)
18. Y. Silberberg, Opt.Lett. 15, 1282 (1990)
19. M. Segev, B. Crosignani, A. Yariv and B. Fisher, Phys.Rev.Lett. 68, 923 (1992)
20. B. Crosignani, M. Segev, D. Engin, P. Di Porto, A. Yariv and G. Salamo, J.Opt.Soc.Am. B 10, 446 (1993)
21. G.C. Duree, J.L. Shultz, G.J. Salamo, M. Segev, A. Yariv, B. Crosignani, P. Di Porto, E.J. Sharp and R.R. Neurgaonkar, Phys. Rev. Lett. 71, 533 (1993)
22. S. Solimeno, B. Crosignani and P. Di Porto, "Guiding, Diffraction and Confinement of Optical Radiation" (Academic, Orlando, Fla.1986).
23. M.D. Feit and J.A. Fleck, Jr., J.Opt.Soc.Am. B 5, 633 (1988)
24. N. Akhmediev, A. Ankiewicz and J.M. Soto-Crespo, Opt.Lett. 18, 411 (1993)
25. B. Crosignani and A. Yariv, J.Opt.Soc.Am. A 1, 1034 (1984)
26. For a more detailed discussion of when the SVA is necessary, see : B. Crosignani, P. Di Porto and A. Yariv, Opt.Commun. 78, 237 (1990)
27. B. Crosignani and P. Di Porto, Opt. Commun. 89, 453 (1992)
28. B. Crosignani, P. Di Porto and S. Piazzolla, Pure Appl.Opt. 1, 7 (1992)
29. B. Crosignani and P. Di Porto, Opt.Lett. 18, 1394 (1993)
30. A. Barthelemy, C. Froehly, S. Maneuf and F. Reynaud, Opt.Lett. 17, 844 (1992)
31. See, e.g., A. Yariv, "Quantum Electronics" 3rd ed. (Wiley, New York, 1989)

Chapter 6

ADVANCES IN SEMICONDUCTOR INTEGRATED OPTICS

A. CARENCO

1. Introduction

The emergence of optical fibers in commercial systems is undoubtedly one of the landmark events of this decade. This application stems naturally from the exceptional transmission performance offered by single-mode silica fibers operating in the near infra-red (1.3 and 1.5 µm), in terms of both loss and bandwidth.

In order to explore and eventually exploit the full potential of single mode fibers, new components exhibiting superior performance are required. Although current and forthcoming systems, comprised primarily of long-haul digital arteries and special-purpose local area networks, rely on today's relatively high-priced components, the use of single-mode fiber in the loop will hinge on our capability to fabricate batches of the necessary optoelectronic circuits on a mass scale.

Research in integrated optics is therefore focussed on generating innovative elements while lowering costs, to enable the development of new architectures based on single-mode fiber. Work on guided waves is aimed not only at improving existing devices (emitters and receivers), but also creating new ones to amplify, modulate, switch or multiplex light signals. These development efforts are spurred by the design of systems featuring increasingly complex architectures. The deployment of such circuits is expected to lead to greatly expanded link capacity, mostly by using :
a) time division multiplexing (TDM) in high data rate systems; and,
b) wavelength division multiplexing (WDM) in multichannel incoherent (optical demultiplex-direct detection) and coherent (heterodyne detection-electrical demultiplex) transmission, networking and switching systems.

All require more complex stations with a large number of interconnected optical devices; and all benefit greatly from optical amplification and the consequent reduction in the number of optical/electrical conversions.

This chapter describes current research efforts conducted in integrated optics using III-V semiconductors (mostly InP), towards the emergence of Opto-Electronic Integrated Circuits (OEICs) which should have a significant impact on system evolution. The main obstacles are discussed as well as the different approaches taken in research laboratories to overcome such problems, in a particularly fast-paced international context.

A. Carenco - FRANCE TELECOM/CNET/Centre PAB 196, Avenue H. Ravera, BP 107, 92225 Bagneux, France

2. Monolithic Integration

To date, many systems have employed discrete components combined in a hybrid form. Single-mode optical circuits can be fabricated from existing discrete components (which are readily available in the laboratories or commercially), assembled in hybrid form using standard or polarisation-maintaining fiber pigtails. In addition to the basic elements-laser and photodetector-there is a wide variety of active and passive single-mode components which have been produced in thin-film technology in many different materials (glass, ferroelectrics, semiconductors,...). Hybrid assemblies have played and are playing a crucial role in the set-up of experimental links.

Hybrid circuits are becoming increasingly complex, particularly in systems combining WDM or High Density WDM (in heterodyne detection). The natural tendency is to contemplate monolithic integration of functions on a single chip to achieve OEICs, or Photonic Integrated Circuits (PICs)[1], which are the subset of OEICs that aims to replace the individually aligned single-mode fiber connections between guided-wave optical devices with lithographically produced waveguides on a single semiconductor substrate. The opportunity to replace discrete optical interconnections with integrated waveguides should yield substantial savings in packaging and also improve system robustness, while providing higher and more stable performances, despite the fact that individual components can no longer be built on the best-suited substrate material.

2.1. The Host Material

Because of their favourable radiation properties, direct bandgap semiconductor materials appear to show the greatest potential for integration. The transmission windows accommodated by existing optical fibers presently limits the choice of materials to III-V semiconductors.

Such materials can be utilised for monolithic structures containing all of the optical elements (light source, detector, switch or modulator,..) as well as the high-speed electronic drive circuitry.

The first attempts at integration focussed on GaAs. Today, InP is being explored since it emits at longer wavelengths which are transmitted more efficiently through silica fibers.

2.2. Semiconductor Waveguides

The optical waveguide is the basic element acting as interface to various active components in the circuit.

Waveguides [2] are fabricated from variably doped GaAs or InP, since the effective mass of carriers (mostly the electrons) in both materials is low enough to have a significant contribution to free-carrier resonance effects. Usually, large refractive index changes (several times 0.1) yielded by single or double heterostructures (SH or DH) based on variations in their composition are preferred over small refractive index changes (several times 0.001) in homostructures (n⁻ InP/ n+ InP).

Among the families of lattice-matched materials from columns III (In, Ga, Al) and V (P,As, Sb), the most commonly employed systems are the following [3] :

(1) for $\lambda < 0.9$ μm, GaAs/Ga$_{1-x}$Al$_x$As, where the refractive index decreases when the parameter x is increased; and,

(2) for λ in the 1 to 1.6 μm range, InP/In$_{1-x}$Ga$_x$As$_y$P$_{1-y}$, which is perfectly matched for

x=0.466 y and InP/In$_{1-x-y}$Ga$_x$Al$_y$As with x+y =0.468 + 0.017 y, where the refractive index of the quaternary is higher than that of InP.

A continuous range of energy gap and refractive index can be varied smoothly by changing x and y. In the first system, GaAs is the waveguide core and the cladding is made up of the ternary AlGaAs. In the second system, the quaternary constitutes the core and InP the cladding. This point is brought up because the optical quality of the waveguide depends on the homogeneity of the core composition, and this is *a priori* more difficult to achieve in a quaternary than in abinary alloy system.

Most devices are fabricated using metal-organic (MOCVD) or molecular beam epitaxy (MBE) growth techniques, which are well suited for growing large areas (several 2 or 3-inch wafers) and enable precise control over layer thickness as well as over composition and doping profiles. Furthermore, the ability to grow extremely thin layers down to the atomic scale is essential for the fabrication of multiquantum well (mqw) and superlattice (sl) structures. Quantum wells exhibit significant electronic and excitonic properties at room temperature, which pave the way to applications such as emission, modulation, optical bistability and photodetection. The possibility of judiciously combining layer thickness and composition of wells and barriers in sl and qw structures has opened ways to "engineer" the bandgap and refractive index in order to design better performing circuits. Moreover, each individual layer can be intentionally grown with a slight mismatch [4]. The induced compressive or tensile strain modifies the semiconductor band structure [5]. It is a very fruitful means to lower the threshold of lasers, to enhance the gain and power saturation of amplifiers and to realize polarisation insensitive modulators and amplifiers [6-8].

Among the variety of possible waveguide structures, the rib and the buried waveguide structures have attracted the most interest (Fig.1). The rib structure offers the advantage of being fabricated after epitaxy; the lateral confinement of light can be modified throughout the circuit simply by varying the etching depth around the rib. Bent guides (Fig.2) with a shorter radius of curvature (< 1 mm) can be made with negligible loss [10,11].

Numerous experimental results have shown that the main causes of propagation losses in straight waveguides are absorption by free carriers and diffusion due to imperfections. In the 1.5 µm region, losses due to free carriers in the n$^+$ and p$^+$ layers used for electrical purposes are roughly estimated to be : α (cm^{-1}) = 1.3 · 10^{-18} n$^+$ (cm^{-3}) and an order of magnitude[2] more for p$^+$.

Adapted technologies based on wet and dry etching techniques have been developed for patterning 2-D waveguide with a smooth process [12]. Dry etching which is less crystal-axis dependent is very useful for realising bent waveguides or totally reflecting corner mirrors providing large directional changes with low additional loss (<1 dB in a 45° Reactive-Ion-Etched DH mirror) [13]. Today, low loss figures are usually achieved for both TE and TM modes in undoped InGaAsP/InP DH waveguides : 0.1- 0.5 dB/cm at 1.5 µm [14]. The

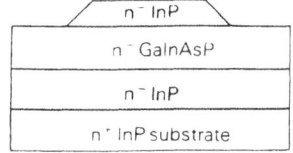
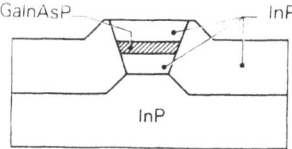

Fig. 1. Cross-section: (Left) of a rib waveguide (typical width = 3 µm, quaternary thickness = 0.5µm); and, (Right) of a buried guide (typical width = 1.5 µm).

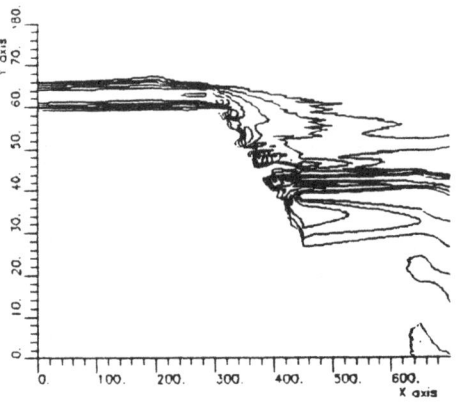

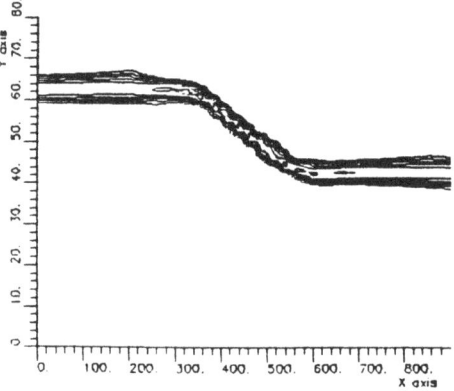

Fig. 2. Simulation by BPM of propagation in an InP/InGaAsP DH rib waveguide laterally-overetched in the bent zone; curvature radius in the symmetric S-shaped guide = 0.5 mm (top), = 1 mm (bottom). [Ref. 9-11]

significant progress registered over recent years is the results of better control over the composition and thickness of the quaternary system, as well as the improvement in the material's morphology.

Tapering is being more and more studied to improve the coupling loss from a tightly-confined small-sized waveguide (typically 1x2 μm^2 mode cross-section) which optimizes most of the active devices into an external fiber which exhibits a large circular mode (diameter 10 μm at $1/e^2$ intensity). In spite of some extra steps in the fabrication process, the adiabatic mode transformation which matches the output mode to the fiber offers the great advantage of relaxing alignment positioning tolerances. Coupling loss figures betwen a lensed-fiber and an AR coated DH waveguide have been reduced to less than 3 dB [15].

2.3. Design Issues

A PIC designer has to face two key issues, an optical one, and an electrical one.
(1) The optical waveguide engineering challenge is: - to design a low loss waveguide with the constraints on doping types and levels (which contribute to the largest part of light attenuation); - to realise efficient coupling between single-mode waveguides which can be

differently shaped (buried, rib, strip-loaded,...) and made of layers having different bandgap energy (less or near the propagation photon energy) according to their function: gain in a laser or amplifier guide, absorption in a photodiode, or high transparency in interconnecting passive waveguides; and, - to tackle the fiber-coupling issue.

(2) The optoelectronic challenge is : - to confine as much as possible the current injection and the electrical field within the mode area to improve the electro-optic overlap; - to provide different electrical polarisations to components which are very closely laid out on the PIC :with a reverse bias on photodiodes, modulators,... and a forward bias on lasers, amplifiers,...; and, - thus, to achieve a high degree of electrical isolation between various active devices of the PIC.

2.4. Tunable Lasers and Laser Arrays

Tunable lasers are key components for WDM networks and switching systems. Indeed, the number of channels is limited directly by the tuning range of the source. Distributed-Feedback-Bragg (DFB) and Distributed-Bragg-Reflector (DBR) lasers are the most developed tunable sources, exhibiting a continuous tunability of about 10 nm [16], and up to 100 nm in step mode with a 40 mA tuning current range [17]. A fast tuning response time (around 0.5 ns) has been measured on DBR lasers [18]. Several new structures based on integrated optic elements which use, like tunable DBR, the carrier-injection tuning effect (i.e., codirectional coupler structures or Y junction structures) have been demonstrated. A 57 nm tuning range has been measured on a InGaAsP/InP laser with a vertical coupler filter [19]. A large effort is being devoted to research on new electro-optic effects which could be more efficient than carrier-injection and thus significantly improve the tuning range while maintaining a narrow laser linewidth. Integrated multi-wavelength sources (laser arrays emitting a stable wavelength comb) are currently attracting a great deal of attention. UV exposure through a phase mask, e-beam, and synchrotron- optical- radiation lithography are the techniques which have been developed to fabricate arrays of submicron Bragg grating mirrors with different pitches. Arrays of 2 nm-spaced 20 DFB lasers have been reported by different groups. An array of 18 multiwavelength- 4 nm spaced- DFB lasers, made up of strained mqw layers (providing a broad gain spectrum), has been integrated with a star coupler and an optical amplifier on a 1x4 mm^2 chip [20]. Thermal dissipation constitutes a critical issue in such devices. Very often, simultaneous cw operation of all the emitters is not possible. Reduction of threshold currents over the wavelength range is a must.

2.5. Laser-modulator Integration

External modulation is a very attractive solution, compared with direct modulation of semiconductor lasers, in term of electrical bandwidth and wavelength "chirp". In particular, the electroabsorption modulators have demonstrated high performance for telecommunication applications. Electroabsorption waveguide modulators are built like semiconductor lasers or amplifiers (Fig.3). The electroabsorption material as a part of the waveguide core is subjected to an electrical field by reverse biasing the pin diode. Three kinds of electroabsorption effects are used : Franz-Keldysh (fk), Quantum-Confined-Stark-Effect (qcse) and Wannier-Stark localisation (ws) [21]. Although based on different physical mechanisms, they all imply a deformation and a red-shift of the absorption edge. The fk concerns bulk material, the qcse exists in mqw and ws localization has been studied only recently for InP-based electroabsorption modulators.

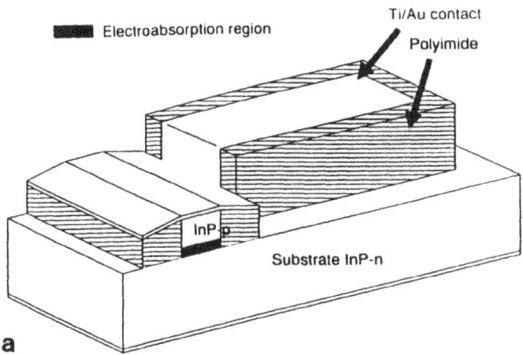

Fig. 3a. Electroabsorption waveguide modulator.

Electroabsorption modulators are well suited to integrated optics : they are very short (< 200 µm), they are intrinsically very fast (< 1 ps), and they are very efficient. A good figure of merit for comparing electroabsorption modulators is the bandwidth-to-drive-voltage ratio, provided that the operating wavelength and device length are such that the insertion loss and extinction ratio are fixed. The best values to date have been obtained for ws and qcse modulators. WS modulators exhibit a 18 GHz bandwidth with a < 1 V drive voltage [22]; qcse modulators a 26 GHz bandwidth with < 2 V [23]. 20 Gbit/s operation with 1.2 V has been recently achieved with a 100 µm long qcse modulator (Fig.3) [24]. Waveguide electroabsorption modulators are attractive also as very short pulse generators (several 10 ps) with 20 GHz repetition rate, in order to transmit solitons at 1.55 µm in optical fibers over thousands of km [25].

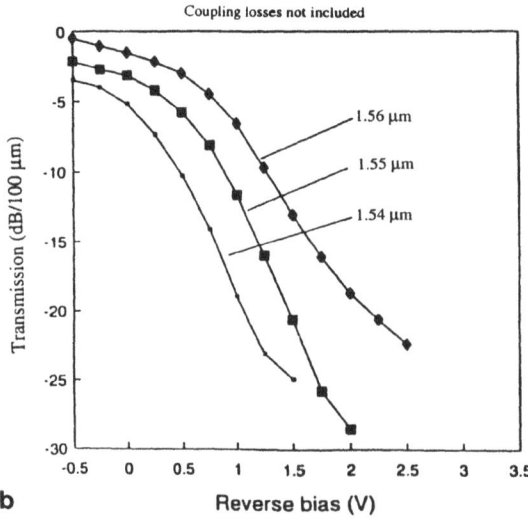

Fig. 3b. Absolute transmission per 100 µm of device length vs reverse bias applied to a modulator based on Quantum Confined Stark effect in InGaAsP/InGaAsP mqw, for several wavelengths in the TE mode.

The prospect now is to integrate electroabsorption modulators with other types of optical components, in particular a tunable single-frequency laser. Basically both structures can be made identical, provided that the number of wells is not too low, as required for a laser, or too high, as required for a modulator. The principal remaining issue is the compatibility of the operating wavelength.

The most prevalent concepts in the fabrication of monolithic laser waveguide structures are:
    (1) butt-coupling, which implies several etch and regrowth steps; or
    (2) evanescent coupling, which decreases the number of regrowths [1].

Most practical realizations have been based so far on bulk-material structures grown in several steps. The use of mqw enables a much more attractive solution, which simplifies the integration process. It is based on bandgap-engineering epitaxy in which the material gap is locally controlled by means of selective deposition through windows of various widths[26] (Fig.4). Using this technique, several realisations of butt-jointless DFB laser-modulators have been made, demonstrating operation at high bit rates (> 10 Gbit/s) under a very low driving voltage (< 2 V) [27].

Generation of very short soliton pulses (around 10 ps) at 4.9 Gbit/s has also been observed in an extended-cavity laser integrated with a modulator [28].

2.6. Guided-wave Receiver Integration

The integration of a photodiode with a waveguide [29] is a relevant OEIC issue, which is very similar to the problem of laser integration. Considerable attention is being paid to waveguide-fed photodiodes, which exhibit a great advantage over conventional top-illuminated photodetectors in their product of bandwidth and internal quantum efficiency, resulting from a distributed thin absorbing layer and a reduced carrier transit time.

The most commonly fabricated waveguide-fed pin diodes are based on the evanescent coupling concept. The structure can be designed to yield either complete

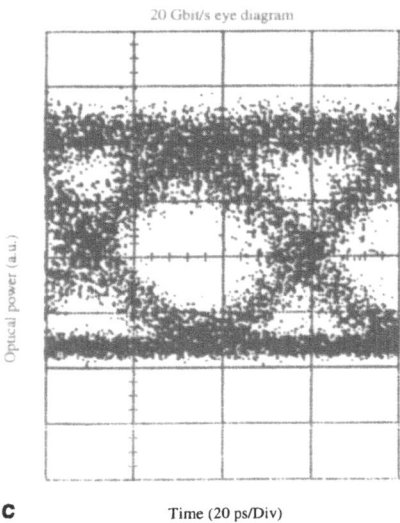

Fig. 3c. 20 Gbit/s eye diagram (sequence length = 2^{23}-1 bits). [Ref. 21-24]

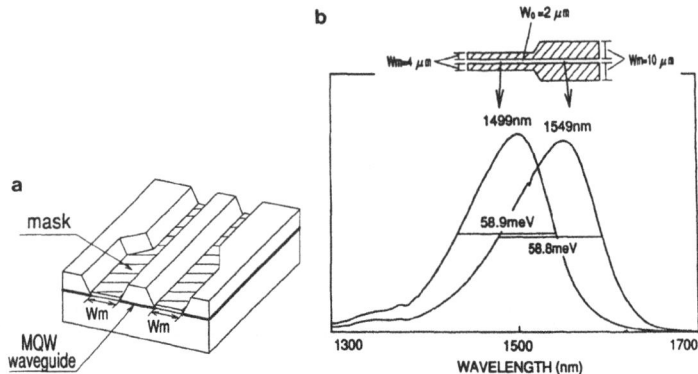

Fig. 4. (a) Schematic structure of a selectively grown mqw waveguide; (b) Photoluminescence spectra of mqw structures with different silica mask stripe widths. [Ref. 26]

absorption with a very short coupling lenght (< 30 µm) [30], or very weak one (i.e. tap detection in self-routing switch nodes [29]).

Waveguide detectors have been considered for WDM and HDWDM receivers :
- grating spectrographs and pin diode arrays, to demultiplex up to 65 channels with a 1 nm spacing [31]; and,
- balanced pin pairs integrated with different building blocks of a heterodyne receiver, as reported by several authors.

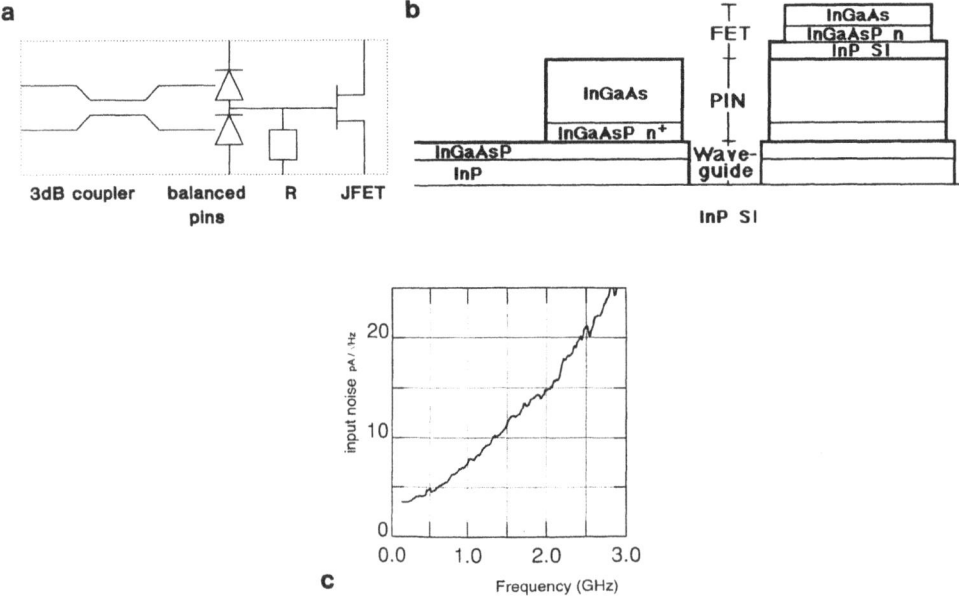

Fig. 5. (a) Layout of an integrated coherent receiver on an Fe:InP substrate, consisting of a 3 dB directional coupler, a pair of balanced pin photodiodes, a load resistor and a JFET transistor; (b) Layer cross-section; (c) Input noise current density measured between 130 MHz and 3 GHz. [Ref. 36]

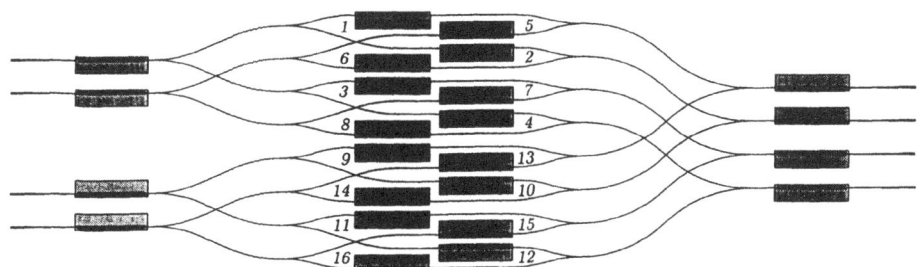

Fig. 6. Mask layout of a 4x4 laser amplifier gate switch array realised on InP. [Ref. 37]

The building blocks to consider in the latter receiver are a broadband-tunable-single frequency laser, polarisation splitters, 3dB TE and TM couplers, pairs of balanced photodiodes and an amplifier front-end. A polarisation matching scheme has to be used to overcome the signal fading due to polarisation fluctuations of the signal transmitted by the fiber. The complexity of a polarisation diversity receiver [10-11] implies monolithic integration, for practical use of coherent optical systems.

So far, different technical approaches have demonstrated several OEICs, including either a local oscillator, a 3 dB coupler and a pair of balanced photodiodes [32,33], or a 3 dB coupler, a pair of balanced photodiodes and a JFET front end (Fig.5) [34-36]. The complete integration of the circuit and improvement of its performance, which is below that of hybrid receivers, remain a great challenge.

2.7. Optical Crossbar Switches

The optical switching matrix is expected to play an important role, for instance in optical cross-connections within future transport networks. The components are required to be able to perform code-and frequency-transparent switching of optical signals without electrical conversion. Efforts are underway to realise basic modules for assembling large switches and thereby, to demonstrate the feasibility of constructing guided-wave switching networks.

Considering III-V materials for realising the switching matrix allows the integration of polarisation insensitive optical amplifiers on the chip. Fig.6 illustrates a tree-structure 4x4 matrix where the basic elements are passive Y splitters/ combiners, the switching function being performed by 16 on/off optical amplifiers switched by the injection current [37]. Passive circuit losses (propagation and splitting) are compensated by the gate switches and by booster amplifiers. The maximum observed fiber-to-fiber gain was 6 dB; the polarisation dependence was rather significant (> 6 dB), but the crosstalk was fairly low (< -40 dB).

3. Conclusions

Integrated optics is strongly driven by fiber-systems requirements. H.Kogelnik claimed in Neuchâtel during the 6th ECIO [38] : "As the complexity of systems increases, the potential for integrated optics applications should increase. Ultimately, we may see sytems that are so complex that they will depend on optical integration." Is Integrated Optics ready to accept this challenge ?

Significant advances resulting from improved III-V epitaxial growth and related processing techniques have been observed. Appropriate fabrication technologies are emerging. Computer-assisted simulation tools for optical propagation ("Beam Propagation Method") and electrooptical interaction modeling are being developed [for example 9,39]. Routine fabrication of phase-shifted DFB mqw lasers and more recently of DFB laser and modulator chips does exist. However, the use of more stabilised processes would be beneficial to developing PICs of greater complexity with a higher yield. The definition of a standard technology is impeded by the large variety of physical configurations always appearing (mqw, strained mqw, sl, alloy mixing [40],..), opening many new ways to explore and many combinations to investigate. However, the ever-decreasing gap between the performance of hybrid circuits and that of PICs confirms the strong potential of InP in integrated optics and is a strong encouragement to further development of the technology.

ACKNOWLEDGEMENTS. Thanks are due to A.Bruno, S.Chelles, F.Devaux, F.Ghirardi and G.Hervé-Gruyer for their helpful contributions.

References

1. T. L. Koch and U. Koren, IEE Proc. J., 138, p.139, (1991)
2. A. Carenco, "Semiconductor waveguides in III-V materials for integrated optics", 4th ECIO, Glasgow, p.1, (1987)
3. M. Allovon and M. Quillec, IEE Proc. J., 139, p.148, (1992)
4. A. Mircea, A. Ougazzaden, J. Barrau, J. C. Bouley, J. Charil, C. Kazmierski and G. Leroux, "Quantum well structures with compensated strain for optoelectronic applications", Proc.IPRM conf. Paris, TuB1, (1993)
5. P. Voisin, Proc.SPIE, 861, p.88, (1988)
6. A. Mathur and P. D. Dapkus, Appl.Phys.Lett., 61, p.2845, (1992)
7. J. E. Zucker, "Quantum effects enhance integrated optics performance", Laser Focus World, p.101, March (1993)
8. S. Chelles, F. Devaux, A. Ougazzaden, A. Mircea, F. Huet and M. Carre, "10 Gbit/s polarisation insensitive electroabsorption modulator using strained InGaAsP/InGaAsP multiquantum wells", ECOC'93, Montreux (Switzerland), We c7.3, p.317 (1993)
9. G. Herve-Gruyer and M. Filoche, ALCOR / BPM software developed and distributed by FT/CNET
10. F. Ghirardi, J. Brandon, M. Carre, A. Bruno, L. Menigaux and A. Carenco, "Polarization-splitter based on modal birefringence in InP/InGaAsP optical waveguides", Photon.Techn.Lett., 5, p. 1047 (1993)
11. F. Ghirardi, J. Brandon, M. Carre, A. Bruno, L. Menigaux, M. Filoche, G. Herve-Gruyer and A. Carenco, Proc. OPTO'93, p.101, May 1993, ESI publications, Paris (France)
12. L. H. Spiekman, F. P. Vanham, M. Kroonwijk, Y. S. Oei, J. J. Vandertol, F. H. Groen and G. Coudenys, Proc. 6th ECIO, Neuchâtel, 2-30, (1993)
13. A. L. Burness, P. H. Loosemore, S. N. Judge, I. D. Henning, S. E. Hicks, G. F. Doughty, M. Asghari and I. White, Electron.Lett. 29, p.520, (1993)
14. A. Carenco, G. Herve-Gruyer, L. Menigaux, A. Mircea and A. Ougazzaden, SPIE Proc., 1141, p.119, (1989)
15. G. Wenger, G. Müller, B. Sauer, D. Seeberger and M. Honsberg, Proc.ECOC, Berlin, p.927, (1992)
16. R. C. Alferness, Proc.OFC/IOOC, San Jose, p.11, (1993)
17. Y. Yoshikuni, Y. Tohmori, T. Tamamura, H. Ishii, Y. Kondo, M. Yamamoto and F. Kano, Proc.OFC/IOOC, San Jose, p.8, (1993)
18. F. Delorme, S. Slempkes, P. Gambini and M. Puleo, Proc.OFC/IOOC, San Jose, p.36, (1993)

19. R. C. Alferness, U. Koren, L. L. Buhl, B. I. Miller, M. G. Young, T. L. Koch, G. Raybon and C. A. Burrus, Electron.Lett., 60, p.3269, (1992)
20. C. E. Zah, F. J. Favire, B. Pathak, R. Bhat, C. Caneau, P. S. D. Lin, A. S. Gozdz, N. C. Andreadakis, M. A. Koza and T. P. Lee, Electron.Lett. 28, p.2361, (1992)
21. F. Devaux and A. Carenco, Proc. 6th ECIO, Neuchâtel, 6-1, (1993)
22. F. Devaux, E. Bigan, M. Allovon, J. C. Harmand, F. Huet, M. Carre and J. Landreau, Appl.Phys.Lett. 61, p.2773, (1992)
23. F. Devaux, E. Bigan, A. Ougazzaden, F. Huet, M. Carre and A. Carenco, Electron.Lett., 28, p.2157, (1992)
24. F. Devaux, F. Dorgeuille, A. Ougazzaden, F. Huet, M. Carre, A. Carenco, M. Henry, Y. Sorel, J. F. Kerdiles and E. Jeanney, "20 Gbit/s operation of high-efficiency InGaAsP/InGaASP mqw electroabsorption modulator with 1.2 V drive voltage", Photon.Techn.Lett., 5, p. 1288 (1993)
25. T. Takaoka, Y. Mayamoto, K. Hagimoto, K. Wakita and I. Kotaka, Electron.Lett., 28, p.897, (1992)
26. T. Sasaki and I. Mito, Proc.OFC/IOOC, San Jose, p.210, (1993)
27. M. Kato, M. Suzuki, M. Takahashi, H. Sano, T. Ido, T. Kawano and A. Takai, Electron.Lett., 28 p.1157, (1992)
28. P. B. Hansen, G. Raybon, U. Koren, B. I. Miller, M. G. Young, M. A. Newkirk, M. D. Chien, B. Tell and C. A. Burrus, Proc.OFC/IOOC, San Jose, PD22, (1993)
29. M. Erman, P. Riglet, P. Jarry, B. G. Martin, M. Renaud, J. F. Vinchant and J. A. Cavailles, IEE Proc. J, 138, p.101, (1991)
30. A. Bruno, A. Carenco, L. Menigaux, J. Thomas, J. Brandon, M. Filoche and A. Scavennec, Proc. ECOC-IOOC, Paris, p.409, (1991)
31. J. B. Soole, A. Scherer, Y. Silberberg, H. P. Leblanc, N. C. Andreadakis, C. Caneau and K. R. Poguntke, Electron.Lett., 29, p.558, (1993)
32. T. L. Koch, U. Koren, R. P. Gnall, F. S. Choa, F. Hernandez-Gil, C. A. Burrus, M. G. Young, M. Oron and B. I. Miller, Electron.Lett. 25, p.1621, (1989)
33. H. Takeuchi, K. Kasaya, Y. Kondo, H. Yasaka, K. Oe and Y. Imamura, IEEE Photon.Techn.Lett. 1, p.398, (1989)
34. R. J. Deri, R. Welter, E. C. M. Pennings, C. Caneau, J. L. Jackel, R. J. Hawkins, J. J. Johnson, H. Gilchrist and C. Gibbons, Electron.Lett. 28, p.2332, (1992)
35. H. Heidrich, Proc. 6th ECIO, Neuchâtel, 2-17, (1993)
36. A. Bruno, L. Giraudet, E. Legros, P. Blanconnier, J. Thomas, M. Carre, M. Billard, F. Ghirardi, L. Menigaux, A. Scavennec and A. Carenco, Proc 6th ECIO, Neuchâtel, 6-8, (1993)
37. M. Gustavsson, B. Largerström, L. Thylen, M. Janson, L. Lundgren, A. C. Mörner, M. Rask and B. Stoltz, Electron.Lett. 28, P.2223, (1992)
38. H. Kogelnik, Proc 6th ECIO, Neuchâtel, 1-3, (1993)
39. G. Herve-Gruyer, M. Filoche and F. Ghirardi, Proc 6th ECIO, Neuchâtel, 14-52, (1993)
40. Y. Chen, J. E. Zucker, B. Tell, N. J. Sauer and T. Y. Chang, Electron.Lett., 29, p.87, (1993)

Chapter 7

SILICA-ON SILICON INTEGRATED OPTICS

R.R.A.SYMS

1. Overview

1.1. Advantages of Silica-on-Silicon Integrated Optics

Silica-on-silicon (SiO_2 / Si) appears the most promising materials system for integrated optics yet developed. It has the following major advantages over previously-developed and competing technologies (namely Ag - Na ion-exchanged glass, Ti: $LiNbO_3$, GaAlAs / GaAs and InGaAsP / InP):
- The fabrication process is cheap, and compatible with that used for Si-based microelectronics.
- Connection to single-mode optical fibres may also be achieved cheaply, and with low loss.
- A wide range of low-loss passive components may easily be produced, on large substrates.
- Slow-speed switching components are possible, based on the thermo-optic effect.
- Hybrid integration may be used to add optical sources.
- Some active components (amplifiers) appear feasible and are currently being developed.

The only major disadvantage is the current lack of an operating mechanism for high-speed switches (e.g. an electro-optic effect). Silica-on-silicon devices should therefore find application in systems for fibre-to-the-home. Immediate requirements are for cheap, passive star components.

1.2. Basic Device Geometry

Fig. 1 shows the essential elements of an SiO_2 / Si integrated optic chip.
This consists of:
1) Substrate: The substrate is single-crystal Si, normally a (100) - oriented, 525 μm thick p-type wafer. Substrate crystallinity is exploited in the fabrication of V-groove alignment features, used to position single-mode fibres for accurate butt-coupling to guides in the integrated optic circuit.
2) Buffer layer: A buffer layer must be used to isolate the waveguide from the high-index substrate. Normally, this layer is a considerable thickness (> 12 μm) of pure SiO_2, and fabrication of such a thick layer represents a considerable technological challenge.
3) Channel guide cores: Waveguides are typically formed as buried channel guides, for compatibility with optical fibres. Guide cores are made from a layer of doped SiO_2, etched into

---

R.R.A.Syms - Optical and Semiconductor Devices Section, Department of Electrical and Electronic Engineering, Imperial College of Science, Technology and Medicine, Exhibition Road, London SW7 2BT, UK

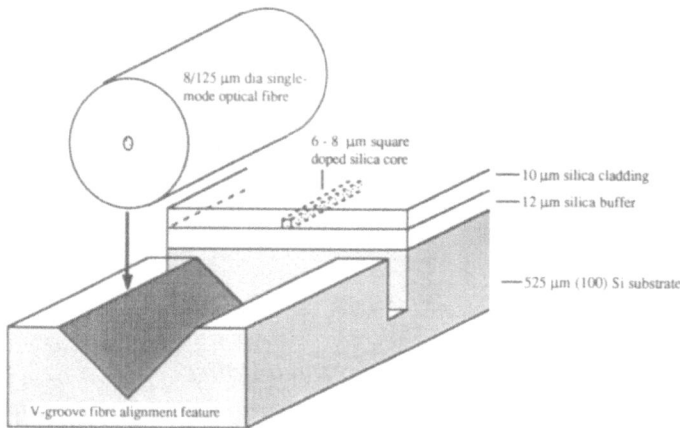

Fig. 1. Basic geometry of silica-on-silicon integrated optical devices.

a rectangular cross-section. Further processing may be used to modify the core shape and reduce sidewall scattering. Two main waveguide variants are currently being investigated, based on large, low-Δn guides, and smaller, medium-Δn guides. Each offers advantages and disadvantages in terms of propagation loss, fibre coupling loss, and the minimum allowable bend radius. These are summarised in Table 1 for TiO_2: SiO_2 devices[1]. Even smaller, high-Δn guides have also been formed in a few materials; however, these suffer very high fibre coupling loss (> 3 dB / facet).

4) Cladding layer: A thick overcladding (typically pure SiO_2, 10 - 20 µm thick.) is used to bury the waveguide cores, to symmetrise the modal fields and isolate them from their surroundings.

In addition, other forms of guide are also currently being investigated. In these, the refractive index changes necessary for waveguiding are obtained by local densification of silica glass by either laser or electron beam irradiation. At present, such guides are less highly developed and stable than topographic guides.

2. Glass Deposition

2.1. Dopants

The basic optical material of silica-on-silicon components is SiO_2. However, the refractive index differences needed to form waveguides are established by controlled doping

Table 1. Characteristics of SiO_2 : TiO_2 single-mode channel guides (after Ref.1).

Core size (µm)	Δn (%)	Propagation loss (dB/cm)	Fibre coupling loss (dB per facet)	Bend radius (mm)
8 x 8	0.25	< 0.1	0.05	25
6 x 6	0.75	0.3	0.5	5

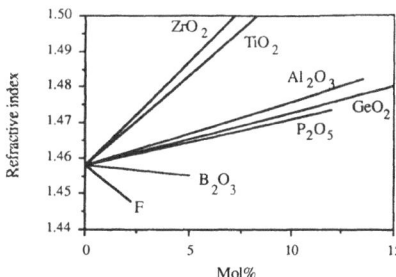

Fig. 2. Variation of refractive index with dopant concentration for SiO_2 doped with various impurities, measured at 0.633 µm wavelength (after Ref. 2).

with impurities:
- Dopants causing a decrease in refractive index include B_2O_3 and F.
- Dopants causing an increase in index include Al_2O_3, As_2O_3, GeO_2, N_2, P_2O_5, TiO_2, and ZrO_2.

The refractive index of pure SiO_2 is 1.458 at 0.633 µm wavelength. Fig. 2 summarises the dependance of refractive index on dopant concentration for some of the materials above[2]. Generally, the dopants are similar to those used in fibre manufacture; however, some have been found to be more suitable than others for the particular application of integrated optics.
For example:
- Since TiO_2 offers a large refractive index change, very high Δn guides may be formed very simply in TiO_2: SiO_2. This materials scheme has therefore been investigated extensively.
- Since phosphorus is pentavalent rather than tetravalent, it is a network modifier. P_2O_5: SiO_2 therefore has a low reflow temperature, making it useful for the formation of cores whose cross-sectional shape will be modified by controlled melting.

Waveguides have also been produced using pure oxide and nitride films thermally compatible with SiO_2. In particular, Si_3N_4 (refractive index 2) has been used to produce extremely small, high-Δn guides. Example waveguide construction schemes are therefore as shown in Table 2.

For each material system, the main fabrication difficulty lies in the formation of thick layers of high-quality silica and doped silica on silicon substrates. Suitable deposition methods are described in the next Section.

Table 2. Construction schemes for silica-on silicon guides based on various dopants.

Buffer layer	Guide layer	Cladding layer	Δn
F : SiO_2	SiO_2	F : SiO_2	low
SiO_2	P_2O_5 : SiO_2	SiO_2	low to medium
SiO_2	TiO_2 : SiO_2	SiO_2	low to high
SiO_2	Si_3N_4	SiO_2	very high

2.2. Deposition Methods

A large number of different deposition methods have now been developed to deposit the thick glass layers required for silica-on-silicon integrated optics. These include:
1) Thermal oxidation and nitridation;
2) Sputtering;
3) Chemical vapour deposition (CVD);
4) Plasma-enhanced chemical vapour deposition (PECVD);
5) Flame hydrolysis deposition (FHD);
6) Sol-gel deposition (SGD).

However, the first two methods are older and somewhat restrictive, and only the last four are being actively developed for fabrication of optical devices. Waveguide technologies based on each of these methods are summarised in Table 3, and details of each process are given in the following Sections.

2.2.1. Thermal Oxidation and Nitridation.

High quality, thick SiO_2 films may be grown on Si substrates by thermal oxidation in a simple tube furnace[3]. Silica may be grown in dry or wet atmospheres, following the reactions:

$$Si + O_2 \rightarrow SiO_2 \quad \text{or} \quad Si + 2H_2O \rightarrow SiO_2 + 2H_2 \tag{1}$$

The growth rate is limited by need for gas molecules to diffuse through the film to the oxide: silicon interface, where they react. For long oxidation times t, the film thickness t_f varies with time as:

$$t_f = \sqrt{(B)} \tag{2}$$

where B is the parabolic rate constant at the oxidation temperature T. Table 4 details the rate constant for growth in dry and wet atmospheres. As can be seen, the main disadvantage of the process is the relatively small value of B, requiring a long immersion in a wet atmosphere at an

Table 3. Fabrication methods for silica-on silicon integrated optics.

Method	Material system	Laboratory	Location
LPCVD	Si_3N_4 / SiO_2	AT & T Bell Labs	Murray Hill, USA
" "	$P_2O_5 : SiO_2 / SiO_2$	"	" "
" "	$As_2O_3 : SiO_2 / SiO_2$	BTRL	Ipswich, UK
PECVD	Si_3N_4 / SiO_2	LETI	Grenoble, France
" "	$P_2O_5 : SiO_2 / SiO_2$	"	" "
" "	$P_2O_5 : SiO_2 / SiO_2$	BNR Europe	Harlow, UK
FHD	$TiO_2 : SiO_2 / SiO_2$	NTT	Ibaraki, Japan
" "	$GeO_2 : SiO_2 / SiO_2$	"	" "
" "	$TiO_2 : SiO_2 / SiO_2$	Furukawa Electric Co.	Chiba, Japan
" "	$TiO_2 : SiO_2 / SiO_2$	PIRI	Columbus, USA

Table 4. Parabolic rate constant B for oxidation of silicon in dry and wet atmospheres (data taken from Ref. 3).

Oxidation temperature (°C)	B (μm^2/hr) (Dry atmosphere)	B (μm^2/hr) (Wet atmosphere)
1200	0.045	0.720
1100	0.027	0.510
1000	0.0.0117	0.287
920	0.0049	0.203

elevated temperature to obtain a thickness suitable for optical applications (e.g. 24 days at 1100°C for ≈16 μm thickness[4]). In addition, the process is not easily modifiable for the formation of doped oxides; consequently, it is most commonly used for the formation of buffer layers. Some particular dopants have been incorporated by further thermal processing. For example, silicon oxynitride guiding layers have been formed by thermal nitridation of SiO_2 layers in NH_3 at similar temperatures[5].

2.2.2. Sputtering. R.F. sputtering can be used to deposit dielectric layers on Si by physical transfer of material from a source known as a target (arranged in a large sheet parallel to the substrate). Coating is performed in a vacuum chamber, into which an inert gas (Ar) is bled at a pressure of ≈ 10^{-3} - 10^{-2} Torr. The gas is then excited into a plasma by an RF field in the space between target and substrate. The former is configured as the cathode, and the latter as the anode, so that positively charged Ar^+ ions strike the target preferentially, and eject atoms from it. Although the sputtered atoms are ejected in random directions, many move towards the nearby substrate. After undergoing further collisions with ions in the plasma, the sputtered atoms strike the substrate from a wide range of angles. The technique can therefore be used to coat non-planar surfaces. Sputtering has been used to deposit many materials for optical waveguide applications (e.g. Corning #7059 glass[6], pure SiO_2[7], pure Al_2O_3[8] etc.). Its main disadvantage for production is the slow deposition rate (around 1 μm/hr).

2.2.3. CVD and PECVD. Chemical vapour deposition (CVD) is a method of depositing glassy films from hot gas mixtures. It originated in the VLSI microelectronics industry, and its further development for optics has been pursued mainly by AT & T Bell Labs [9,10] following earlier work[11]. The physical layout of the equipment used can take a variety of forms, depending on the direction of the gas flow (horizontal or vertical) and the form of heating used. For example, a horizontal-tube hot-wall reactor might consist of a quartz tube with an external resistive heater. The substrate is placed on a quartz support, and loaded into the tube through a removable end-cap. Process gases are injected from one end, and heated prior to arrival at the substrate by contact with the walls. They flow over the substrate as a laminar boundary layer, and are adsorbed on the surface where they react to form the desired film. By-products of the reaction and unreacted gases are exhausted at the downstream end of the tube.

The temperature and pressure are both highly dependent on the process used[3]. One major branch of CVD involves the thermally-activated oxidation of hydrogen compounds, such as silane (SiH_4), phosphine (PH_3) and arsine (AsH_3). For example, pure silica layers may be deposited at atmospheric pressure by a mixture of silane and oxygen, at a temperature of 400 - 450°C, following the reaction:

$$SiH_4 + O_2 \rightarrow SiO_2 + 2H_2 \tag{3}$$

Doped silica layers can also be formed easily. For example, phosphosilicate glass (P_2O_5: SiO_2) can be produced following the two simultaneous reactions:

$$SiH_4 + 2O_2 \rightarrow SiO_2 + 2H_2O, \quad \text{and} \quad 2PH_3 + 4O_2 \rightarrow P_2O_5 + 3H_2O \tag{4}$$

A silica-on-silicon process based on Si / SiO_2 / P_2O_5: SiO_2 / SiO_2 may therefore be implemented in a single reactor equipped with controllable sources of SiH_4, PH_3 and O_2.

Similarly, silicon nitride can be produced from dichlorosilane and ammonia at $\approx 750°C$, or from silane and ammonia at $\approx 900°C$, following:

$$3SiCl_2H_2 + 4NH_3 \rightarrow Si_3N_4 + 6HCl + 6H_2, \quad \text{or} \quad 3SiH_4 + 4NH_3 \rightarrow Si3N_4 + 12H_2 \tag{5}$$

Though CVD is conceptually a simple process, in reality it is complicated by two aspects. The first is the difficulty of ensuring uniform film deposition since the temperature, concentration, chemical composition and velocity of the gas mixture can vary considerably along the length of the tube and across its diameter. The second is the need to handle gases that are often toxic, explosive or corrosive (or a combination of all three). Some gases are highly dangerous; silane is toxic, and explodes in contact with air. Double-layer stainless steel pipework is therefore a standard safety precaution.

Using modified CVD process, it is possible to deposit films at lower substrate temperatures (200 - 350°C), by supplying the necessary energy from an electrically-excited plasma. This variant is known as plasma-enhanced chemical vapour deposition (PECVD); its development for optics has mainly been pursued by LETI[12,13] and BNR Europe Ltd.[14] Fig. 3 shows a typical parallel-plate PECVD rig.

The substrate is placed on a heated susceptor, which is arranged to act as one of a pair of R.F. electrodes. A plasma is then established between the electrodes, and the process gases are then bled into this region (which may have a very high electron temperature). The plasma might be derived from an inert gas; for example, SiO_2 films may be deposited by reacting SiH_4 and N_2O in an argon plasma. Alternatively, it might itself form one of the reactants; SiON may be deposited using SiH_4, N_2O and NH_3, with the resulting film composition being determined by the N_2O: NH_3 ratio[15].

2.2.4. Flame Hydrolysis Deposition. Flame hydrolysis deposition (FHD) is a method of depositing glasses in the form of a powdery 'soot'. It was originally developed for the

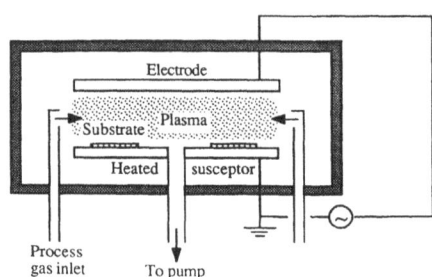

Fig. 3. Arrangement for plasma-enhanced chemical vapour deposition.

fabrication of optical fibre preforms in the VAD process, and its further development for integrated optics has been performed mainly by NTT[1,16-18]. In FHD, silica is produced by hydrolysing a mixture of $SiCl_4$ and O_2 in a glassburner, following the reaction:

$$SiCl_4 + 2H_2O \rightarrow SiO_2 + 4HCl \tag{6}$$

The result is a stream of small silica particles, or 'soot', which are blown onto the substrate by the oxy-hydrogen flame, where they stick. During deposition, large numbers of substrates (e.g.[16] 50) are rotated on a turntable, and the soot stream is traversed to ensure uniform coverage (see Fig. 4).

A variety of dopants may also be incorporated in the films very simply. For example, TiO_2 and GeO_2 may be included by adding $TiCl_4$ or $GeCl_4$ to the gas stream, while P_2O_5 and B_2O_3 dopants may be obtained from BCl_3 and PCl_3 sources. Because the composition of each layer may be varied merely by adjusting the flow of the relevant gas, graded-index guides and tapered structures may be built up by flame hydrolysis.

Thick films may be built up very quickly by FHD. However, after the deposition of sufficient material, the porous soot must be compacted or 'sintered' in a furnace at $\approx 1300°C$ to produce glass of optical quality. This results in considerable (90%) film shrinkage, making the final thickness uncertain.

2.2.5. Sol-gel Deposition. The sol-gel process[19] is an alternative method of depositing glassy films, based on the hydrolysis and polycondensation of metal alkoxides. Three stages are involved: first, the formation of a stable suspension of particles within a liquid (a sol); second, further processing of the sol to form a continuous solid network permeated by liquid (a gel); and third, drying of the gel to form a porous glass. Usually, the sol is applied to a surface to be coated by spin-coating, so sol-gel material is also referred to as spin-on-glass. One commonly-used process involves the fabrication of silica (SiO_2) films from tetraethylorthosilicate ($Si(OC_2H_5)_4$, or TEOS). However, a wide range of other metal alkoxides can be used, allowing a corresponding range of dielectric films. For example, refractive index variations can be introduced by variations in the sol composition (e.g. by the use of TiO_2, derived from a tetrapropylorthotitanate or TPOT precurser, as a dopant).

For silica, the basic sol-gel process involves the reaction:

$$Si(OR)_4 + 2H_2O \rightarrow SiO_2 + 4ROH \tag{7}$$

However, this proceeds in a number of steps. The alcoxide is first hydrolysed, so that a number of OR groups on each of the TEOS molecules are replaced by OH. For example, mixing TEOS with a quarter the stochiometric ratio of water and refluxing in the presence of a catalyst (e.g.

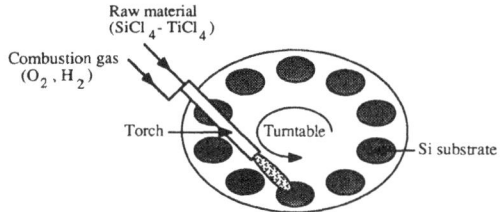

Fig. 4. Arrangement for flame hydrolysis deposition (after Ref. 1).

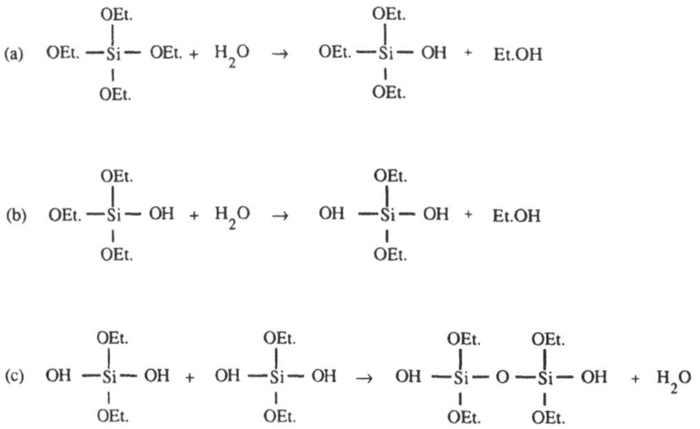

(a) $OEt.-\underset{\underset{OEt.}{|}}{\overset{\overset{OEt.}{|}}{Si}}-OEt. + H_2O \rightarrow OEt.-\underset{\underset{OEt.}{|}}{\overset{\overset{OEt.}{|}}{Si}}-OH + Et.OH$

(b) $OEt.-\underset{\underset{OEt.}{|}}{\overset{\overset{OEt.}{|}}{Si}}-OH + H_2O \rightarrow OH-\underset{\underset{OEt.}{|}}{\overset{\overset{OEt.}{|}}{Si}}-OH + Et.OH$

(c) $OH-\underset{\underset{OEt.}{|}}{\overset{\overset{OEt.}{|}}{Si}}-OH + OH-\underset{\underset{OEt.}{|}}{\overset{\overset{OEt.}{|}}{Si}}-OH \rightarrow OH-\underset{\underset{OEt.}{|}}{\overset{\overset{OEt.}{|}}{Si}}-O-\underset{\underset{OEt.}{|}}{\overset{\overset{OEt.}{|}}{Si}}-OH + H_2O$

Fig. 5. The sol-gel process for production of silica from tetraethylorthosilicate: a) hydrolysis of one ligand group, b) hydrolysis of two groups, and c) polymerisation by condensation.

HCl), will replace on average one group, as shown in Fig. 5a; repeating the process will result in the replacement of two groups (Fig. 5b). These reactions may be slowed by dilution with solvent (ethanol) so that the sol is stable enough for use.

When some groups have been hydrolysed, polymerization can take place as shown in Fig. 5c. Here the reaction by-product is water, so this causes further hydrolysis and polymerization in a chain reaction until a connected network (known as a gel) is built up. Again, this reaction can be slowed by dilution with solvent. However, when the sol is coated onto the substrate (e.g. by spin-coating) the solvent is driven off. This accelerates the polymerisation reaction, resulting in rapid gelation on the wafer surface. Gelation and final solvent evaporation are normally completed by a baking step.

Drying involves a reduction in volume of the spun layer, which results in tensile stresses in the film. After a certain deposited thickness (normally a few µm), this generally results in failure of the film through cracking. Recently, a new process has been developed for the deposition of thick (20 µm) sol-gel films on silicon substrates, based on repetitive spin-coating followed by densification by rapid thermal annealing[20]. This has overcome the problem of stress-cracking, and enabled waveguides to be constructed using both the TiO_2: SiO_2 / SiO_2 and the P_2O_5: SiO_2 / SiO_2 systems. Fig. 6 shows a typical configuration for annealing, based on a conventional rapid thermal processor.

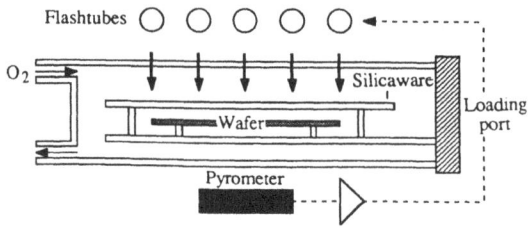

Fig. 6. Arrangement for rapid thermal annealing of sol-gel films.

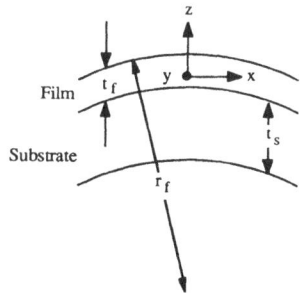

Fig. 7. Geometry for calculation of film stress and birefringence.

2.3. Properties of Deposited Films

All the processes used for one-sided deposition of thick silica films yield bowed Si wafers, implying that the films will be stressed. Stress and film curvature are related by Stoney's formula[21,22]:

$$\sigma_f = \{E_s t_s^2 / [6 (1 - v_s) t_f]\} \{1/r_s - 1/r_f\} \tag{8}$$

where $1/r_s$ is the curvature of the substrate without film, $1/r_f$ is the curvature with film, t_s, E_s and v_s are the thickness, Young's modulus and Poisson's ratio of the substrate, and t_f is the thickness of the film (Fig. 7). Following the standard convention, tensile stresses are positive, and compressive stresses negative. In the latter case, the film surface is typically convex and r_f is positive.

Mechanical constants for Si and SiO$_2$ are given in Table 5. Assuming a wafer thickness of 525 µm, a film thickness of 10 µm and a 3 m radius of curvature (say), we obtain a stress of $\sigma_f = -275$ MN/m². Much of this can be ascribed to differential thermal expansion as the wafer is cooled from the process temperature T_p to room temperature T. Assuming roughly constant thermal expansion coefficients - true to reasonable approximation - the stress caused by thermal effects alone is given by:

$$\sigma_{th} = E_f (\alpha_s - \alpha_f) (T - T_p) / (1 - v_f) \tag{9}$$

Table 5. Mechanical and optical properties of Si and SiO$_2$.

Property	Units	Si	SiO$_2$
α	/°C	2.6 x 10⁻⁶	0.55 x 10⁻⁶
E	N/m²	1.08 x 10¹¹	7.2 x 10¹⁰
v		0.42	0.19
n @ λ = 0.633 µm		3.5	1.458
dn/dT	/°C		1.1 x 10⁻⁵
q$_{12}$ - q$_{11}$	m²/N		2 x 10⁻¹²

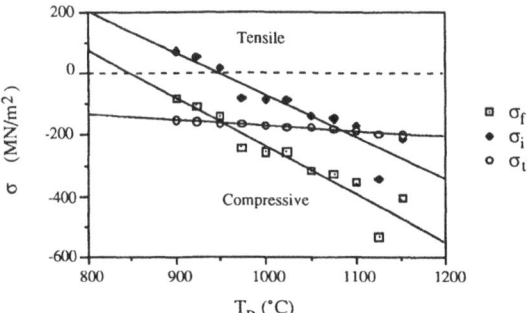

Fig. 8. Variation of film stress, thermal stress and intrinsic stress with rapid thermal annealing temperature for sol-gel phosphosilicate glass (after Ref. 23).

where E_f, v_f and α_f are the Young's modulus, Poisson's ratio and thermal expansion coefficient of the film, and α_s is the expansion coefficient of the substrate. Assuming a process temperature of (say) 1000°C, we then obtain σ_{th} = -180 MN/m². In this example, thermal stresses account for the bulk of the observed stress; the remaining contribution is intrinsic stress (σ = -95 MN/m²), introduced by the deposition process itself. Thermal stresses might be reduced by using substrates with lower expansion coefficients (e.g. SiO_2), but Si has so far proved overwhelmingly popular.

All stresses depend heavily on the process parameters. For example, Fig. 8 shows the variation of σ_f, σ_{th} and σ_i with annealing temperature for films of sol-gel phosphosilicate glass deposited by repetitive spin-coating and rapid thermal annealing[23]. This shows an increase in σ_f and σ_i with T_p. Above a certain temperature ($\approx$ 950°C), σ_i is always compressive; this condition appears to be necessary to prevent film cracking.

The presence of a large biaxial film stress causes refractive index changes through the photoelastic effect. In the geometry of Fig. 7, the changes in the refractive index seen by fields polarized in the x, y and z directions due to equal stresses σ_f applied in the x and y directions are[24]:

$$\Delta n_{x,y} = -n^3/2 \cdot \{ q_{11} + q_{12}\} \sigma_f; \qquad \Delta n_z = -n^3/2 \cdot 2q_{12} \sigma_f \qquad (10)$$

where q_{11} and q_{12} are components of the stress-optical tensor [q]. Since $q_{11} \neq q_{12}$, the material becomes birefringent, with:

$$n_e - n_o = \Delta n_z - \Delta n_{x,y} = -n^3/2 \cdot \{q_{12} - q_{11}\} \sigma_f \qquad (11)$$

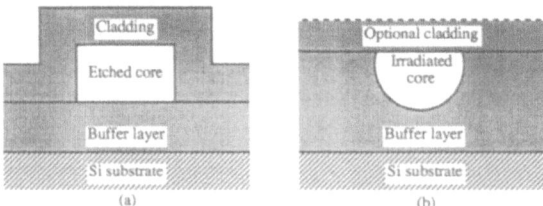

Fig. 9. Silica-on-silicon channel guides, formed with a) topographic and b) irradiated cores.

In silica, $q_{12} > q_{11}$, so for compressive film stresses $n_e - n_o$ is positive. Assuming $\sigma_f = -275$ MN/m^2 as before, we obtain $n_e - n_o = 8.8 \times 10^{-4}$. A similar birefringence $n_{TM} - n_{TE}$ is observed between the effective indices of waveguide modes. Since the magnitude of the birefringence is significant compared with the index changes used to form the guide, steps may be taken to control it in some applications.

3. Channel Waveguide Fabrication

3.1. Waveguide Variants

Each of the processes described in the previous Section may be used to deposit suitably thick layers of silica or doped silica on silicon substrates. Buried channel waveguides may then be formed in these layers by a variety of different methods. Fig. 9 shows the two most important guide types.

In topographic guides, a core of material of raised refractive index is formed on a buffer layer and then buried by a further cladding layer. Cores of rectangular cross-section may be formed simply by etching (Fig. 9a), while alternative cross-sections may be obtained by further processing (e.g. reflow). Both methods can yield low-loss guides, but cause significant surface modulation. A surface flat enough for further processing can be achieved if the guide is formed by local modification of a single layer of material. Processes conventionally used to form guides in bulk glasses (e.g. silver-sodium ion exchange) have not yet been applied to deposited films, which typically consist of pure SiO_2 or SiO_2 with inappropriate dopants. Instead, the emphasis has been placed on the use of local annealing to increase the refractive index by densification. Both laser and electron beam irradiation have been used, and the resulting core can be buried using an optional cladding layer (Fig. 9b)). Details of each of these processes are given in the following Sections.

3.2. Topographic Waveguide Fabrication

Etched channel waveguide fabrication begins with the deposition of a bilayer glass structure (normally consisting of a doped SiO_2 guiding layer on an undoped buffer layer) on an Si substrate. The wafer is then coated with a mask material, which is patterned lithographically with waveguide features and then wet etched. The mask is used as an etch stop during reactive ion etching (RIE) or reactive ion beam etching (RIBE), which is used to remove most of the upper layer material, leaving the waveguides as rectangular cores (Fig. 10a).

Reactive ion etching can be based on either fluorine or chlorine chemistry. For example, in the former case, a C_2F_6 - C_2H_4 plasma has been used with an amorphous silicon mask[1]; in the

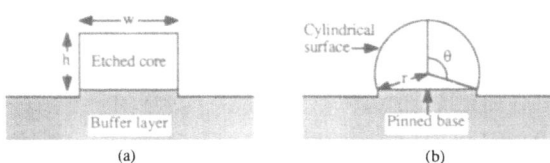

Fig. 10. Topographic waveguides, after a) reactive ion etching, and b) reflow.

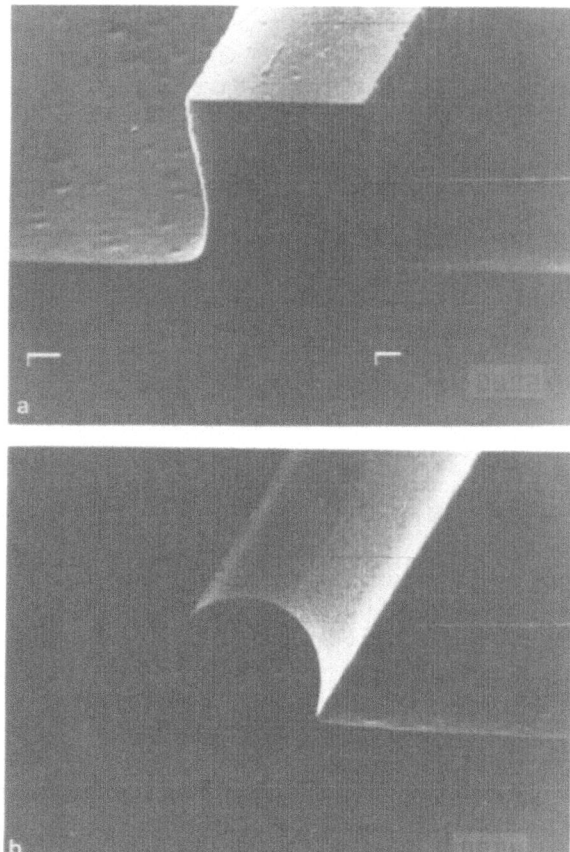

Fig. 11. SEM photographs of topographic waveguide cores formed in sol-gel phosphosilicate glass a) after reactive ion etching and b) after reflow at 1100 °C.

latter, a CHF_3 - Ar - O_2 plasma with a Cr mask[20]. Etch rates are typically in the range 500 - 1500 Å/min. After RIE, the core walls are normally rough. For example, Fig. 11a) shows an SEM photograph of a channel waveguide core formed in sol-gel phosphosilicate glass, immediately after RIE.

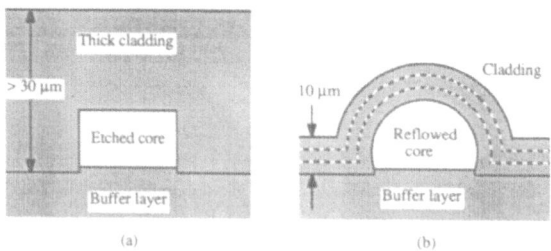

Fig. 12. Topographic waveguides, after burial by a) FHD and b) CVD.

Fig. 13. Topographic waveguide after spin-coating of sol-gel glass (after Ref. 20).

Depending on the materials used, the core may be reflowed to lower the surface roughness and hence reduce scattering losses in the final guides. This may be achieved very simply if the core glass has a lower melting point than the buffer layer material (achieved through, for example, differences in phosphorous content). At the core melt temperature, surface tension forces pull the core into a cylindrical cross-section[25] (Fig. 10b). The final core shape may then be found by simple geometry, equating the initial and final cross-sectional areas. For example, the core radius r is the solution to:

$$w \times h = r^2 \{\theta - \sin(\theta)\cos(\theta)\} \tag{12}$$

where the angle θ is defined by the pinned base, so that $r \sin(\theta) = w/2$. Fig. 11b shows an SEM photograph of the waveguide core of Fig. 11a, immediately after reflow at 1100°C. If glasses of high melt temperature (e.g. TiO_2: SiO_2) are used for both core and cladding, melting may still be used to reduce sidewall roughness. However, in this case a cylindrical core cross-section is not obtained[26].

After reflow, guides are buried beneath a thick cladding layer. Using FHD, an extremely thick cladding (30 μm!) can be deposited very simply, so that the resulting guide has an ideal symmetric core as shown in Fig. 12a. Using CVD, deposition of a thick cladding is more time-consuming. Furthermore, material is deposited on the core and on the exposed buffer layer at a uniform rate, so that the shape of the growing surface gradually evolves as shown in Fig. 12b. Only a limited degree of local planarization is therefore obtained by this method. Using the sol-gel process, the waveguide is buried using further spin-coating. In this case, more global planarization is achieved. For example, Fig. 13 shows the surface profile of sol-gel TiO_2: SiO_2 channel guides buried by spin-coating; here the radius of curvature above the guide is ≈ 5 mm[26].

3.3. Propagation Losses of Topographic Waveguides

The propagation losses of topographic waveguides fabricated by most of the methods described are now extremely low. For example, Table 6 gives examples to illustrate the rapid

Table 6. Historical reduction of propagation loss in silica-on silicon topographic channel waveguides.

Material system	Method	Laboratory	Propagation loss (dB/cm)	Wavelength (μm)	Date	Ref.
SiO_2	Oxidation	Cincinnati Univ.	< 1	0.633	1983	4
$TiO_2 : SiO_2$	FHD	NTT	0.5	1.3	1986	17
$Si_3N_4 - SiO_2$	CVD	AT & T	< 0.3	1.3 - 1.6	1987	9
$TiO_2 : SiO_2$	FHD	NTT	0.1	1.55	1990	1
$P_2O_5 : SiO_2$	PECVD	LETI	0.1	1.55	1990	12
$GeO_2 : SiO_2$	FHD	NTT	0.01	1.55	1990	18

progress achieved in loss reduction. The main contributions to loss arise from:
- Sidewall scattering: this is largely eliminated by reflow of the guide core and cladding;
- Extrinsic absorption caused by the presence of hydrogen, primarily near 1.39 μm (excitation of 2nd harmonic of stretching vibrations of O-H bonds) and 1.52 μm (3rd harmonic of Si-H bond and 2nd harmonic of N-H bond): this can be dramatically reduced by thermal annealing[9,12]; and,
- Intrinsic absorption caused by the tails of the mid-infrared absorption bands of S-O and similar bonds. The most important of these lie at 7.3 μm (B-O), 8.0 μm (P-O), 9.0 μm (Si-O) and 11.0 μm (Ge-O)[2]. For operation at near-infrared wavelengths, germania is therefore the most favourable dopant, as suggested by the data of Table 6.

3.4. Waveguides Fommed by Local Densification

Waveguides may be formed by local densification of planar silica layers, which are often not fully densified in their as-deposited state. Two methods are currently being investigated: 1) Laser irradiation; and, 2) Electron beam irradiation. Both technologies are in their infancy compared with topographic waveguide fabrication, but offer a number of potential advantages. Only a single layer is required to form a buried channel guide, and the slight degree of densification involved leaves the surface flat enough for further processing. However, questions concerning propagation loss and guide stability have yet to be addressed.

Fig. 14a shows densification by laser irradiation[27,28]. Here a focussed beam from a high power laser is scanned mechanically over the silica layer. The radiation can be absorbed directly if the laser wavelength lies in one of the mid-IR absorption bands; high-power CO_2 lasers (λ = 10.6 μm) are therefore appropriate. Other wavelengths may be used with overcoatings (e.g. metal films) having different absorption characteristics. Densification is caused primarily by local heating; consequently, most effort has concentrated on porous sol-gel material rather than fully consolidated films. Energy densities of 4 - 6 J/cm^2 are required for waveguide formation. However, little work has been performed on channel waveguide formation, and propagation losses are quoted as high[27,28]. Fig. 14b shows electron beam irradiation. In this case, the effects are caused by a combination of heating and bond-breaking, although the latter dominates when the substrate is cooled. In the past, high-dose irradiation of bulk silica[29,31] and chalcogenide glass[32,33] has been used to cause local refractive index changes. More recently, the technique has been applied to glasses deposited by CVD and PECVD. The method is more highly developed than laser irradiation, and complete processes for fabrication

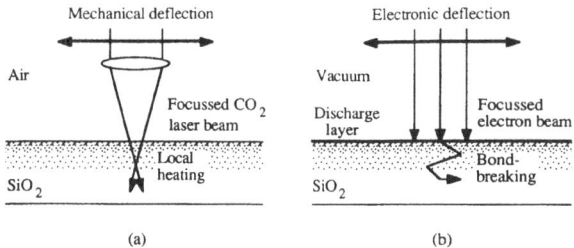

Fig. 14. Waveguide formation by a) laser and b) electron beam irradiation.

of single-mode channel guide devices have already been demonstrated.

The irradiation-induced changes are graded (with a depth profile similar to the concentration variation obtained in ion implantation) as the electrons gradually loose energy through multiple scatterings. The irradiation depth is characterised by the Grün range R_g, the depth at which all the electron energy has been dissipated. Using experimental data for pure silica, Everhart and Hoff [34] obtained the following relation between R_g and the initial electron energy E:

$$R_g (\mu m) = 0.0181 \, E \, (keV)^{1.75} \qquad (13)$$

Fig. 15 shows the variation of R_g with E for $0 < E < 40$ keV; over this range, R_g varies between 0 and 10 µm, so that the depth of the induced changes can be made compatible with single-mode integrated optics. An energy of 25 keV yields guides suitable for operation at 1.5 µm wavelength.

Normally, electron beam irradiation results in slight compaction of the glass, accompanied by an increase in refractive index. Planar and channel guide formation by irradiation through narrow strip apertures has therefore already been demonstrated in PECVD SiO_2 supplied by LETI, France[35,36]. Direct electron-beam writing of channel guides has also recently been performed[37]. However, the reverse effect (irradiation-induced expansion, accompanied by a decrease in refractive index) has been observed in some other PECVD-deposited materials[38].

Buried channel guides can be fabricated very simply in materials having a positive irradiation-induced refractive index change. Irradiation is performed through a gold mask deposited on the surface of the wafer. The gold acts as a barrier to electrons, allowing selective irradiation of areas using a flood beam. In practice, a relatively complicated tri-level metallisation is required[39]. This consists of:

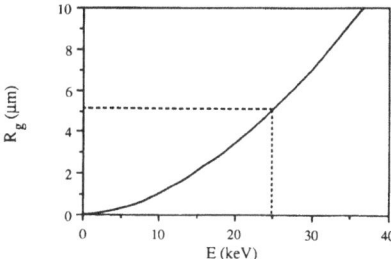

Fig. 15. Variation of the Grün range R_g with electron energy, for silica.

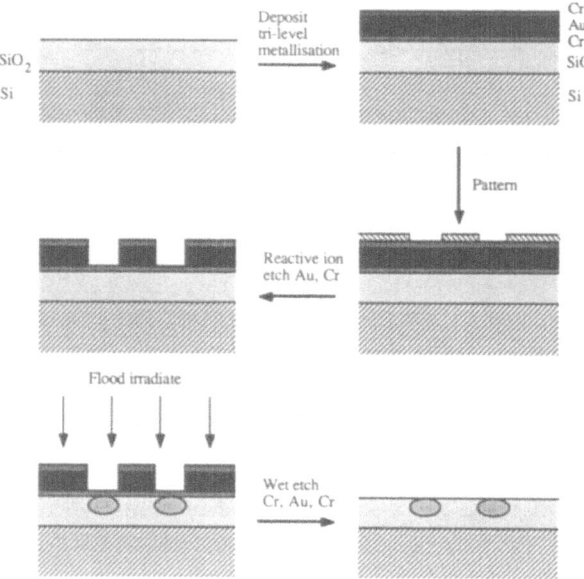

Fig. 16. Masking sequence for channel waveguide formation by electron beam irradiation.

Layer 1 - a 400 Å Cr adhesion layer, which also serves to prevent charge accumulation.
Layer 2 - a 7000 Å Au layer, which acts as the irradiation mask; this is patterned by reactive ion etching using Ar and O_2 to maintain dimensionality and to allow narrow stripe features to be formed.

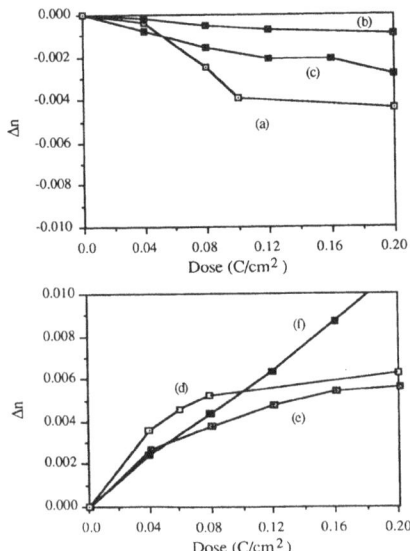

Fig. 17. Variation of the irradiation-induced refractive index change with irradiation dose for: a) BNR SiON; b) BNR FSG; c) BNR SiO_2; d) BNR PSG; e) BT SiO_2; and, f) BT ASG. Irradiations: 1a) - 1e) performed at 25 keV; 1f) at 13 keV (after Ref. 39).

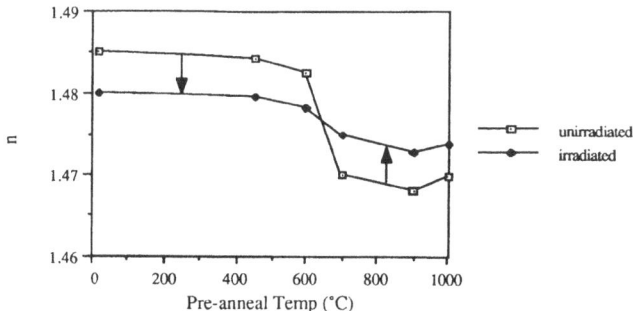

Fig. 18. Refractive index of BNR SiON before and after electron irradiation at 25 keV with a dose of 0.2 C/cm^2, after a 60 minute pre-irradiation annealing step at various temperatures (after Ref. 39).

Layer 3 - a 2000 Å Cr layer, which serves as a mask for the RIE step used to pattern Layer 2.

These layers are removed after irradiation, leaving the glass surface smooth and featureless. Fig. 16 shows the fabrication sequence for materials whose refractive index is increased by irradiation. Here, guides are fomed simply by irradiation through a strip aperture.

Fig. 17 shows the variation of the irradiation-induced refractive index change with charge dose for a number of PECVD and CVD-deposited materials supplied by BNR Europe Limited (BNR) and British Telecom Research Laboratories (BT)[39]. As can be seen, there is a wide variation in sensitivity and in maximum achievable refractive index change. Most importantly, the induced change is negative for some materials and positive for others.

These effects have been linked to variations in the OH content. Materials with a high OH content typically expand on irradiation, so that the refractive index falls, while materials with a low OH content compact. Consequently, the sign of the induced changes may be reversed by thermal annealing before irradiation. For example, Fig. 18 shows the effect of annealing PECVD-deposited silicon-oxynitride for 60 minutes at different temperatures before irradiation at a fixed dose (0.2 C/cm²). For low anneal temperatures, the refractive index falls on irradiation, while above 600°C it rises[39].

Propagation losses of channel guides fabricated by this technique are still relatively high; for example, 1.6 dB/cm has been achieved at 1.5 µm wavelength in PECVD-deposited SiO$_2$[39].

4. Coupling to Optical Fibres

4.1. Mode-Matching

Besides the possibility of low-loss channel waveguides, silica-on-silicon integrated optics offers cheap, low-loss connection to single-mode optical fibres. Now, the coupling efficiency η between an input transverse field E_{in} and the μth transverse modal field E_μ of a waveguide is given by[40]:

$$\eta = |<E_{in}, E_\mu>|^2 / \{<E_{in}, E_{in}><E_\mu, E_\mu>\} \tag{14}$$

where the inner product $<E_a, E_b>$ between two fields $E_a(x, y)$ and $E_b(x, y)$ is defined as:

$$<E_a, E_b> = \int_{-\infty}^{+\infty} \int_{-\infty}^{+\infty} E_a E_b^* dx\, dy \qquad (15)$$

For high coupling efficiency, the two fields should be accurately matched in size and shape, and coincide spatially. For connection to standard 1.5 μm single-mode fibre (which has a mode-field diameter of ≈ 8 μm), SiO_2 / Si channel guides should therefore have large core size and low Δn. As shown in Table 1, 8 x 8 μm cores with Δn of 0.25% gives η = 98.8%, or 0.05 dB loss[1].

However, with such a large modal field, tight bending in the planar optical circuit is impossible; the minimum tolerable bend radius in this example was 25 mm. Some trade-off between bend and coupling loss is possible; decreasing the core size and increasing the Δn to 6 x 6 μm and 0.75%, respectively, reduces the minimum bend radius to 5 mm, but increases the coupling loss to 0.5 dB. However, the bend radii needed in high density circuits or delay lines[41] cannot be achieved without much smaller, higher-Δn waveguides (e.g. 3 x 3 μm GeO_2-doped cores with Δn of 2%, which offer usable bends of 2 mm radius[42]). These cause excessive coupling loss. Attention has therefore been given to adiabatic tapers capable of transforming the guide from large core, low Δn type at the fibre input to small core, high Δn type in the optical circuit. Two methods have been demonstrated. The first is a modification of flame hydrolysis deposition; in this, the gas composition and torch scan pattern are gradually varied to deposit a thinner, higher Δn guiding layer near the centre of the circuit[43]. The waveguide width is then varied by lithography. In the second, a small-core, high Δn guide is first fabricated with uniform parameters, and then modified by local diffusion of dopants. This is done by clamping one end of the substrate in a cooled holder, and heating the other to a high temperature (Fig. 19). For example, heating at 1300°C for 5 hours to form a 25 mm long taper decreased the coupling loss of a 6.5 x 6.5 μm GeO_2-doped guide with a Δn of 0.75% from 0.43 to 0.08 dB[44]).

4.2. Mechanical Alignment

Even with good mode matching, efficient coupling will not be obtained if the relative positions of fibre and waveguide are inaccurately specified. For low cost, alignment must be passive and achieved through integrated mechanical alignment features rather than by active optimisation of position. Some early attempts were made to use the thick silica layers themselves to fabricate suitable features[45]; however, the Si substrate is now used almost exclusively.

The technology used is anisotropic etching down Si crystal planes[46], based on earlier work on fibre connectors[47] and components for pigtailing $Ti:LiNbO_3$ devices[48]. Two etch mixtures are commonly used, based on KOH and EDP (a mixture of ethylenediamine, pyrocatechol and water) respectively. KOH requires a Si_3N_4 surface mask, because it attacks SiO_2; EDP can be used directly with silica, and is therefore compatible with silica-on-silicon integrated optical devices

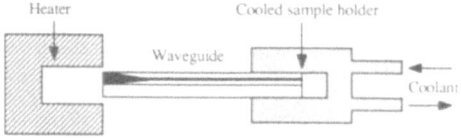

Fig. 19. Experimental set-up for taper formation by local diffusion of dopants (after Ref. 44).

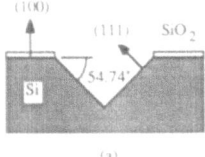 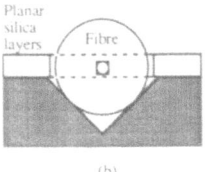

Fig. 20. V-groove formation by anisotropic etching: a) basic geometry, and b) fibre location.

In the most commonly used orientation, etching of (100) oriented Si through rectangular openings in a silica mask layer results in the formation of V-grooves, which can act as kinematic mounts for optical fibre. The topology of fully-formed V-grooves is shown in Fig. 20a. The desired shape is generated because of the large difference between the etch rates R_{100} normal to the surface and R_{111} normal to the (111) planes. Typically, this difference is of the order of 40: 1. Anisotropically etched silicon can be used merely as a precision mount for ribbon optical fibre, locating the individual cores on a precise spacing. A prepared fibre end may then be polished and glued to the cleaved end facet of a waveguide device[48,49]. However, it is advantageous if the waveguide and alignment features are fabricated on a common substrate; in this case, the sloped ends of the grooves must be removed (for example, by precision sawing[50]); Fig. 20b shows the location of the fibre within an integrated groove. Fully pigtailed devices (e.g. 1 x 16 power splitters[51]) have now been demonstrated. Fig. 21 shows a pigtailed channel waveguide formed from TiO_2: SiO_2 sol-gel glass.

5. Passive Waveguide Devices

5.1. Device Types

The main device variants of SiO_2 / Si integrated optics are broadly similar to those

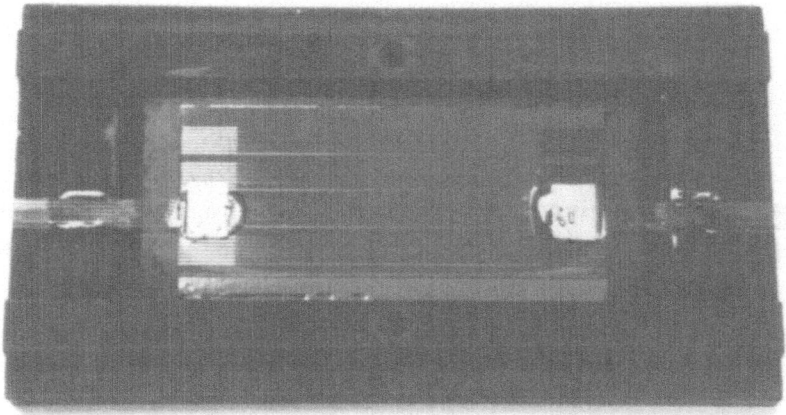

Fig. 21. SiO_2/Si waveguide chip with ribbon fibre pigtails held in internal V-grooves (photo courtesy A.S. Holmes).

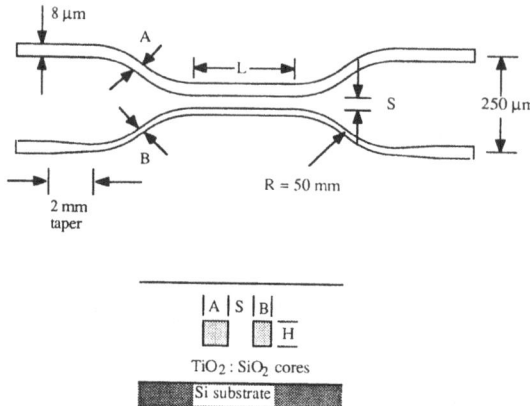

Fig. 22. Broad-band power splitter based on asymmetric directional coupler (after Ref. 53).

previously developed for Ti: LiNbO$_3$[40]. However, due to the larger substrate sizes available, far more complex circuits have become possible. Key components are: 1) Power splitters based on Y-junctions and directional couplers; 2) Stars based on tree-structures and radiative distributors; and, 3) MUX/DEMUX components based on Mach-Zehnder interferometers and gratings. Since it is amorphous, silica is not electro-optic. However, a number of slow-speed controllable devices operating via the thermo-optic effect have also been developed. These include: a) Phase modulators; and, b) Switches and switch arrays.

5.2. Power Splitters and Stars

1 x 2 and 2 x 2 power splitting functions are performed by Y-junctions and directional couplers. The former are waveguide forks with branch angles of 1° or less, and provide broadband operation; they may also be cascaded into binary tree structures to provided higher-order splitting. Directional couplers consist of two parallel waveguides which exchange power by evanescent coupling. According to standard theory[40], the coupling efficiency of a directional coupler is given by:

$$\eta = \sin^2\left\{\sqrt{(v^2 + \xi^2)}\right\} / \left(1 + \xi^2 / v^2\right) \tag{16}$$

where $v = \kappa L$ is the normalised coupling length, κ is the coupling coefficient, L is the device length, $\xi = \Delta\beta L/2$, and $\Delta\beta$ is the difference between the propagation constants of the two guides. κ is given by:

$$\kappa = k_o^2 \langle \Delta n^2 E_1, E_2 \rangle \tag{17}$$

where $k_o = 2\pi/\lambda$, $E_1(x, y)$ and $E_2(x, y)$ are normalised transverse fields of the modes in the two guides, and $\Delta n^2(x, y)$ is the perturbation in relative dielectric constant seen by each guide due to the presence of its neighbour. Symmetric couplers ($\Delta\beta = 0$) having $\kappa L = \pi/4$ can act as 50% (3 dB) power splitters. The coupling coefficient depends exponentially on the guide separation; however, in most SiO$_2$ / Si systems, usable lengths lie in the range 0.1 mm < L < 5 mm.

For broadband networks, wavelength-independent characteristics analogous to those of fused tapered couplers[52] are important. However, variations in coupling efficiency arise in symmetric couplers because of the dependence of κ on wavelength. One structure providing good broadband performance is shown in Fig. 22. This is an asymmetric coupler, constructed with cores of similar height H but differing widths A, B. Through careful choice of these parameters, and of the guide separation S and device length L, variations in κ and $\Delta\beta$ can be traded off against each other to obtain wavelength-flattened coupling efficiencies of 50% ± 5% over the range[53] 1.2 - 1.6 µm.

Using components of this type, highly efficient low-order star couplers for broadcast systems may be constructed. For example, Fig. 23a shows a 2 x 8 star based on a single directional coupler and two Y-junction trees. Fig. 23b shows a combinatorial[54] 8 x 8 star[55] based entirely on directional couplers, which divides the power from a single input equally among the 8 outputs. In this case, waveguide intersections are required; these have both low loss and low crosstalk provided the angle of intersection is chosen suitably[56].

When N is large, the combinatorial approach is impractical. In this case, an alternative method is used, in which the power is coupled between input and output ports by radiation[57]. Fig. 23c shows a typical structure; here the input and output guides are arranged in regular fans, and connected via a section of planar guide.

Light emerging into this region from the end of an input guide will spread by diffraction, so that the far-field pattern covers the entrance to the output fan, coupling to the individual waveguide modes with good uniformity. Theoretical predictions for total efficiency are as high as 35%, independent of N. High efficiencies have been achieved in practise, and the number of ports has risen very rapidly. For example, the sizes of 1 x N devices have increased from 1 x 16[51] to 1 x 128[58], while the sizes of N x N devices have risen correspondingly from 19 x 19[49,59] to 144 x 144[60].

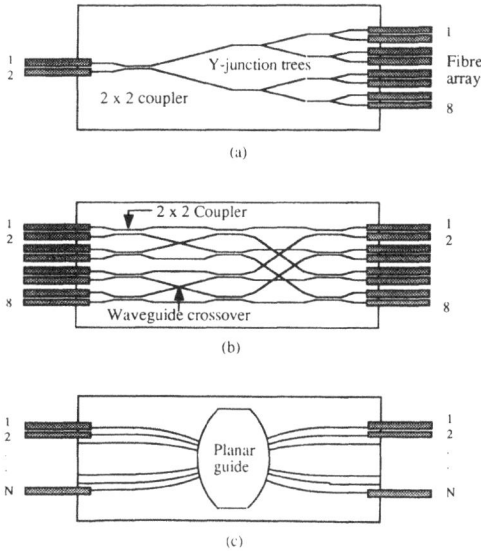

Fig. 23. a) 2 x N star based on directional coupler and Y-junction tree, b) N x N star based on directional couplers (after Ref. 55), and c) N x N radiative star.

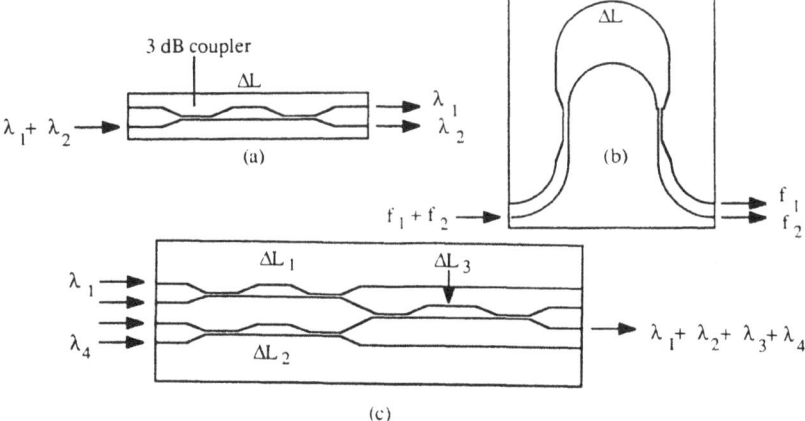

Fig. 24. Mach Zehnder interferometers for a) wavelenght demultiplexing and b) frequency demultiplexing (after Ref. 62); c) 4-channel wavelength multiplexer (after Ref. 10).

5.3. Multiplexers and Demultiplexers

Efficient wavelength-division multiplexers and demultiplexers are required for WDM systems. Versatile filters may be constructed from Mach-Zehnder interferometers (MZIs), as shown in Fig. 24[61,62]. Fig. 24a shows the basic principle. Two 3 dB couplers are combined into an MZI with an imbalance of ΔL between the two arms. According to standard theory[40], the transfer efficiencies for the two output ports for an input to the lower port are:

$$\eta_{upper} = \cos^2(\beta\Delta L/2) \qquad \eta_{lower} = \sin^2(\beta\Delta L/2) \qquad (18)$$

where $\beta = 2\pi n_{eff}/\lambda$, and n_{eff} is the effective index. Thus, by appropriate choice of ΔL, η_{upper} and η_{lower} may be made unity at two different wavelengths λ_1 and λ_2. A multiplexed input may then emerge with the individual components spatially separated. As ΔL increases, λ_1 and λ_2 approach each other, so devices with large imbalance (Fig. 24b) may be used for frequency-division multiplexing.

More complicated networks containing multiple interferometers can be constructed to improve the filter response or increase the number of channels. Fig. 24c shows a 1.5 µm MUX based on three MZIs with different imbalances. This combined four channels separated by 8 nm, with an average fibre-to-fibre loss of 2.6 dB and -16 dB crosstalk[10]. Similar devices can act as matched filters[63].

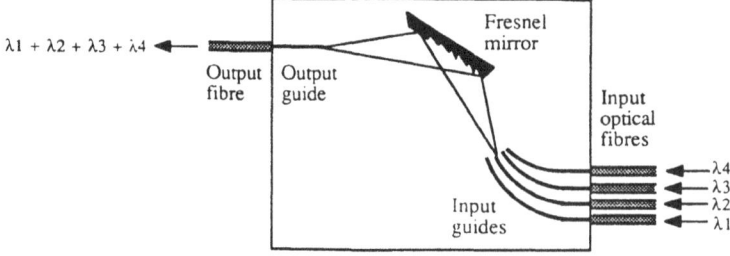

Fig. 25. Dispersive four-channel wavelength multiplexer (after LETI Technical Data Sheet DOPT 9053).

Multiplexers may also be based on a single dispersive component such as a grating, which effectively provides all the required filters in parallel. For example, Fig. 25 shows a four-channel MUX based on a curved, blazed reflection grating[64]. After leaving the input channel guides, the input beams propagate as diverging cylindrical waves in a planar waveguide. They are combined into a single cylindrical beam converging on the output channel guide by a Fresnel mirror (the reflective equivalent of a Fresnel lens), which provides a wavelength-dependent, reflective focusing action. The mirror is formed by etching a blazed pattern through the silica overlayers to the substrate, so that total internal reflection is obtained at the silica/air interface. This 1.5 µm MUX had a channel separation of 20 nm, an insertion loss of < 6 dB, and crosstalk better than -20 dB.

Relatively little work has been performed on Bragg grating devices (although these promise higher selectivity) because the simplest grating structures do not provide spatially-separated outputs. Corrugated gratings with 1.54 µm centre wavelength and 1.5 nm bandwidth have been fabricated in Si_3N_4 / SiO_2[65], and grating/resonator combinations have been used to achieve bandwidths as low as 0.2 nm[66]. More recently, gratings have been directly photoinduced in GeO_2-doped waveguide cores by holographic exposure to UV radiation[67], following the method previously demonstrated for fibres[68]. This method is potentially extremely attractive; however, the resulting gratings have only limited refractive index modulation, although increased modulation has been reported recently for material consolidated in a reducing atmosphere[69].

5.4. Device Tuning

Fabrication tolerances often result in devices failing to fulfill their design specifications. Two techniques have been investigated for device tuning after fabrication. Both are based on irradiation, and are broadly similar to methods previously described for waveguide fabrication:
- Spot irradiation with a CO_2 laser has been used to adjust the passband of MZI-based filters[70];
- Flood irradiation with an electron beam has been used to adjust the coupling coefficient of directional couplers[71].

For example, Fig. 26 shows the variation in coupling efficiency with length L for symmetric directional couplers fabricated using arsenosilicate glass (ASG) cores and silica cladding. As predicted by Eq. 16, the variation of η with L follows a sin^2 curve. After flood irradiation with doses of 0.04 and then 0.14 C/cm² using an electron beam of sufficient energy to reach the cores (25 keV), the period of the oscillations increases, corresponding to a reduction in coupling coefficient. This is due to an increase in modal confinement (which reduces the overlap integral of Eq. 17), caused by a rise in the core refractive index on irradiation.

5.5. Switchable Components

Slow-speed optical phase shifters and modulators may be constructed very simply, based on the thermo-optic effect. Fig. 27 shows a phase modulator, which consists of a Ti metal thin-film heater of length L located above a channel waveguide.

Under normal conditions, the optical path accumulated in the device length is $\phi = \beta L = 2\pi n_{eff} L/\lambda$. When a current is passed through the heater, a temperature gradient is established across the silica layers that is sufficient to remove the heat generated, by conduction to the substrate. Some heat is also removed by convection. Assuming the guide temperature rises by ΔT, the resulting phase change is:

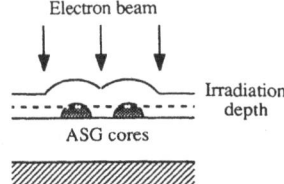

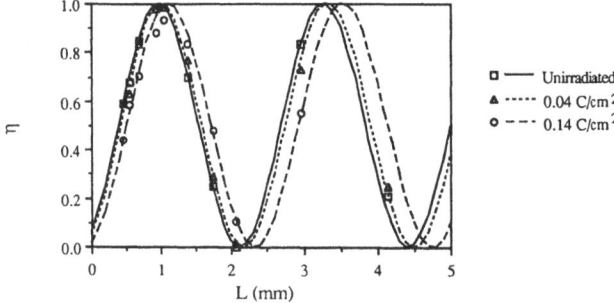

Fig. 26. Tuning of directional couplers by electron beam irradiation: coupling efficiency versus length for unirradiated devices, and devices irradiated with 0.04 and 0.14 C/cm² charge dose. Points are experimental data; lines theoretical best fit (after Ref. 71).

$$\Delta\phi = 2\pi/\lambda\{dL/dT\, n_{eff} + L\, dn_{eff}/dT\}\, \Delta T = 2\pi/\lambda\, \{\alpha_f\, n_{eff} + dn_{eff}/dT\}\, L\Delta T \tag{19}$$

where $\alpha_f = 1/L\, dL/dT$ is the thermal expansion coefficient of the guide. Assuming that $n_{eff} \approx n_{silica}$, Table 5 suggests that $dn_{eff}/dT \gg \alpha n_{eff}$, so that:

$$\Delta\phi \approx 2\pi/\lambda\, dn_{eff}/dT\, L\Delta T \tag{20}$$

A temperature rise of 6.8°C is therefore required to obtain a phase change of π radians in a device of length 10 mm at 1.5 μm wavelength. This can be achieved with an electrical power of ≈ 0.5 W, with a response time of 2 msec[1].

Thermo-optic devices are polarization-insensitive, and their operation does not depend on absolute temperature. However, they do suffer some disadvantages. In particular, since the heat they generate is not well localised, they cannot be used to control a directional coupler by

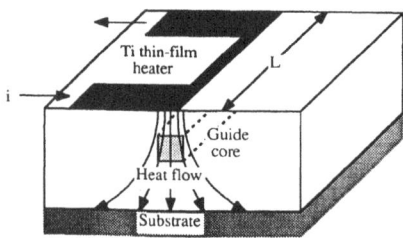

Fig. 27. Construction of thermo-optic phase modulator.

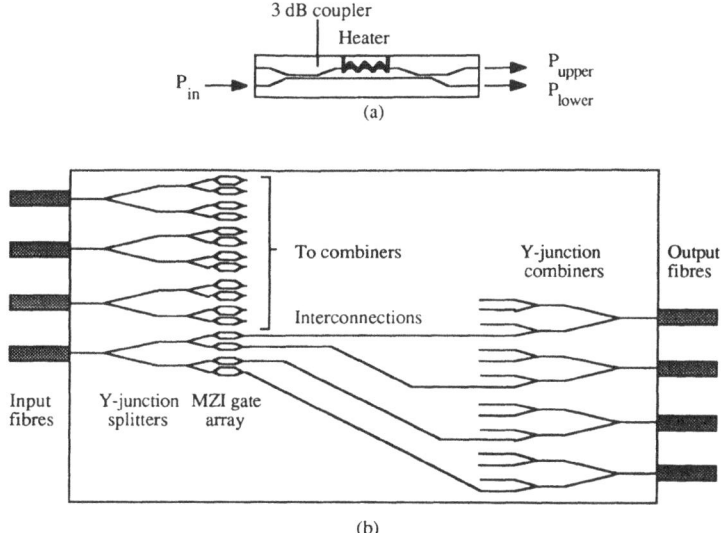

Fig. 28. a) Interferometric switch based on thermo-optic phase modulator; b) optical gate matrix switch using Mach-Zehnder interferometer gates (after Ref. 75).

desynchronising the two guides. The building-block most commonly used in Ti: LiNbO$_3$ integrated optics for switching and tunable filtering is therefore inappropriate for SiO$_2$ / Si at present.

Thermo-optic phase modulators can be used to provide modulation or switching when they are incorporated into a Mach-Zehnder interferometer as shown in Fig. 28a[72,73]. In this case, the outputs from the two arms are given by:

$$\eta_{upper} = \cos^2(\Delta\phi/2) \qquad \eta_{lower} = \sin^2(\Delta\phi/2) \qquad (21)$$

where $\Delta\phi$ is as in Eq. 20. Two-way switches of this type can be formed into gate matrix switching arrays. Unfortunately, the required architecture is more complex than for comparable coupler-based Ti: LiNbO$_3$ devices.

Fig. 28b shows the construction of a typical gate matrix switch array, which has already been demonstrated in 4 x 4 form[74,75]. This is constructed from Y-junction trees and thermo-optically-driven interferometric modulators. Effectively, it is composed of four cross-connected 1 x 4 power splitters and sixteen modulators, and therefore functions as a strictly non-blocking broadcast switch[76]. Total insertion losses and crosstalk reported are 27 dB and -12 dB, respectively.

6. Active Waveguide Devices

In addition to the range of passive components previously described, active devices such as optical amplifiers are now being constructed using SiO$_2$ / Si integrated optics. The principles involved are very similar to those of existing fibre-based amplifiers, but integration offers a cost advantage.

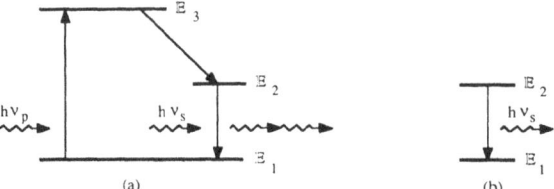

Fig. 29. Energy levels of a three level optical amplifier.

Fig. 29a shows the energy levels of a three-level optical amplifier, which might be used as a signal booster. Gain is achieved through the process of stimulated emission, in which a signal photon of frequency v_s such that $hv_s = E_2 - E_1$ triggers a downward transition of a bound electron in the active material from state 2 to the ground state (state 1), generating a second photon of the same frequency. This process can be made to dominate over absorption of the signal (which promotes electrons from state 1 to state 2) only if the population of state 2 is made to exceed that of the ground state. This is achieved by pumping the material with a source of photons of frequency v_p and energy $E_p = hv_p$. Absorption of these photons results in the promotion of electrons from the ground state to a third state with energy E_3 such that $E_p = E_3 - E_1$. The electrons then decay non-radiatively to state 2. If the pump beam is strong enough and the lifetime of state 2 is long enough, population inversion is obtained. However, spontaneous emission of photons of the signal frequency also occurs, through the decay of electrons from state 2 to state 1 (Fig. 29b). This acts as a source of optical noise.

Silica may be transformed into an active material by doping with rare earth ions. For operation at ≈1530 nm, doping with Er^{3+} ions is used. Optical gain can be provided at this wavelength by pumping at 980 nm, which promotes electrons from the $^4I_{15/2}$ ground state to the $^4I_{11/2}$ state. Electrons in this state quickly decay to the $^4I_{13/2}$ metastable state. Stimulated transition from this state back to the ground state then provides the gain[77-79]. Two possible sources can currently be used to generate the pump wavelength: for high powers, Ar^+-pumped Ti: Al_2O_3 lasers are used, and for lower powers, specially developed semiconductor laser diodes[80]. The main problem with erbium doping is the difficulty of obtaining high concentrations of
uniformly-dispersed ions in the host material[79]. Specifically, Er is known to form concentration clusters in a pure silica host, which limit the dopant concentration to ≈ 100 ppm in SiO_2 or GeO_2: SiO_2. Already, this has been identified as a key difficulty in fibre-based amplifiers, necessitating the use of long fibre lengths (e.g. 90 metres[81]) to achieve the required gain. In an integrated device, the problem is even more acute, and very careful attention must be paid to the host material. It has already been demonstrated in fibre amplifiers that the use of alumina[82,83] or phosphorus pentoxide[84] co-doping can drastically reduce the clustering problem in silica. Clustering occurs because the rare-earth oxides (e.g. Er_2O_3) are generally insoluble in SiO_2. However, Er_2O_3 is soluble in Al_2O_3, which in turn is soluble in SiO_2. The Al_2O_3 can then form a solvation shell around the rare earth ion, and the resulting complex is readily incorporated into the silica network. The same principle is assumed to apply in P_2O_5 co-doping[79]. Waveguide devices can then be constructed using a four-component glass. For example, fibre-based amplifiers have been constructed using Er_2O_3: Al_2O_3: GeO_2: SiO_2. Here, two of the dopants - germania and alumina - are used to control the optical refractive index and the dispersion of the third dopant (the erbium ions) independently.

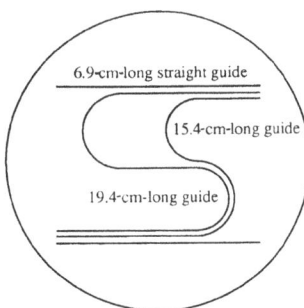

Fig. 30. Folded path of integrated optical amplifier (after Ref. 87).

Similar principles have already been applied to planar devices. NTT have already demonstrated Nd- and Er-doped silica waveguide lasers[85,86], and have recently constructed a 20 cm-long Er-doped amplifier with 15 dB gain at a pump power of 500 mW [87]. Fig. 30 shows the folded layout used to achieve such a path length in integrated device dimensions.

In each case, a three-component glass was used. The waveguide material was SiO_2 co-doped with P_2O_5, formed by flame hydrolysis. The active materials were then introduced by soaking the soot in an alcohol solution containing a hydrated rare earth chloride before sintering, and Er concentrations of 5500 - 8000 ppm were achieved. Other fabrication methods (e.g. PECVD[88] and sol-gel[89]) are now being adapted to allow the incorporation of erbium in deposited glasses.

ACKNOWLEDGEMENTS. The Author is indebted to co-workers at Imperial College (A.S.Holmes, J.Lewandowski), British Telecom Research Laboratories (K.Cooper and M.Nields) and BNR Europe Limited (M.Grant and G.Cannell) for the provision of some of the results presented here. These were obtained from collaborative programs supported by SERC/DTI and the Furukawa Electric Co. Ltd.

References

1. M. Kawachi, Opt. Quant. Elect. 22, 391-416 (1990)
2. J. Gowar, "Optical Communication Systems", Prentice Hall International (1984)
3. S.M. Sze, "VLSI Technology", McGraw Hill International Book Company (1983)
4. D.E. Zelmon, H.E. Jackson, J.T. Boyd, A. Naumaan, and D.B. Anderson, Appl. Phys Lett. 42, 565-566 (1983)
5. D.E. Zelmon, J.T. Boyd, and H.E. Jackson, Appl. Phys. Lett. 47, 353-355 (1985)
6. J.T. Boyd, R.W. Wu, D.E. Zelmon, A. Naumaan,and H.A. Timlin, Opt. Engng. 24, 230-234 (1985)
7. N. Imoto, N. Shimizu, H. Mori, and M. Ikeda, IEEE J. Lightwave Tech. LT-1, 289-293 (1983)
8. M.K. Smit, G.A. Acket and van der Laan, Thin Solid Films 138, 171-181 (1986)
9. C.H. Henry, R.F. Kazarinov, H.J. Lee, K.J. Orlowsky and L.E. Katz, Appl. Opt. 26, 2621-2624 (1987)
10. B.H. Verbeek, C.H. Henry, N.A. Olsson, K.J. Orlowsky, R.F. Kazarinov and B.H. Johnson, IEEE J. Lightwave Tech. LT-6, 1011-1015 (1988)
11. W. Stutius and W. Streifer, Appl.Opt. 16, 3218-3222 (1977)
12. G. Grand, J.P. Jadot, H. Denis, S. Valette, A. Fournier and A.M. Grouillet, Elect. Lett. 26, 2135-2137 (1990)

13. S. Valette, S. Renard, P. Jadot, P. Gidon and C. Erbeia, Sensors and Actuators A21-A23, 1087-1091 (1990)
14. S. Day, R. Bellerby, G. Cannell and M. Grant, Elect. Lett. 28, 920-922 (1992)
15. D.E. Bossi, J.M. Hammer and J.M. Shaw, Appl. Opt. 26, 609-611(1987)
16. M. Kawachi, M. Yasu and T. Edahiro, Elect. Lett. 19, 583-584 (1983)
17. N. Takato, M. Yasu and M. Kawachi, Elect. Lett. 22, 321-322 (1986)
18. T. Kominato, Y. Ohmori, H. Okazaki and M. Yasu, Elect. Lett. 26, 327-329 (1990)
19. L.L. Hench and J.K. West, Chem. Rev. 90, 33-72 (1990)
20. A.S. Holmes, R.R.A. Syms, Li Ming and M. Green,"Fabrication of buried channel waveguides on silicon substrates using spin-on glass and rapid thermal annealing", Appl. Opt. 32, 4916-4921 (1993)
21. L.M. Mack, A. Reisman and P.K. Bhattacharya, J. Electrochem. Soc. 136, 3433-3437 (1989)
22. T.H.T. Wu and R.S. Rosler, Solid State Technol., May Issue, 65-71 (1992)
23. R.R.A. Syms, "Stress in thick sol-gel phosphosilicate glass films formed on Si substrates" J. Non-Cryst. Solids 167, 16-20 (1994)
24. R. Guenther, "Modern Optics", John Wiley and Sons (1990)
25. C.J. Sun, W.M. Myers, K.M. Schmidt and S. Sumida, IEEE Photon. Technol. Lett. 3, 238-240 (1991)
26. R.R.A. Syms and A.S. Holmes, "Reflow and burial of channel waveguides formed in sol-gel glass on Si substrates" IEEE Photon. Technol. Lett. 5, 1077-1079 (1993)
27. M. Guglielmi, P. Colombo, L. Mancinelli Degli Espositi, G.C. Righini and S. Pelli, Proc. SPIE 1513, 44-49 (1990)
28. D. J. Shaw and T. A. King, Proc. SPIE 1328, 474-481 (1990)
29. A. J. Houghton and P.D. Townsend, Appl. Phys. Lett. 22, 565-566 (1976)
30. C.B. Norris and E.P. Eer Nisse, J. Appl. Phys. 45, 3876-3882 (1974)
31. T.A. Dellin, D.A. Tichenor and E.H. Barsis, J. Appl. Phys. 48,1131-1138 (1977)
32. H. Nishihara, Y. Handa, T. Suhara and J. Koyama, Appl. Opt. 17, 2342-2345 (1978)
33. Y. Handa, T. Suhara, H. Nishihara and J. Koyama, Appl. Opt. 18, 248-252 (1979)
34. T.E. Everhart and P.H. Hoff, J. Appl. Phys. 42, 5837-5846 (1971)
35. S.J. Madden, M. Green and D. Barbier, Appl. Phys. Lett. 57, 2902-2903 (1990)
36. D. Barbier, M. Green and S.J. Madden, IEEE J. Lightwave Tech. 2. 715-720 (1991)
37. J. Bell and C.N. Ironside, Elect. Lett. 27, 448-450 (1991)
38. J. Lewandowski, R.R.A. Syms, S. Madden and M. Green, Opt. Quant. Elect. 23, 703-711(1991)
39. J. Lewandowski, R.R.A. Syms, M. Grant and S.Bailey, "Controlled formation of buried channel waveguides by electron beam irradiation of glassy layers on silicon" Int. J. Optoelectronics, May/June Issue (1994)
40. R.R.A. Syms and J.R. Cozens, "Optical Guided Waves and Devices", McGraw-Hill Book Company (1992)
41. C.J. Beaumont, S.A. Cassidy, D. Welbourn, M. Nield and A. Thurlow, "Integrated silica optical delay line" Proc. ECOC '91, paper TuB5-6 (1991)
42. S. Suzuki, K. Shuto, H. Takahashi and Y. Hibino, Elect. Lett. 28, 1863-1864 (1992)
43. H. Yanagawa, T. Shimizu, S. Nakamura and I. Ohyama, IEEE J. Lightwave Tech. LT-10, 587-592 (1992)
44. M. Yanagisawa, Y. Yamada and M. Kobayashi, Elect. Lett. 28,1958-1959 (1992)
45. H. Terui, Y. Yamada, M. Kawachi and M. Kobayashi, Elect. Lett. 21, 646-647 (1985)
46. D.B. Lee, J. Appl. Phys. 40, 4569-4574 (1969)
47. C.M. Schroeder, Bell. Syst. Tech. J. 57, 91-97 (1977)
48. E.J. Murphy, T.C. Rice, L. McCaughan, G.T. Harvey and P.H. Read, IEEE J. Lightwave Tech. LT-3, 795-799 (1985)
49. H.M. Presby and C. A. Edwards, Opt. Engng. 31, 141-143 (1992)
50. M.F. Grant, R. Bellerby, S. Day, G.J. Cannell and M. Nelson, "Self-aligned multiple coupling for silica-on-silicon integrated optics" Proc. EFOC-LAN, London, June 19-21, 269-292 (1991)

51. S. Day, R. Bellerby, Cannell and M. Grant, Elect. Lett. 28, 920-922 (1992)
52. D.B. Mortimore, Elect. Lett. 21, 742-743 (1985)
53. A. Takagi, K. Jinguji and M. Kawachi, Elect. Lett. 26,132-133 (1990)
54. M.E. Marhic, Opt. Lett. 2, 368-370 (1984)
55. H. Yanagawa, S. Nakamura, I. Ohyama and K. Ueki, IEEE J. Lightwave Tech. LT-8, 1292-1297 (1990)
56. A. Himeno, M. Kobayashi and H. Terui, Elect. Lett. 21,1020-1021(1985)
57. C. Dragone, IEEE J. Lightwave Tech. LT-7, 479-488 (1989)
58. H. Takahashi, K. Okamoto and Y. Ohmori, IEEE Photon. Technol. Lett. 5, 58-60 (1993)
59. C. Dragone, C.H. Henry, I.P. Kaminow and R.C. Kistler, EEE Photon. Technol. Lett. 1, 241-243 (1989)
60. K. Kato, K. Okamoto, H. Okazaki, Y. Ohmori and I. Nishi, IEEE Photon. Technol. Lett. 4, 348-351 (1993)
61. N. Takato, K. Jinguji, M. Yasu, H. Toba and M. Kawachi, EEE J. Lightwave Tech. LT-6, 1003-1010 (1988)
62. T. Ikegami and M. Kawachi, "Passive paths for networks" Physics World, Sept. Issue, 50-54 (1991)
63. J. Nishikido, M. Okuno and A. Himeno, Elect. Lett. 26,1766-1767 (1990)
64. S. Valette, P. Gidon and J.P. Jadot, "New integrated optical multiplexer-demultiplexer realised on silicon substrate" Proc. ECIO '87, May 11-13, Glasgow (1987) Lee H.J., Henry C.H., Kazarinov R.F., Orlowsky K.J. Appl. Opt. 26, 2618-2620 (1987)
66. C.H. Henry, R.F. Kazarinov, H.J. Lee, N.A. Olsson and K.J. Orlowsky, IEEE J. Quant. Elect. OE-23, 1426-1428 (1987)
67. K. O. Hill, Y. Fujii, D.C. Johnson and B.S. Kawasaki, Appl. Phys. Lett. 32, 647-649 (1978)
68. G.D. Maxwell, R. Kashyap, B.J. Ainslie, D.L. Williams and J.R. Armitage, Elect. Lett. 28, 2106-2107 (1992)
69. Y. Hibino, M. Abe, H. Yamada, Y. Ohmori, F. Bilodeau, B. Malo and K.O. Hill, Elect. Lett. 22, 621-623 (1993)
70. H. Uetsuka, M. Kurosawa and K. Imoto, Elect. Lett. 26, 251-253 (1990)
71. R.R.A. Syms, J. Lewandowski, M. Nield and F. Mackenzie, Tuning of silica-on-silicon directional couplers by electron beam irradiation" IEEE Photon. Tech. Lett., April Iusse (1994)
72. W.E. Martin, Appl. Phys. Lett. 26, 562-564 (1975)
73. M. Haruna and J. Koyama, IEE Proc. H (Microwaves, Optics & Antennas) 131, 322-324 (1984)
74. A. Himeno and M. Kobayashi, IEEE J. Lightwave Tech. LT-3, 230-235 (1985)
75. A. Himeno and M. Kobayashi, Elect. Lett. 23, 887-888 (1987)
76. K. Habara and K. Kikuchi, Elect. Lett. 23, 376-377 (1987)
77. J.B. Ainslie, S.P. Craig and S.T. Davey, IEEE J. Lightwave Tech. LT-6, 287-292 (1988)
78. J.B. Ainslie, IEEE J. Lightwave Tech. LT-9, 220-227 (1991)
79. W.J. Miniscalco, IEEE J. Lightwave Tech. 2, 234-250 (1991)
80. M. Shimizu, M. Horiguchi, M. Yamada, I. Nishi, J. Noda, T. Takeshita, M. Okayasu, S. Uehara, E. Sugita, IEEE J. Lightwave Tech. LT-9, 291-295 (1991)
81. K. Suzuki, Y. Kimura and M. Nakazawa, Elect. Lett. 26, 948-949 (1990)
82. J.F. Massicott, R. Wyatt, B.J. Ainslie and S.P. Craig-Ryan, Elect. Lett. 26,1038-1039 (1990)
83. Y. Kimura and M. Nakazawa, Elect. Lett. 28, 1420-1422 (1992)
84. P.R. Morkel, G.J. Cowle and D.N. Payne, Elect. Lett. 26, 632-634 (1990)
85. Y. Hibino, T. Kitagawa, M. Shimizu, F. Hanawa and A. Sugita, IEEE Photon. Technol. Lett. 1, 349-350 (1989)
86. T. Kitagawa, K. Hattori, Y. Shimizu, Y. Ohmori and M. Kobayashi, Elect. Lett. 27, 334-335 (1991)
87. T. Kitagawa, K. Hattori, K. Shuto, M. Yasu, M. Kobayashi and M. Horiguchi, Elect. Lett. 28, 1818-1819 (1992)
88. K. Shuto, K. Hattori, T. Kitagawa, Y. Ohmori and M. Horiguchi, Elect. Lett. 29, 139-141(1993)

89. D. Moutonnet, R. Chaplain, M. Gauneau, Y. Pelous and J.L. Rehspringer, Mat. Sci. Engng. B9, 455-457 (1991)

Chapter 8

INTEGRATED OPTICS ON SILICON: IOS TECHNOLOGIES

S. VALETTE

1. Introduction

Integrated Optics is a very promising technology for the near future, taking into account the expected development of optoelectronics and the availability at low cost of all elements required for advanced optoelectronic systems.

To the well known advantages of optics, integrated optics brings a greater miniaturisation of optical circuits, suppression of most alignment problems and an opening of the way to mass production and low cost optoelectronic devices.

However unlike microelectronics where the monolithic approach has been the key to the success, a hybrid approach now appears, better adapted to the industrial development of optoelectronics[1,2]

This situation arises from fundamental differences between microelectronics and optoelectronics:

- a greater number of different components to be integrated, including light sources, electronic drivers, photodetectors with electrical signal processing, amplifiers, etc.;

- the predominance in optoelectronic architectures of passive elements, which don't require the use of a semiconductor material;

- a lack of real technological unity in the monolithic approach using indium phosphide, leading to a large number of processing steps and mask levels even for the integration of a small number of components;

- a smaller integration density in optoelectronics devices, which drastically modifies the economic advantages potentially brought by a monolithic solution ; and,

- finally, a better flexibility of hybrid approaches, allowing the achievement of optimized devices and configurations.

Of course, the advantages of such a hybrid approach is effective only if interconnection problems between hybrid components can be solved with technical approaches compatible with the objectives of low cost and mass production.

From this point of view, silicon based optoelectronics offers an attractive solution thanks to the mechanical properties of the substrate, its high thermal conductivity which allows direct laser diode hybridization, and the various etching methods, wet or dry, which give great freedom to achieve precise grooves and cavities for passive alignment between optical elements...

S. Valette - LETI (CEA - Technologies Avancées) Centre d'Etudes Nucléaires de Grenoble, Av. de Martyrs - 85 X F 38041 Grenoble Cedex - France

Furthermore, the broad variety of technologies and processes associated with the principal optical materials involved in silicon based integrated optics, silica and silicon nitride, gives great flexibility to build new integrated optics components unrealizable with competing approaches.

2. Description of Silicon Integrated Optics On Silicon (IOS) Technologies[3÷6]

The path to silicon integrated optics was shown about twelve years ago by LETI in France[3,6,7,8,9] and NTT in Japan[4,5,10,11] using respectively CVD and flame hydrolysis deposition (FHD). Nowadays, ATT[2,12] in USA and more recently a lot of new laboratories have choosen to investigate this integrated optics approach.

Regardless the technical method which is used, we can classify the integrated optics structures achieved on silicon into two classes:

1) A "high contrast structure" in which the refractive index difference between the core and the surrounding media ΔN exceeds 10^{-1}, yielding intrinsically high birefringence waveguides having attractive special properties in some specific applications. The core material is generally silicon nitride, silicon oxynitride or alumina whereas the surrounding media are silica. At LETI, such structures using a silicon nitride core have been called IOS1 (for Integrated Optics on Silicon version 1 : see Fig. 1a).

2) A "low contrast structure" in which the ΔN value is low, typically in the range 3 times 10^{-3} to few times 10^{-2}. The waveguide is formed with doped silica and the optical components can be insensitive to the polarization state of the light. Such structures have been called "high silica guides" at NTT and IOS2 (Integrated Optics on Silicon version 2 : see Fig. 1b) at LETI.

In both cases the guided light is isolated from the silicon substrate by a silica buffer layer of greater or lesser thickness, depending of the effective index value of the guided mode.

A more complex structure combining both waveguide types in the same multilayer arrangment was first proposed by LETI[3,14], then by ATT[15,16]. It is the so-called IOS3 (Integrated Optics on Silicon version 3 : see Fig. 1c) which combines silicon nitride and doped silica core waveguides which may or may not be separated by a suitable silica buffer.

A structure employing gratings[14] or adiabatic transistions[15,16] allows efficient controled exchange of the guided light between both types of waveguides. This can be very useful for achieving special optical functions[17,18] such as polarization splitting[16] that are difficult to realize with low contrast structures, or to efficiently connect integrated optics components having different mode profile dimensions.

The performance thus far demonstrated for IOS structures is very good, including low propagation losses (less than 3 dB/m) and low in-plane scattering (lower than 50dB and 43dB measured at 1° off axis for IOS2 and IOS1 respectively). The lower values for propagation losses have been measured using ring resonator configurations at different wavelenths with slightly different IOS2 type structures[19,20,21,22].

Such performance requires a post-annealing operation when CVD is used, in order to suppress undesirable absorption peaks due to the overtones of Si-H, N-H or O-H bond mainly in the 1.4 - 1.5 µm range. The use of a reaction gas without hydrogen or nitrogen could be certainly an attractive way to avoid the need for such annealing. Some solutions of this type are being investigated. With the FHD approach, no annealing is required because of the high temperature consolidation phase during the waveguide formation.

In low contrast structures, the residual birefringence due to internal stresses is on the order of 10^{-4} and limits slightly the polarization insensitivity of integrated components.

Integrated Optics on Silicon: IOS Technologies

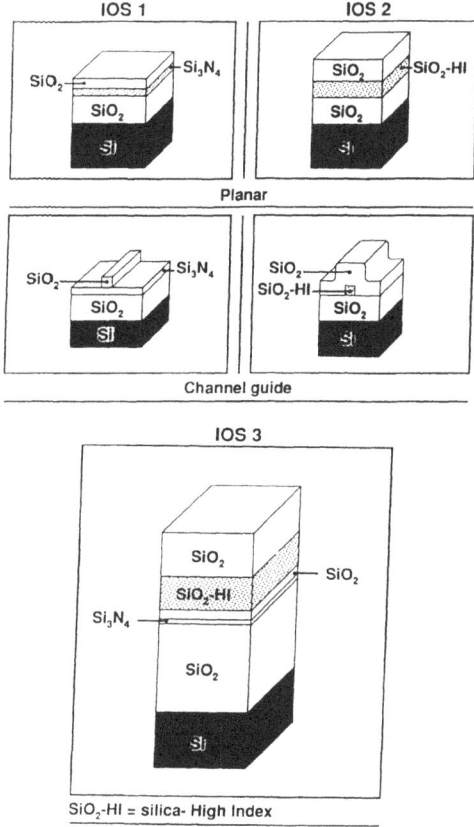

Fig. 1. a),b), and c): Schematic diagram of IOS structures.

However, this value depends strongly on the deposition conditions and can reach a few times 10^{-4} in the worst cases. Various methods have been proposed to reduce or to control this parameter[23, 24].

In high contrast waveguides, the material birefringence can be neglected compared to the modal birefringence, especially when a silicon nitride core is used.

3. Connection of IOS Structures and Other Optoelectronic Components: The Silicon Mother Board Concept

Two problems need to be considered when interconnecting IOS and other optoelectronic components:

1) The mode profile overlap between both structures to be connected: this determines the maximum coupling efficiency which can be reached using butt coupling.

2) The alignment accuracy given by the connection method, which must be generally lower than 1µm when coupling losses lower than 1 dB are required. This rough estimate illustrates the challenge when completely passive alignment is desired.

3.1. Connection with Optical Fibers[2, 25, 26, 27]

With IOS2 type structures the channel guide mode profile can be very suitable for efficient connections with optical fibers because of the similar optical characteristics of both components.

In many cases, however, one desires to increase slightly the refractive index difference ΔN in order to have more freedom to achieve bent microguides with small radii of curvature. This trade-off leads to a slight mismatch between mode fields and to a theoretical fibre-waveguide coupling loss of about 0.1 - 0.2 dB. Because of the rather large field diameter of optical communication fibers, alignment tolerances are relaxed to about 0.5 µm in the perpendicular directions in order to attain a coupling loss below about 0.4 dB. Axial tolerances are greater, especially if an index matching material is used between fibre and guide (typically a gap of 10 µm leads to 0.1 - 0.2 dB excess loss).

Of course, problems are different with high contrast structures where the mode mismatch is generally the principal cause of loss. For instance, with an IOS1 structure and a silicon nitride core 0.16 µm thick, the minimum loss is about 8 - 10 dB, very similar to that obtained for coupling with a standard laser diode. In this case, the use of suitable tapers to provide a mode transformation appears as an attractive solution. The possibility of adiabatic transitions is one of the advantages of IOS3 structures. Such configurations have been used by ATT[15] to demonstrate efficient coupling between laser diodes and doped silica waveguides, and by LETI to connect a reading-writing magneto-optic head with an input single-mode fiber[28].

Very accurate alignment grooves have been already been demonstrated using the well known "V" shape directly produced on a silicon substrate by chemical preferential etching[26, 27]. Results published by STC technology[26] (now BNR Europe) are very impressive: in the case of IOS2 type guides, 50% of passively aligned fibre to waveguide couplings exhibited a loss less than 0.2 dB and 90% less than 0.4 dB. Less than 0.1 dB additional loss was introduced through the use of passive alignment relative to loss measured using active alignment. However, the "V" groove technique required a wet etching process, which can be a drawback in manufacturing stage. Moreover, the direction of the grooves has to follow the 111 crystal plane and this limits the architectural freedom. A new technique using "U" grooves[27, 30] made by microwave dry etching was demonstrated three years ago by LETI. In this approach the positioning of the fibre laterally is provided by the silica structure itself, which acts as a mask for silicon etching with a 1 to 100 selectivity. Vertical alignment is provided by the depth of the "U" groove, which can be controlled by in situ interferometry. Most of the connections using this method exhibit excess loss lower than 0.6 dB with an average value of 0.4 dB. Because this process can be used for self alignment between microguides and optical fibres, additional losses are only due to vertical misalignment. At present the limitation of this technique is mainly given by the etching uniformity over the silicon wafer, which is about 3%. An improved uniformity (1% over wafer seems a reasonable objective) will provide very attractive coupling performance and yield for industrial development.

3.2. Connection between Laser Diodes and III-V Optoelectronic Components

These connections[2,31,32,33] are certainly the key to success of silicon based optoelectronics. The high thermal conductivity of silicon (only twice smaller than copper) is indeed the main parameter. It allows hybridization of laser diodes on the silicon substrate without any degradation of their optical performance, as demonstrated by several commercially available semiconductor lasers mounted on silicon. Of course, in the real case of hybridization with an integrated optical circuit, the problem becomes more complicated because of additional alignment requirements. However, different solutions seem able to overcome these difficulties.

The first technique is the well known flip-chip technique[34, 35, 36, 37, 38, 39, 40, 41] using solder bumps, which has been already industrially developed for infrared imaging[37] and microelectronic Multi-Chip Modules (MCM). Currently, developments in advanced laboratories are concentrating on the development of Optoelectronic Multi-chip Modules (OMCM) and on the concept of the optoelectronic silicon mother board. This approach is especially interesting because the basic processes required for efficient hybridization of components on silicon substrates are well known and have now reached an industrial development stage. That means that the technical effort to develop efficient OMCMs appears reasonable and that flip-chip techniques can offer a near-term route to high functionality optoelectronic modules combining reduced size, improved performance and low manufacturing cost. Although alignment tolerances are particularly severe in specific cases (laser diodes, light modulators) due to the high confinement of the light in such III-V devices, the strong potential of silicon micromachining already mentioned in other Chapters of this volume, combined with the use of surface tension effects during reflow in flip-chip techniques, should allow reaching these objectives.

Another approach could be the development of advanced Epitaxial-Lift-Off (ELO)[42]. Hybridization of different optoelectronic components including laser diodes and light modulators has already been demonstrated using different technical approaches[43, 44]. Published results are good, and highly compatible with the performance requirements. However, although epitaxial lift-off is quite an old technique, no information concerning its suitability for industrial development is readily available. The fact that the ELO technique requires significant technical changes in III-V component manufacturing is certainly a gap to overcome for its future industrial development.

3.3. Connection with Micro-optic Components[45 ÷ 49]

Integrated optics is very suitable and powerful for optical information processing. However, with planar optics, one dimension is lost. This is a drawback for specific applications in the field of sensors[49, 50] or optical memories[45, 46, 47, 48, 49] for which the use of a three dimensional optical beam is required in one part of the device. In such cases, efficient 2D - 3D optical converters are needed. This is also the case for hybridization of photodetectors. Although butt coupling can be used, it is better for industrial development to have the sensitive area of the detector parallel to the waveguide plane: that also requires components which turn the planar guided beam into a free space optical beam. Such functions can be achieved by 45° mirrors or by gratings as demonstrated by Osaka University[45, 46, 47, 48, 49]. However in the latter case, the required periodicity of the grating structures is very small and the efficiency is limited. A solution could be the development of integrated blazed gratings. This last possibility could be one of the more attractive in the near future to achieve optical microsystems associating integrated and micro-optics.

4. Integrated Components and Devices in Silicon Based Integrated Optics

A number of components have been already demonstrated with IOS technologies, mainly in the field of optical communications, which presents the big challenge with the expected development of local area networks.

However, advanced devices have also been achieved in the fields of sensors and optical memories, demonstrating the huge potential of silicon based approaches for low cost applications[3, 4, 5, 6, 51, 52, 53, 54, 55, 56, 57, 58, 59, 60].

4.1. IOS Technologies for Optical Communications

All the basic elements of communications networks have been achieved and published using both FHD and CVD deposition methods.

For conventional components such as couplers, duplexers and beam dividers, the performance obtained with silica waveguides is comparable to that obtained with other approaches and for instance ion exchange waveguides on glass.

For more complex components such as multiplexers[27, 58, 59, 60], polarization diversity receivers including polarization splitters, etc.[16, 17, 18], the flexibility of IOS technologies and the quality of technical processes associated with the basic materials like silica give new possibilities that are thus far forbidden for competing approaches.

The achievement of integrated catadioptric components in silica waveguides perfectly illustrates this flexibility[26, 27, 58, 59, 60].

Such components require very deep etching of the entire waveguiding structure, e.g., etching depths of 20 µm to 30 µm with IOS2 at the 1.55 µm wavelength. Moreover, the smoothness of the etching walls and their verticality with regard to the waveguide plane are very critical parameters when high optical performance is required[61].

Because of the high degree of development of dry etching processes in silica, such requirements can be fulfilled, and several devices using plane or curved integrated mirrors have been already reported. A wavelength multiplexer having four channels and using a dispersive Fresnel-mirror was described some years ago[26, 27, 59]. The improvement of etching processes makes likely the future achievement of similar devices having 16 channels or more working in Littrow configuration and with excess losses on the order of 3 dB.

Although channel waveguide phased arrays[62] are serious competitors to dispersive mirrors, the availibity of an efficient anisotropic etching process is one of the key techniques for advanced integrated optics using silicon based technology.

Another example of the technical advantage of the silicon based integrated optics approach can be illustrated by the so-called IOS3 structure associating both high contrast and low contrast guides on the same silicon substrate.

These structures, as already mentionned, can be very useful to solve connection problems between guided wave elements which have mismatched mode profiles; they can be also used to achieve specific components[16, 17, 18] such as polarization splitters, a basic component for polarization diversity receivers, or multiplexers.

Such optical functions are very difficult to achieve with low contrast waveguides because of the very weak modal birefringence between TE and TM modes. This degeneracy of orthogonal polarized waves can be avoided by using IOS3 structures exhibiting suitable thicknesses of the high contrast guide, associated with adiabatic transitions.

Very promising results have been reported by ATT for a structure combining silicon nitride and phosphorus doped silica guides. Very low insertion loss and high isolation between TE and TM outputs were demonstrated with short length devices[16].

Finally, in order to complete this short review of devices which could have a strong impact on future optoelectronic architectures, we should mention results concerning light emission in silica waveguides by suitable doping: erbium (around 1.54 µm), praseodynium or neodymium[63, 64, 65, 66, 67, 68, 69].

The publication by NTT of silicon based integrated optics amplifiers[67] having a gain of about 15 dB for a waveguide length of 20 cm is certainly one of the more important reports of recent months.

Many laboratories are now involved in this research. Several doping methods have been already tried:

1) Direct introduction of the dopant during silica deposition with structures made by CVD-FHD or more recently by sputtering[69] (the highest gain per unit length has been obtained using this method by ATT: 20 dB over 3 cm).

2) Dipping in a suitable solution (erbium chloride) before the consolidation phase with FHD.

3) Ion implantation at high energy in low contrast silica guides[63, 64].

4) Ion implantation at low energy (< 300 keV) in high contrast guides (silicon nitride core). This last approach allows the use of conventional ion implantation for the industrial development of amplifier devices, in association with appropriate adiabatic transitions for suitable connections with optical fibres[68].

4.2. IOS Technologies for Sensors[50]

The first commercially available device in the field of sensors was the interferometric displacement sensor developed by the french company CSO associated with LETI[70].

This sensor uses a very conventional scheme based on an integrated Michelson interferometer, which gives two interference patterns phase shifted by about 90° in order to obtain information both on the displacement amplitude (by counting fringes) and on the displacement direction. The most critical parameter to achieve such a sensor is the long term wavelength stability of the light source.

This is especially difficult because conventional Fabry-Perot laser diodes exhibiting the well-known mode hopping phenomenon were used. These diodes were chosen because of their low cost and the good availability of these light sources at 0.78 µm wavelength.

Wavelength stabilities $\Delta\lambda/\lambda$ better than 10^{-6} are currently obtained over more than one year by an appropriate choice of the laser diode, a suitable mounting of the light source in front of the chip, and the use of the external cavity effect.

This sensor, using a 7x7 mm chip, exhibits resolution of 10 nm and an accuracy of 0.1 µm over 10 cm dynamic range. The full dynamic range can reach more than 1 meter, limited by the quality of the output lens which is used to collimate the output optical beam.

Associated with a rubidium cell and a feedback loop to control the laser diode wavelength, stability higher than 10^{-9} has been recently observed [71].

This sensor opens the way for new integrated devices in the field of dimensional metrology, which could combine miniaturization and very low cost.

In the field of gyrometry, the recent publication of low loss ring resonators[19, 20, 21] leads us to anticipate low cost integrated gyrometers on silicon chips for low cost, low performance (100 to 1000°/hour) applications.

The use of a spiral scheme[72] and a tricoupler (Fig. 2) to analyse the rotation phase shift appears the best way to succeed, because it relaxes the technical difficulties arising from the need for a high coherence light source in ring resonator schemes[72, 73] (for high wavelength stability and to avoid parasitic interference effects).

For the higher performance range (10 to 100°/hour), the association of a silicon based integrated chip and a fiber loop in an advanced optoelectronic multichip module presented recently by GEC could be a viable approach for a lower cost fiber gyro[74].

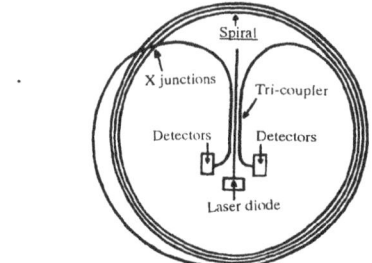

Fig. 2. Spiral configuration for low cost integrated gyrometers.

Finally, two new very attractive kinds of sensors seem well matched to the basic capabilities of silicon integrated optics:

1) The first one concerns chemical sensors for which optics can bring a better selectivity and reliability compared with competing approaches[72, 75, 76]. Integrated spectrophotometers operating near the absorption peaks of chemical substances, offer certainly one of the best ways to achieve efficient chemical sensors. Infortunately, because of the spectral transmission of silicon based waveguides (up to 1.8 µm) and because of the availibility of semiconductor laser diodes, this approach requires working with the overtones of absorption peaks, for which absorption coefficients are very weak. However, losses observed now with silica guides in the range of 1.3 to 1.7 µm allow the achievement of propagation lengths greater than 1 meter without to much attenuation. Such lengths are suitable for detection of the most interesting gases by using the well-known evanescent wave interaction. Such sensors can be realized in practical devices using a double spirale scheme[72] (Fig. 3) which offers a long interaction length together with a suitable reference on a small size chip.

2) The second type of sensors concerns the combination of integrated optics and micromechanics to realize attractive new optomechanical devices[72, 77, 78, 79]. Silicon based integrated optics is perfectly fitted to this approach because of the ability offered by IOS structures to realize waveguiding cantilevers or membranes which can convert an external parameter (pressure, sound, etc.) into an optical signal. Many laboratories are now involved in this promising field[80, 81]. Because of the advanced state of the art of both integrated optics and micromechanisms on silicon, the first commercialized devices of this type can be expected very soon.

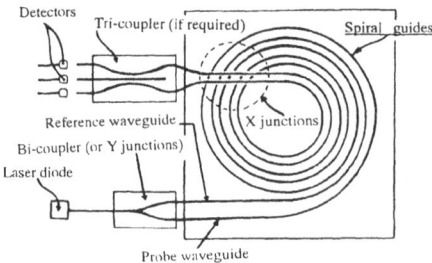

Fig. 3. Double spiral scheme for integrated optics spectrophotometers.

4.3. IOS Technologies and Optical Memories[45 ÷ 48, 82 ÷ 84]

Optical memory heads require small size at low weight for both economical and technical reasons. If we ignore the cost of hybridization and packaging, the cost of an optoelectronic chip alone is directly proportional to its size because of its batch fabrication. For very low cost applications (a few dollars per device) it is therefore very important to reduce the size of a given optical circuit. That challenge is particularly acute for optical memories. On the other hand, small size and low weight optical heads are also needed to attain low information access times.

Integrated optics is therefore well fitted to such applications and several laboratories, mainly in Japan, are working in this direction.

The problem is however difficult, because the optical performance required is very high and because such devices need to use high quality optical 2D to 3D converters in order to obtain a diffraction limited reading or writing spot.

Osaka University was the first to report such integrated devices on silicon substrates for compact-disc applications. The 2D to 3D converter was formed by a focusing grating made by direct electron beam writing[45, 46, 47, 48, 49].

However, the spot size given by the optical converter which is convenient for a feasibility demonstration is still too large for practical use. More recently, a magneto-optic reading writing head has been studied at LETI[28, 82]. This device used an interferometric integrated circuit to analyse the polarization rotation produced by the magnetic media. Associated with a thin film magnetic coil fabricated directly above the integrated optics circuit, the miniaturized chip can be attached to a flying slider. It realizes the optical reading-writing function without any optical projection lens because the gap between the integrated optic circuit output and the disk surface remains very small (typically 0.2 µm).

However, progress in conventional magnetic heads limits the interest in this technical approach for fixed magneto-optic discs.

For removable discs, an approach using a quite similar integrated optics chip associated with an external projection lens is also under study in a European ESPRIT project; it requires additional components integrated on the chip in order to provide auto-focus and tracking function.

Such solutions, combining integrated optics to analyse the optical signal and micro-optics to realize a diffraction limited optical spot, seem now one of the more pragmatic ways to achieve efficient miniaturized optical memory heads. These approaches are being pursued by several laboratories in the world, mainly in Japan.

5. Conclusion

Silicon based technologies have reached a high level of maturity and many components are now available with the required performance for different fields of applications including communications, sensors and optical memories.

The potential cost of these integrated devices could be very low, provided that suitable processes for connections between the silicon mother board and other optoelectronics components like III-V elements and optical fibers can be developed. Various technical solutions are available to meet all the expected requirements. They could be developed in the next two or three years with a reasonable effort because of the available state of the art of silicon technologies.

In the longer term, the development of complete microsystems associating integrated optics and micro-mechanics could be an exciting new challenge[72, 77,78, 79].

In parallel, the association of IOS technologies and suitable materials could lead to efficient active devices using acousto-optic [85, 86] or electrooptic effects by combining silicon based technologies and polymeric materials[87, 88, 89, 90, 91].

References

1. S. Valette, P. Gidon, S. Renard and J.P. Jadot "Silicon based integrated optics technology: an attractive hybrid approach for optoelectronics", Proceedings SPIE, Vol. 1128, Glasses for optoelectronics, pp. 179-185, 1989.
2. C.H. Henry, G.E. Blonder and R.F. Kazarinov, Journal of lightwave technology, 7, 10, pp. 1530-1539, (1989).
3. S. Valette, P. Mottier, J. Lizet, P. Gidon, J.P. Jadot and D. Villani, "Integrated optics on Si substrates: a way to achieve complex optical circuits", Proceedings SPIE, Vol. 651, Integrated Optical Circuit Engineering III, pp. 94-101, Innsbruck, Mars 1986.
4. M. Kawachi, Opt. and Quantum. Electronics, 22, pp. 391-416 (1990).
5. T. Miyashita, S. Sumida and S. Sakaguchi, "Integrated optical devices based on silica waveguide technologies", Proceedings SPIE, Vol. 993, Integrated Optical Circuits Engineering VI, pp. 288-294, 1988.
6. S. Valette, J.P. Jadot, P. Gidon, S. Renard, A. Fournier, A.M. Grouillet, H. Denis, P. Philippe and E. Desgranges, Solid State Technology, pp. 69-75, (1989).
7. P. Mottier and S. Valette, Appl. Optics, 20, pp. 1630-1634 (1981).
8. S. Valette, A. Morque and P. Mottier, Electron. Lett., 18, pp. 13-14 (1982).
9. S. Valette, J.P. Jadot, P. Gidon, S. Renard, G. Grand, A. Fournier, A.M. Grouillet, P. Philippe, H. Denis, E. Desgranges, L. Mulatier and C. Erbeia, "Integrated Photonic circuits on silicon", in Proceedings of NATO Ad. S. In.:" Novel Silicon Based Technologies", pp. 173-240, Klüwer Academic Publishers, 1991.
10. M. Kawachi, M. Yasi and T. Edahiro, Electron Lett. 19, pp. 1583-1584 (1983).
11. N. Takato, M. Yasu and K. Kawachi, Electron. Lett., 22, pp. 321-322 (1986).
12. R. Kazarinov, "Silicon based materials for integrated optics" in NATO A.S.I.: "Novel Silicon based Technologies", Boca Raton (Florida), July 17-19 1989, Klüwer Academic Publishers, 1991.
13. S. Valette and J. Lizet, "Dispositif de multiplexage de plusieurs signaux lumineux en optique intégrée", Patent n°85 03681, March 1985.
14. G. Grand, "Contribution à l'étude de dispositifs optiques intégrés couplés à des capteurs à fibre optique", PHD thesis, Grenoble University, pp. 80-96, 24 September 1985.
15. Y. Shani, C.H. Henry, R.C. Kistler, K.J. Orlowski and D.A. Ackerman, Appl. Phys. Lett., 55, pp. 2389-2391, (1989).
16. Y. Shani, C.H. Henry, R.C. Kistler, R.F. Kazarinov and K.J. Orlowski, Appl. Phys. Lett. 56, pp. 120-121, (1990).
17. Y. Shani, C.H. Henry, R.C. Kistler, R.F. Kazarinov and K.J. Orlowski, IEEE Journal of Quantum Electronics, 27, 3, pp. 556-566, (91).
18. R. Adar, C.H. Henry, R.F. Kazarinov, R.C. Kistler and G.R. Weber, Journal of Lightwave Technology, 10, 1, pp. 46-50, (1992).
19. J. Bismuth, P. Gidon, F. Revol and S. Valette, "Low losses ring resonators fabricated from silicon based integrated optics technics", Proceedings of 7th Optical Fibre Sensors Conference, OFS 90, Sydney, p. 105-108, December 2-6, 1990.
20. J. Bismuth, P. Gidon, F. Revol and S. Valette , Electronics Letters, 27, pp. 722-724, April 1991.

21. R. Adar, Y. Shani, C.H. Henry, R.C. Kistler, G.E. Blonder and N.A. Olsson, Appl. Phys. Lett., 58, pp. 444-445, (1991).
22. Y. Inoue, T. Diminato, Y. Tachirawa and O. Ishida, Electronics Letters, 28, 7, pp. 684-685, (1992).
23. K. Kawachi, N. Takato, K. Jingusi and M. Yasu, "Birefringence control in high-silica single mode channel waveguides in silicon", Proceedings of OFC/IOOC 87 Conference, pp. 125-127, 1987.
24. A. Sugita, K. Jinguji, N. Takato and M. Kawachi, IEEE Journal on selected areas in communication, 8, 6, pp. 1128-1131, (1990).
25. E.J. Murphy and T.C. Rice, IEEE J. Quantum Electronics, QE-22, 6, pp. 928-933, (1986).
26. G. Grand, S. Valette, G.J. Cannell, J. Aarnio and M. Del Giudice, "Fibre pigtailed silicon based low cost passive optical component", Proceedings of 16th European Conference on Optical Communication, ECOC 90, p. 525-528, Amsterdam, 16-20 Sept. 1990.
27. G. Grand, J.P. Jadot, S. Valette, H. Denis, A. Fournier and A.M. Grouillet, "Fiber pigtailed wavelength multiplexer/demultiplexer at 1.55 microns integrated on silicon substrate", Proceedings of EFOC-LAN conference, pp. 108-113, Munich, 25-29 June 1990.
28. S. Renard and S. Valette, "Integrated reading and writing optical heads: a way to a multigigabyte multi-rigid-disk drive", Proceedings of the SPIE conference on Optical Data Storage 91, vol. 1499, pp. 238-246, Colorado Springs, 25-27 February 1991.
29. H.M. Bresby, "Connectorized integrated star coupler on silicon", Proceedings of the 42nd Electronic Components and Technology Conference 1992, pp. 630-632, San Diego, May 1992.
30. G. Grand, H. Denis and S. Valette, Electronics Letters, 27, 1, pp. 16-17, (1991).
31. H. Terui, Y. Yamada, M. Kawachi and M. Kobayashi, Electronics Letters, 21, pp. 646-648, (1985).
32. E.E.L. Friedrich, M.G. Oberg, B. Borberg, S. Nilsson and S. Valette, Journal of Lightwave Technology, 10, 3, pp. 336-340, (1992).
33. J.P. Loppe, A.J.T. De Krijger and O.J.J. Noordman, Electron. Lett., 27, 2, pp. 162-163, (1991).
34. M.J. Wale and C. Edge, IEEE Trans. Comp. Hyb. Manuf. Tech., 13, 4, pp. 780-786, (1990).
35. See, for instance, A.D. Trigg, GEC Journal of Research, 7, 7, pp. 16-17, (1989) and references within.
36. M.J. Wale, C. Edge, F.A. Randle and D.J. Pedder, "A new self-aligned technique for the assembly on integrated optical devices with optical fibres and electrical interfaces", Proceedings of European Conference on Optical Communication, ECOC 89, paper ThA 19-7, pp. 368-371, 1989.
37. G.L. Destefanis, Semicond. Sci. Technol., 6, pp. C88-C92 (1991).
38. K.P. Jackson, E.B. Flint, M.F. Cina, D. Lacey, J.M. Trewhella, T. Caulfield and S. Sibley, "A compact multichannel transceiver module using planar-processed optical waveguides and flip-chip optoelectronic components", Proceeding of the 42nd Electronic Components and Technology Conference 1992, pp. 93-97, San Diego, May 1992.
39. M.S. Cohen, M.F. Cina, E. Bassous, M.M. Oprysko, J.L. Speidelli, F.J. Canora Jr. and M. J. Defranza, "Packaging of high-density fiber/laser modules using passive alignment techniques", Proceedings of the 42nd Electronic Components and Technology Conference 1992, pp. 98-107, San Diego, May 1992.
40. C.A. Armiento, A.J. Negri, M.J. Tabasky, R.A. Boudreau, M.A. Rothman, T.W. Fitzgerald and P.O. Haugsaa, "Four channel, long wavelength transmitter arrays incorporating passive laser/singlemode-fiber alignment on silicon waferboard", Proceedings of the 42nd Electronic Components and Technology Conference 1992, pp. 108-114, San Diego, May 1992.
41. B. Imler, K. Scholz, M. Cobarruvias, R. Haitz, V.K. Nagesh and C. Chao, "Precision flip-chip solder bump interconnects for optical packaging", Proceedings of the 42nd Electronic Components and Technology Conference 1992, pp. 508-512, San Diego, May 1992.
42. P. Demeester, J. Pollentier, L. Buydens and P. Van Daele, "Novel optoelectronic devices and integrated circuits using epitaxial-lift-off", Proceedings of the SPIE Conference on Optoelectronic devices and IC's, Aachen, Oct. 90.

43. J. Pollentier, L. Buydens, P. Van Daele and P. Demeester, IEEE Photonics Technology Letters, 3, 2, pp. 115-117 (1991).
44. M. Yanagisawa, H. Terui, K. Gutd, T. Miya and M. Kobayashi, IEEE Photonics Technology Letters, 4, 1, pp. 21-23, (1992).
45. H. Nishihara "Recent studies of miniaturization of optical disk pick-ups in Japan", Proceedings of the SPIE, Vol. 1248, Storage and retrieval Systems and Applications, pp. 88-95, (Santa Clara, 13-15 February 1990).
46. T. Suhara and H. Nishihara "Integrated - optic disk pick-up devices using waveguide holographic components", Proceedingsw of SPIE, Vol. 1136, pp. 92-99, 1989.
47. K. Yokomori, S. Fujita, S. Misawa, T. Kihara, M. Aoki, A. Hiroe, A. Takaura, Y. Nakayama and H. Funato, Japanese Journal of Applied Physics, 31, 2B, pp. 548-550, (1992).
48. S. Ura, Y. Furukawa, T. Suhara and H. Nishihara, J. Opt Soc. Am., 7, 9, pp. 1759-1763, (1990).
49. T. Suhara and H. Nishihara, IEEE J. of Quantum Electronics, QE-2, 6, pp. 845-867, (1986).
50. S. Valette, S. Renard, J.P. Jadot, P. Gidon and C. Erbeia, Sensor and Actuators, A21-A23, pp. 1087-1091, (1990).
51. K. Imoto, H. Sano and M. Miyazaki, Applied Optics, 26, 19, pp. 4214-4219 (1987).
52. A. Takagi, K. Jinguji and M. Kawachi, Electronics Letters, 26, 2, pp. 131-133 (1990).
53. C. Dragone, Electronics Letters, 24, n°15, July 1988.
54. C. Dragone, C.H. Henry, I.P. Raminiw and R.C. Kistler, IEEE Photonics Technology Letter, 1, 8, pp. 241-243, (1989).
55. N. Takato, T. Kominato, A. Sugita, K. Jinguji, H. Toba and M. Kawachi, IEEE Journal of selected areas in communications, 8, 6, pp. 1120-1127 (1990).
56. B.H. Verbeek, C.H. Henry, N.A. Olsson, N.J. Orlowsky, R.F. Kazarimov and B.H. Johnson, J. Lightwave Technology, 6, pp. 1011-1015 (1988).
57. A. Sugita, M. Okuno, T. Matsunaga, M. Kawachi and Y. Ohmori, "Strictly non-blocking 8x8 integrated optical matrix switch with silica based waveguides on silicon substrate", Proceedings of European Conference on Optical Communication ECOC-90, pp. 545-549, Amsterdam, 16-20 Sept. 1990.
58. S. Valette, Journal of Modern Optics, 35, 6, pp. 993-1005 (1988).
59. S. Valette, P. Gidon and J.P. Jadot, "New Integrated Optical Multi/Demultiplexer realized on Si substrate", Proceedings of the fourth European Conference on Integrated Optics ECIO 87, p. 145-149 Glasgow May 11-13, 1987, C.D.W. Wilkinson, J. Lamb editors, 1987.
60. S. Valette, J.P. Jadot, P. Gidon and S. Renard "New integrated optics structure on silicon substrate application to optical communications and optical interconnects", Proceedings of SPIE 862: Optical Interconnexions, pp. 20-26, Cannes, 17-18 November 1987.
61. A. Kimeno, A. Terui and M. Kobayachi, J. of lightwave technology, 6, 1, pp 41-46, (1988).
62. M. Zirngibl, G. Dragone and J.H. Joyner, IEEE Phot. Tech. Lett, 4, 11, pp 1250-1253 (1992).
63. A. Polman, A. Lidgard, D.C. Jacobson, C. Becker, R.C. Kistler, G.E. Blonder and J.M. Poate, Applied Physics Letters, 57, 26, pp. 2859-2861 (1990).
64. A. Polman, D.C. Jacobson, D.J. Eaglesham, R.C. Kistler and. J.M. Poate, J. Appl. Phys., 70, 7 (1991).
65. Y. Hibino, T. Kitagaw, M. Shimizu, F. Hanawa and A. Sugita, IEEE Photonics Technology Letters, 1, 11, pp. 349-350 (1989).
66. T. Kitagawa, K. Hattori, M. Shimizu, Y. Ohmori and M. Kobayachi, Electronics Letters, 27, 4, pp. 334-335 (1991).
67. T. Kitagawa, K. Hattori, K. Shuto, M. Yasu and M. Kobayashi, Electronics Letters, 28, 19 pp. 1818-1819 (1992).
68. O. Lumholt, H. Bernas, A. Chabli, J. Chaumont, G. Grand and S. Valette, Electronic letters, 28, 34, pp 2242-2243 (1992).

69. J. Shmulovich, A. Wang, Y.H. Wang, P.C. Becker, A.J. Bruce and R. Adar, Electronics Letters, 28, 13, pp. 1181-1182 (1992).
70. J. Lizet, P. Gidon and S. Valette, "Integrated Displacement Sensor Achieved on Si Substrate", in Proceedings of the Fourth European Conference on Integrated Optics, ECIO 87, Glasgow, 206-211, May 11-13, 1987, C.D. W. Wilkinson and J. Lamb, Eds.
71. L. Pujol, D. Bouteaud and M. Achtenhagen, "Interféromètre intégré stabilisé par absorption atomique", Proceedings of Opto 92, pp. 254-257, Paris 14-16 Avril 1992.
72. S. Valette "Integrated Optical Sensors", Proceedings of the European Conference on Integrated Optics ECIO 93, Neuchatel, 18-22 April 1993, pp. 12-1 to 12-3.
73. R.A. Bergh, H.C. Lefevre and H.J. Shan, Journal of Lighwave Technology, LT2, n°2, pp. 91-107 (1984).
74. G.N. Blackie and I.R. Croston, GEC Journal of Research, 10, 2, pp. 106-110 (1993).
75. J.P. Dakin and W.F. Croydon, "Application of fibre optics in gas sensing", Proceedings of the EFOC/LAN 88 Conference, pp. 238-239, Amsterdam, June 29th, July 1st 1988.
76. W. Weldon, P. Phelan and J. Hegarty, Electronics letters, 29, 6, pp. 560-561 (1993).
77. S. Valette, "Active components in silicon based integrated optics technologies", Proceedings of OPTO 91, pp. 263-271, Paris 26-28 Mars 1991.
78. H. Bezzaoui and E. Voges, Sensors and Actuators A., 29, pp. 219-223 (1991).
79. E. Voges, H. Bezzaoui and M Hoffmann, "Integrated Optics and microstructures on silicon", Proceedings of the European Conference on Integrated Optics ECIO 93, Neuchatel, 18-22 April 1993, pp. 12-4 to 12-6.
80. K. Fisher, D. Zurhelle, R. Hoffmann, F. Wasse and J. Muller, "Fully integrated optical force and pressure sensor based on SiON layers", Proceedings of the European Conference on Integrated Optics, ECIO 93, Neuchatel, 18-22 April 1993, pp. 12-7 to 12-9.
81. C. Wagner, J. Frankenberger and P.P. Deimel, "Optical pressure sensor based on a Mach Zehnder interferometer integrated with a lateral a-Si pin photodiode", Proceedings of ECIO 93, Neuchatel, 16-22 April 1993, pp 12-10 to 12-11.
82. V. Lapras, P. Labeye and P. Gidon, "Development of a reading-integrated optical circuit for a magneto-optical head", Proceedings of ECIO 93, Neuchatel, 16-22 April 1993, pp 12-34 to 12-35.
83. B.N. Kurdi, "Integrated optics for optical data storage", Proceedings of ECIO 93, Neuchatel; 16-22 April 1993, pp 12-17 to 12-19.
84. T. Suhara, H. Ishimaru, S. Ura and S. Nishihara, Transactions of the IEICE, E73, n°1, pp. 110-115 (1990).
85. See, for example, F.S. Hickernell, "Zinc-oxyde thin film surface wave transducers", Proceedings of the IEEE, 64, 5, pp. 631-635, (1976), and references therein.
86. P. Mottier, S. Valette and J.P. Jadot, Optics and laser technology, 18, 2, pp. 89-92 (1986).
87. J. Zyss, J. Molecular Electron., 1, pp. 25-45 (1985).
88. R. Lytel, G.F. Lipscomb, M. Stiller, J.I. Thackara and A.J. Ticknor "Organic integrated otical devices in nonlinear optical effects in organic polymers", Edited by J. Messier, F. Kajzar, P. Prasad, D. Ulrich, NATO Series, Vol. 162, Academic Publishers Dordrecht, Boston, London, 1989.
89. G.R. Mohlmann, "Polymer electro-optic devices", Proceedings of ECOC 90, pp. 833-840, Amsterdam, Sept. 1990
90. D.G. Girton, S.L. Kwiatkowski, G.F. Lipscomb and R.S. Lytel, Appl. Phys. Lett., 58, 16, pp. 1730-1732 (1991).
91. C.C. Teng, Appl. Phys. Lett., 60, 13, pp. 1538-1540 (1992).

Chapter 9

ARE GLASSES SUITABLE FOR OPTOELECTRONICS ?

A. MONTENERO

In a paper published in 1992, Dorn et al.[1] reviewed five classes of materials suitable for use in nonlinear optics (see Table I).

As one can see, the authors state that among the classes considered, glasses are the least and fibers (together with semiconductors) the most suitable. In spite of the ideas of the previous authors, it may be worthwhile to spend some time discussing this very large class of materials. Of course this paper does not pretend to be exhaustive and to review all the work done in this field. I hope only that it can give some ideas about how to use glasses in nonlinear optics.

Glasses are well known for their linear optical properties: they are or can be made highly transparent, isotropic, formable in any shape and, by changing the composition, they can be tailored to achieve any intermediate value, in a wide range of possible values, of almost all the properties. For example, the refractive index n of a glass could be considered as due to contributions of all the atoms involved. So we can use the following formula, where each component contributes its own part:

$$n \propto P_1 n_1 + P_2 n_2 + ... + P_i n_i \qquad (1)$$

where n = resulting refractive index, Pi = percentage of component i, n_i = refractive index of component i. All this can be achieved with relatively easy and cheap manufacturing techniques.

Table I. Materials suitability in non-linear optics[(Ref. 1] (l = well suited, m = suitability to be verified).

Function	Semicon.	Polymers	Ferroelec.	IO-Glass	Fibres
E/O	l				
O/E	l	m			
Amplification	l	m	m	m	l
Space Switch	l	l	l	m	l
Time Switch					l
λ-Switch	l	m	m	m	l
Logic	l	m	m	m	m

Abbreviations: E/O = Electro-to-optic conversion; O/E = Optic-to-electro conversion; IO-Glass = Glasses for Integrated Optics

A. Montenero - Istituto di Strutturistica Chimica, Viale delle Scienze, Parma, Italy

But glasses also show nonlinear properties. When an electromagnetic field interacts with an atomic system, in addition to a linear induced polarization, there is a polarization proportional to higher order terms of the applied electric field. The polarization (P), thus induced in a medium by external optical electric fields (E) of frequency ω, is expressed by the power series:

$$P(\omega)=\chi^{(1)}E(\omega)+\chi^{(2)}(-\omega, \omega_1, \omega_2)E(\omega_1)E(\omega_2)+\chi^{(3)}(-\omega, \omega_1, \omega_2, \omega_3)E(\omega_1)E(\omega_2)E(\omega_3)+... \quad (2)$$

where $\chi^{(n)}$, the complex dielectric susceptibilities, are tensors and are related to the microscopic structure of the material. Classical or linear optics depends on $\chi^{(1)}$, the first order linear term; the second order non-linear susceptibility $\chi^{(2)}$ is the origin of second harmonic generation, parametric mixing, etc.. The second order non-linearity $\chi^{(2)}$ is important in non centrosymmetric materials, e.g. lithium niobate $LiNbO_3$. In isotropic materials these contributions should vanish; thus the discovery of efficient second harmonic generation in optical fibers is surprising.

The third-order contributions to the total polarization are given by:

$$\chi_i^{(3)}(\omega_4)=\Sigma \, \chi_{jkl}^{(3)}(-\omega_4,\omega_1,\omega_2,\omega_3) \cdot E_j(\omega_1) \, E_k(\omega_2) \, E_l(\omega_3)$$

where $\omega_4 = \omega_1 + \omega_2 + \omega_3$ and E_j, E_k, E_l are three separately applied electric fields with their own frequency ($\omega_1, \omega_2, \omega_3$) and polarization direction. The presence of three electric fields makes many processes possible. They are: third harmonic generation; $\chi^{(3)}$ (-3ω, ω, ω, ω); Raman and Brillouin scattering; (-ω$_1$, ω$_2$, -ω$_2$, ω$_1$); many three or four wave mixing processes such as degenerate four wave mixing; $\chi^{(3)}$ (-ω, ω, ω, -ω); and, refractive index and absorption changes. We want here to focus on refractive index changes.

The total refractive index is:

$$n = n_0 + n_2 <E^2> \quad (3)$$

with n_0 the linear index of refraction, $<E^2>$ the time averaged square of electric field of the incident light (in esu) and n_2 the non-linear refraction index. In SI units $n = n_0 + n_2 I$ with I the intensity of the incident beam.

When a monochromatic beam of frequency ω is linearly polarized and the medium is isotropic, n_2 (esu) is related to the real part of $\chi^{(3)}$ by

$$n_2 = 12\pi/n_0 \, \text{Re} \, \chi^{(3)} (-\omega,\omega,\omega,-\omega) \quad (4)$$

The non-linear properties depend of course on the frequency of the incident beam, and in particular whether or not ω is close to absorption bands, i.e., whether transitions are resonant or nonresonant. When resonant, the effects can be large and observable with small optical fields. Examples are absorption edges or exciton resonances of many semiconductors, organic polymers, chalcogenide glasses, and glasses doped with semiconductors or organic dyes. There is high power dissipation and slow response. In the nonresonant case, the optical nonlinearity is predominantly of electronic origin and needs strong optical fields. The response time is very short and heating can be minimum.

There are many ways to measure the optical nonlinearities. For what concerns the third order nonlinearity, the most frequently used techniques are: interferometry; third harmonic generation; three wave mixing; four wave mixing; and, Z-scan.

The first glasses[3,4] studied for their nonlinear properties were those having high linear refractive index. These glasses are usually transparent in the visible region and so the values obtained are limited to the nonresonant third order optical nonlinearity n_2:

	$\chi^{(3)}$	Method	λ(nm)
SiO_2	$.36 \cdot 10^{-14}$	TRI	1060
SF59	$7.5 \cdot 10^{-14}$	DFWM	1060
PbO-SiO_2	$8 \cdot 10^{-14}$	DFWM	1060
PbO-GeO_2	$11 \cdot 10^{-14}$	DFWM	1060
PbO-Ga_2O_3-Bi_2O_3	$42 \cdot 10^{-14}$	DFWM	1060

From the first empirical approaches for calculating n_2 good predictions were obtained by a better knowledge of the structure of glasses and in particular of their bonds:

$$n_2(10^{-13} \text{ esu}) = K(n_d-1)(n_d^2+2)^2 / v_d(1.52+(n_d^2+2)(n_d+1) \, v_d/6n_d)^{1/2} \qquad (5)$$

n_2 being the nonlinear refractive index, n_d the linear refractive index at the d line of He, v_d the Abbe number (the reciprocal of the wavelength dispersion of the linear refractive index of the material at this wavelength), and K a constant obtained by fitting experimental data for several oxide glasses and crystals. For simple glasses this formula gives a good approximation, but when polarizable elements are involved the deviation becomes very large.

As is known, glasses are homogeneous materials which can contain other particles conferring special properties. The heating process makes the formation of a two phase glass possible: a vitreous homogeneous matrix and a crystalline ordered phase. When the separated crystals are semiconductors or metals, the glasses show very interesting features. Thus, if we heat borosilicate glasses containing Cd ions together with sulphur or selenium, nucleation and crystallization processes start[6], with the formation of crystals like CdS, CdSe or $CdS_xSe_{(1-x)}$:

	$\chi^{(3)}$ (e.s.u.)	Method	λ(nm)
CdSSe-doped borosilicate	$9 \cdot 10^{-12}$	THG	1090
Corning CS3-68	$13000 \cdot 10^{-12}$	DFWM	530
Schott RG 695	$3000 \cdot 10^{-12}$	DFWM	690

The semiconductor containing glasses show large optical nonlinearity at photon energies near the band gap. The $\chi^{(3)}$ values usually are in the range of 10^{-8} to 10^{-9} esu and the response time is 10^{-11} sec. The nonlinear effect depends on the particle size. This fact, together with the possibility of quite easily modifying the matrix and the crystal composition, gave origin to a large research effort on this kind of glasses. We may consider two limiting cases: considering R, the particle radius, and a_{ex}, the effective bulk exciton Bohr radius, we can have: $R < 2a_{ex}$ where a strong confinement effect is present; $R > 4a_{ex}$ where a weak confinement effect is present.

By modifying the experimental procedures it is possible to obtain R less than, equal to or greater than a_{ex}. The semiconductor crystals included in the glassy matrix could be copper and silver halides, but the greatest effort has been put on cadmium sulphides, selenides and tellurides. But, as had previously occured in the '70s for the so called TMO glasses, an unsolved problem is the consistent control of the properties of glasses made under supposedly

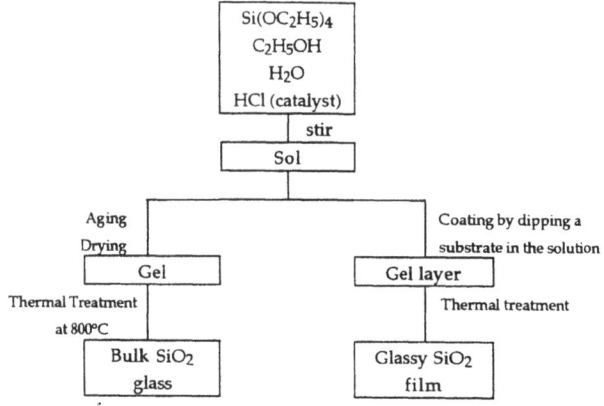

Fig. 1. Schematic preparation of silica glass and glassy films by the sol-gel method.

identical conditions. Awareness of this problem was heightened by Schanne-Klein et al.[7], who found significant differences in experimental and commercial glasses.

A similar problem is present in another class of nonhomogeneous glasses with NLO properties: glasses containing very small metallic particles. In this case it is very important to control the particle size, but because the metallic clusters tend to aggregate, becoming too large, the nonlinear properties are less than expected.

When we think of glasses, we think of materials obtained at high temperature and by melting. The high temperature needed for melting glass batches is often a problem, because the higher the temperature, the more difficult is process control. One way to partly solve this problem is to use low-melting-point glasses. Up to now only a few efforts have been made, but one exception is the work at Corning Incorporated. They used low temperature glasses, which have low glass transition temperature, good chemical resistance, and convenient optical quality. They can incorporate organic dyes which show large $\chi^{(3)}$ and are excited by easily available lasers. In particular, the dyes incorporated were Rhodamines, Cresyl violet, Coronene, Pyrene, Phtalocyanines, 1,2,5,6-Dibenzoanthracene[8].

In recent years one particularly promising technique for obtaining glasses has been developed, which has not yet shown all its potentialities: this method is called Sol-Gel. The technique is based on the formation of a glass network starting from organometals like $Si(OR)_4$ with R = -CH_3, -C_2H_5, etc. or $RSi(OR)_3$, etc. through chemical reactions of hydrolysis and condensation.

Hydrolysis: $M(OR)_n + xH_2O \rightarrow M(OH)_x(OR)_{n-x} + xROH$
Polycondensation: dehydration: $-M-OH + H-O-M \rightarrow -M-O-M- + H_2O$
dealcoholation: $-M-OH + R-O-M \rightarrow -M-O-M- + ROH$

By means of these reactions, conducted at ambient or low temperature, a material is obtained which is like a solid inorganic sponge embedding a lot of reaction liquid. By thermal treatments the pores collapse and finally a glass is formed identical to that prepared by conventional methods, but with the advantage of being fabricated at low temperature. Starting from organosilicon compounds we can obtain a pure silica glass at about 800°C. It must be emphasized that not only silicon compounds give these reactions: other metals can also be used

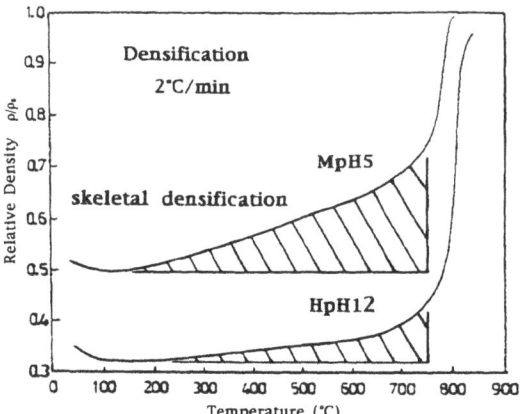

Fig. 2. Changes in the relative density as a function of temperature [Ref. 9].

as alkoxides (Fig. 1). If we stop heating at any intermediate temperature, different structures are obtained, mainly due to different pore size (Fig. 2)[9].

Another attractive characteristic of this technique is that we can easily prepare thin films by dipping the substrate in the solution. For example, Hashimoto. et al. [10] obtained iron oxide films deposited on silica glass with a thickness t = 0.07 µm. The measured $\chi^{(3)}$ was equal to $58 \cdot 10^{-11}$ esu, three orders of magnitude higher than that of Al_2O_3 single crystal.

At UCLA in Los Angeles Mackenzie and co-workers[11] prepared thin ferroelectric glassy films by sol-gel with interesting results, using the experimental procedure shown in Fig. 3.

It is easy to see that it is possible to include in this open structure any kind of material, whether by incorporating it into the starting solution, or by embedding it after the "sponge" is ready.

Matsuoka et al. [12] also dispersed gold particles in TEOS (tetraethylorthosilicate). The average particle diameter was in the range 14-60 Å with $\chi^{(3)} = 7.7 \cdot 10^{-9}$ esu, which is a very

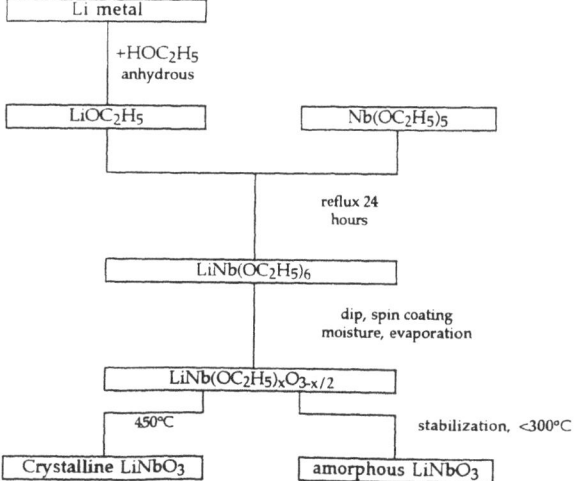

Fig. 3. Schematic representation of the sol-gel process for $LiNbO_3$[Ref. 11].

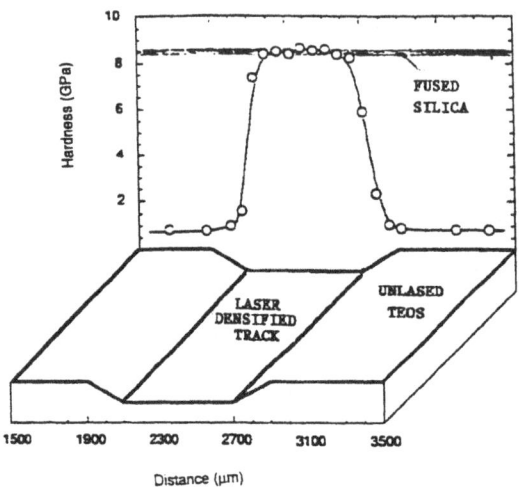

Fig. 4. Hardness profile of a laser-densified channel [Ref. 17].

good result. When heating the sample at 1000°C the $\chi^{(3)}$ value decreased by a factor of two, probably due to crystal growth. Takada et al.[13,14] prepared borosilicate glasses doped with CdS and with PbS (r = 2.5 - 7.5 nm). Spanhel et al.[15] obtained good samples containing CdS quantum dots by using a multifunctional inorganic-organic sol-gel processing.

Another way to obtain chalcogenides was proposed by Tohge et al.[16], who used cadmium-thiourea or zinc-thiourea complexes doping a silicon alkoxide solution; another route they proposed was the immersion of porous sol-gel derived glasses in alcoholic solutions of selenourea.

But maybe the sol-gel technique shows its biggest advantages in the formation of waveguides. Traditional waveguide fabrication methods are based on physical depositions or ion chemical exchange. However Taylor and Fabes [17] (Fig. 4) demonstrated that laser densification of a sol-gel derived planar waveguide leads to an increase of the refractive index, as expected. In comparison with the usual furnace densification process, the laser treatment retains more water molecules in the network, which contributes to the increase in the refraction index. Guglielmi et al.[18] by laser densification of SiO_2-TiO_2 films also obtained waveguides and found that the residual carbon in the film enhances the optical losses. The wide possibilities offered by the sol-gel technique are greatly enhanced by the possibility of inserting organic

Table II. Absorption intensity of Rhodamine B in SiO_2 matrix [Ref. 19].

% TiO_2	n_f (λ = 632.8 nm)	t [μm]	α_{peak} [cm^{-1}]	$\Delta n_f/\Delta \lambda$ [μm^{-1}]
25	1.60	0.44	$1.2 \cdot 10^{-4}$	0.06
30	1.62	0.33	$2.7 \cdot 10^{-4}$	0.12
50	1.70	0.40	$2.4 \cdot 10^{-4}$	0.18
70	1.77	0.29	$1.3 \cdot 10^{-4}$	0.21
90	1.87	0.33	$3.0 \cdot 10^{-4}$	0.33

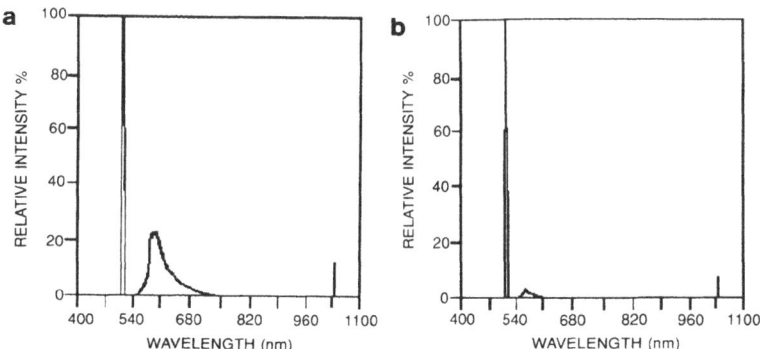

Fig. 5. Luminescence spectra of: (a) starting solution containing Rhodamine, (b) film [Ref. 19].

dyes directly in the solution or after the gelification. In this way, it is possible to add the advantage of using an organic dye, with its wavelength tunability due to a large fluorescence spectrum, to the good mechanical properties of glass (compared to dyes incorporated in polymers), and without any problem due to transport of liquid (compared to dye solutions).

Probably the most extensively studied dyes trapped in gel-derived glasses are Rhodamine B and 6G. In our lab we prepared[19] planar optical waveguides doped with Rhodamine B in a matrix of SiO_2. The first parameter checked was thermal stability. The peak centered at $\lambda = 562$ nm shifts towards the blue, its intensity increases and decreases with the temperature and disappears at $T = 210°C$. The peak centered at $\lambda = 527$ also shifts to the blue and reaches a maximum intensity, but at 210°C it is still very intense and disappears only at $T > 350°C$. The third peak, at $\lambda = 355$ nm, also moves towards the blue, but its intensity remains almost constant until $T = 240°C$, when it disappears (Table II,). The fluorescence spectrum of the deposited film was lower than that of the solution (Fig. 5 *a* and *b*). Rhodamine 6G is very similar to Rhodamine B, so we prepared films by using this other dye but in the system SiO_2-TiO_2, where the titania percentage ranged from 24 to 87 %. The first measurements in this case were done in order to determine the influence of the composition on the refractive index.

Due to the thin film thickness (about 0.4 µm), it was impossible to measure the refractive index of samples containing less than 87% of titania by means of the one-prism coupling technique. In this case the value obtained is $n_f = 1.886$ and the thickness $t = 0.4$ µm. By analyzing absorption spectra it was possible[20] to obtain information on all the other samples (Table III). Measurements are in progress in order to estimate the third order susceptivity and to design channel waveguide in these films.

Table III. Optical parameters of TiO_2-SiO_2 films calculated by absorption spectra [Ref. 20].

$T_{treat.}$ (°C)	λ_1 (nm)	I_1	λ_2 (nm)	I_2	λ_3 (nm)	I_3
30 (sol)	562	2,821	513	2,723	352	2,195
70	562	0,846	527	0,735	355	0,122
150	544	1,633	513	1,854	343	0,480
180	549	1,300	513	1,545	346	0,392
210	--	--	518	1,864	343	0,465
240	--	--	518	1,569	343	0,464
270	--	--	505	0,465	--	--
300	--	--	505	0,465	--	--
350	--	--	505	0,146	--	--
500	--	--	--	--	--	--

As a final comment I would like to quote what Mackenzie wrote in one of his papers[21]: "... Areas of Sol-Gel research which should be fruitful in the future for optics are: a) Organically modified ceramics in which not only one phase is optically active but that both phases are active and that the organic constituent interacts with the inorganic constituent; b) Amorphous phases obtained from sol-gel solutions which are "active" rather than "passive", for instance amorphous magnetics and amorphous ferroelectrics; c) New nonlinear optical materials with high $\chi^{(3)}$ and $\chi^{(2)}$; d) New and improved oxide glasses; e) Multifunctional materials based on new nanocomposites; and, f) Porous gels impregnated with optically active materials."

ACKNOWLEDGEMENTS. Finally I would like to thank: M. Bertolotti, S. Curziotti, E. Fazio, G. Gnappi, F. Michelotti, F. Ricciardiello, C. Sibilia with whom this work has been a pleasure.

References

1. R. Dorn, D. Baums, P. Kersten and R. Regener, Adv. Mater. 4, 464-473 (1992).
2. W.L. Smith, "Handbook of laser Science and Technology" Vol. 3, p. 259, M.J. Weber Ed. CRC Press - Boca Raton (1986).
3. R. Adair, L. Chase and S.A. Payne, J. Opt. Soc. Am. B, 4, 875, (1989).
4. D.W. Hall, M.A. Newhouse, N.F. Borrelli, W.H. Dumbaugh and D.L. Weidman, Appl. Phys. Lett, 54, 1293 (1989).
5. H. Nasu, Y. Ibara and K. Kubodera, J. Non-Cryst. Solids, 110, 229, (1989).
6. R.K. Jain and R.C. Lind, J. Opt. Soc. Am., 73, 647 (1989).
7. M.C. Schanne-Klein, F. Hache, D. Ricard and C. Flytzanis, J. Opt. Soc. Am. B, 9, 2234-2239 (1992).
8. R. Lu, "Nonlinear Optical Effects in Organically Doped Low Temperature Glasses and Heavy Metal Oxides", presented at "Topical Meeting on Intelligent Glasses", Venezia (Italy), 13-14 September 1991.
9. C.J. Brinker, G.W. Scherer, E.P. Roth, J. Non-Cryst. Solids, 72, 345-368 (1985).
10. T Hashimoto, T. Yoko and S. Sakka, J. Ceram. Soc. Japan 101, 64-68 (1993).
11. Y. Xu and J.D. Mackenzie, Integrated Ferroelectrics, 1, 17-42 (1992).
12. J. Matsuoka, R. Mizutani, S. Kaneko, H. Nasu, K. Kamiya, K. Kadono, T. Sakaguchi and M. Miya, J. Cer. Soc. Japan 101, 53-58 (1993).
13. T. Takada, T. Yano, A. Yasumori, M. Yamane and J.D. Mackenzie, J. Non-Cryst. Solids 147&148, 631-635 (1992).
14. T. Takada, T. Yano, A. Yasumori and M. Yamane, J. Ceram. Soc. Japan 101, 73-75 (1993).
15. L. Spanhel, E. Arpac and H. Schmidt, J. Non-Cryst. Solids 147&148, 657-662 (1992).
16. N. Tohge, M. Asuka and T. Minami, J. Non-Cryst. Solids 147&148, 652-656 (1992).
17. D.J. Taylor and B.D. Fabes, J. Non-Cryst. Solids 147&148, 457-462 (1992).
18. M. Guglielmi, P. Colombo, L. Mancinelli Degli Esposti, G.C. Righini, S. Pelli and V. Rigato, J. Non-Cryst. Solids 147&148, 641-645 (1992).
19. A. Montenero, G. Gnappi, S. Curziotti, M. Bertolotti, C. Sibilia, F. Michelotti and F. Ricciardiello, Proc. Congress "Omaggio Scientifico a Renato Turriziani", Vol. 2, 2.357-2.376 Rome (1992).
20. M. Bertolotti, P. Di Francesco, E, Fazio, G. Gnappi, M. Gressani, F. Michelotti, A. Montenegro, G. Nicolao and C. Sibilia, PAC-RIM Conference, Honolulu 7-10 Nov. 1993.
21. J.D. Mackenzie, J. Ceram. Soc. Japan 101, 1-10 (1993).

Chapter 10

LINEAR AND NONLINEAR OPTICAL PROPERTIES OF POLYMER WAVEGUIDES

F. MICHELOTTI

1. Introduction

Experimental evidence that organic semiconductors show some of the highest figures of merit among all the nonlinear optical materials of $\chi^{(3)}$ type has increased the interest in these materials during the last few years[1]. Due to charge conjugation along the polymer backbone or to the presence of side chains constituted of organic molecules with high hyperpolarizabilities, the third order nonlinear response can be strongly enhanced.

We can define the figure of merit of a material characterizing its use in a directional coupler[1] as $W=\Delta n_{sat}/(\alpha\lambda)$, where Δn_{sat} is the saturated refractive index change, α is the linear absorption coefficient and λ is the operating wavelength. In the case of polydiacetylene PTS for example, the out of resonance W is larger than 100. The same figure of merit is about 10 in the case for example of GaAlAs. Moreover, in the case of PTS we have a response time of the order of 10^{-12} s, while it is of the order of 10^{-8} s for GaAlAs.

Polymeric materials are also very convenient for waveguide fabrication, due to their easy processability. The spin-coating technique is widely used in semiconductor device technology because of its versatility and simplicity. A polymer solution in a suitable solvent is dropped on an optical quality glass substrate that is immediately spinned at constant speed around the axis orthogonal to the surface to be coated. The film thickness obtained at the end of this operation depends on the viscosity of the solution, on the spinning speed and on the spinning time. The typical thickness which can be obtained [2], without compromising the optical uniformity of the films ranges between 0.2 µm and 10 µm.

In this chapter we describe the production and characterization of Poly-α-Methyl-Styrene (PAMS) and Poly-Phenyl-Acetylene (PPA) thin films. Thin films of the two polymers were spun over fused silica and pyrex test substrates in order to characterize the deposition process. Then they were spun over fused silica substrates which had been previously etched, to obtain two grating couplers with a period of $\Lambda=0.55$ µm. Thickness and refractive index of the waveguides were investigated by m-line spectroscopy while the propagation losses were evaluated through the analysis of the intensity of the scattered light along the propagation direction.

F. Michelotti - Dipartimento di Energetica, Università di Roma "La Sapienza", Via A. Scarpa 16, 00161 Roma, Italy

The first polymer (PAMS), which exhibits no significant nonlinear optical properties, has been studied in order to obtain a preliminary characterization of the deposition process, and because it can be used as a host matrix for other nonlinear optical materials (semiconductor microcrystallites or dye molecules) or for rare earth ions ($Er^{3+}, Nd^{3+}, Pr^{3+}$).

The second polymer (PPA) was investigated because of its nonlinear optical properties. The third order nonlinear susceptibility of PPA was studied by D.Neher et al., who measured $\chi^{(3)}(-3\omega;\omega,\omega,\omega) = 7.0 \cdot 10^{-12}$ esu at $\lambda = 1.064$ µm, by third harmonic generation[3].

The results of a nonlinear grating coupling experiment are reported, showing a negative thermal change of the refractive index of PPA films under irradiation with a CW modelocked Nd:YAG source at $\lambda = 1.064$ µm, with pulse duration $\tau = 70$ ps and high repetition rate $f_{rep} = 76$ MHz. The effective nonlinear refractive index has been found to be $n_2^{eff} = -7.8 \cdot 10^{-12}$ cm^2/W.

2. Polymer Preparation

Poly-α-Methyl-Styrene was prepared by cationic polymerization of the monomer at low temperature. 50.0 g of α-MethylStyrene were put into the reactor and cooled externally at -80°C with a mixture of acetone and solid carbon dioxide. Then a solution of 7.0 g of anhydrous aluminium trichloride in 700 ml of 1-chloropropane at about 0°C were added dropwise to the solution of the monomer. The reaction mixture was mantained at -80°C by adding further amounts of solid CO_2 when needed in the external Dewar. The mixture was then allowed to react into the Dewar flask overnight; then it was heated to room temperature and concentrated by distilling off part of the solvent. Then a large excess of methanol was added, the precipitated polymer was recovered by filtration, and the polymer was again dissolved and reprecipitated to eliminate traces of catalyst still present. The yield relative to the initial monomer was 76%. Phenyl-Acetylene (PA) was polymerized in basic solvents such as triethylamine, or in bulk under the action of Rh(I) complexes[4]. When polymerization was carried out in bulk a co-catalyst of sodium ethoxide was added to the reaction mixture. If a suitable concentration of co-catalyst is used, it is possible to achieve the polymerization of PA at room temperature. The resulting polymer is readily soluble in organic solvents.

3. Waveguides Preparation

Optical quality fused silica and pyrex substrates were first cleaned with acetone then put into an ultrasonic bath in a chloroform solution and washed for 60 minutes. The substrates were then rinsed several times in deionized water, always in an ultrasonic bath (each bath lasted 60 minutes).

The substrates were dried in a nitrogen atmosphere. Before performing the spin coating operation they were baked for 30 minutes in a oven whose temperature was kept fixed at 90°C. This last operation is necessary in order to eliminate residual water traces on and inside the glass surface. Before spinning, the samples were allowed to cool at room temperature for at least 30 minutes.

In the case of PAMS four solutions in chlorobenzene, with polymer concentrations ranging from 0.1 gr/ml to 0.4 gr/ml, were prepared in four different containers. In the case of PPA three solutions in chlorobenzene, with polymer concentrations ranging from 0.05 gr/ml to 1.5 gr/ml, were prepared. The saturation concentration of the solution was lower than that for

PAMS and was strongly dependent on the temperature at which the polymer had been synthesized. Polymers which have been synthesized at lower temperature give, for the same concentration, a more viscous solution with a lower saturation concentration. For the highest concentrations the solution process was accelerated by putting the container in an ultrasonic bath for 20 minutes. In each case the solutions were filtered through a Millipore 0.22 µm filter in order to eliminate dust particles and bubbles.

The solutions were used to perform the spin coating procedure. We spun at several fixed spin speeds without changing the spinning time, which was fixed at 20 seconds. The angular speed was mantained between 600 and 3000 rpm. After spinning each sample was baked in an oven for 30 minutes at 90°C, below the glass transition temperature T_g, in order to achieve complete evaporation of the solvent.

The waveguides obtained were studied in order to obtain the refractive index and thickness of the films. Two techniques were used: optical transmission spectra analysis and interferometric microscope observation.

In the first case, measuring the transmission spectrum in a transverse configuration we obtain oscillations in the transmittance T versus wavelength λ, due to interference in the film. The transmittance curves can be fitted with the theoretical expression[5]:

$$T = \frac{n_s}{n_a} \frac{\tau_{12}^2 \tau_{23}^2 e^{-2k_2 \eta}}{1 + \rho_{12}^2 \rho_{23}^2 e^{-4k_2 \eta} + 2\rho_{12}\rho_{23} e^{-2k_2 \eta} \cos(\phi_{23} + \pi + \phi_{12} + 2n_f \eta)} \cdot \left\{ 1 - \left(\frac{1 - n_s}{1 + n_s} \right)^2 \right\} \quad (1)$$

where n_f, n_s, n_a are the refractive indices of the film, air and substrate, $\eta = 2\pi h/\lambda$, h is the film thickness, k_2 is the extinction coefficient of the film and τ_{12}^2, τ_{23}^2, ρ_{12}^2, ρ_{23}^2, ϕ_{12} and ϕ_{23} are given by the following expressions:

$$\rho_{12}^2 = \frac{(n_1 - n_2)^2 + k_2^2}{(n_1 + n_2)^2 + k_2^2} \qquad \rho_{23}^2 = \frac{(n_3 - n_2)^2 + k_2^2}{(n_3 + n_2)^2 + k_2^2} \quad (2)$$

$$\tau_{12}^2 = \frac{4n_1^2}{(n_1 + n_2)^2 + k_2^2} \qquad \tau_{23}^2 = \frac{4(n_2^2 + k_2^2)}{(n_3 + n_2)^2 + k_2^2} \quad (3)$$

$$\tan \phi_{12} = \frac{2k_2 n_1}{n_2^2 + k_2^2 - n_1^2} \qquad \tan \phi_{23} = \frac{2k_2 n_3}{n_2^2 + k_2^2 - n_3^2} \quad (4)$$

From curve fitting in a weakly absorbing part of the spectrum, we can evaluate the film refractive index n_f and thickness h. These data can then be utilized, in conjunction with reflection spectra, to obtain the absorption coefficient in the absorbing part of the spectrum. As an example, in Fig. 1, we show the spectrum obtained for a 1.9 µm thick PAMS film (solution concentration 0.4 gr/ml, spinning speed 1400 rpm) over a fused silica substrate. The quality of the spectrum indicates a great uniformity of the deposited film. For the case considered in Fig. 1, fitting data with the expression (4.1), we have found n_f=1.60 at λ=0.630 µm and h=1.80 µm.

The second measurement technique, using the interferential microscope, takes less time. In this case, we produce a scratch in the film and measure the fringe shift across the scratch[6]. Using the value of the refractive index measured with the preceding technique at the wavelength of the measurement and measuring the fringe shift $\Delta\phi$ with respect to the interfringe $\Delta\Phi$ we have:

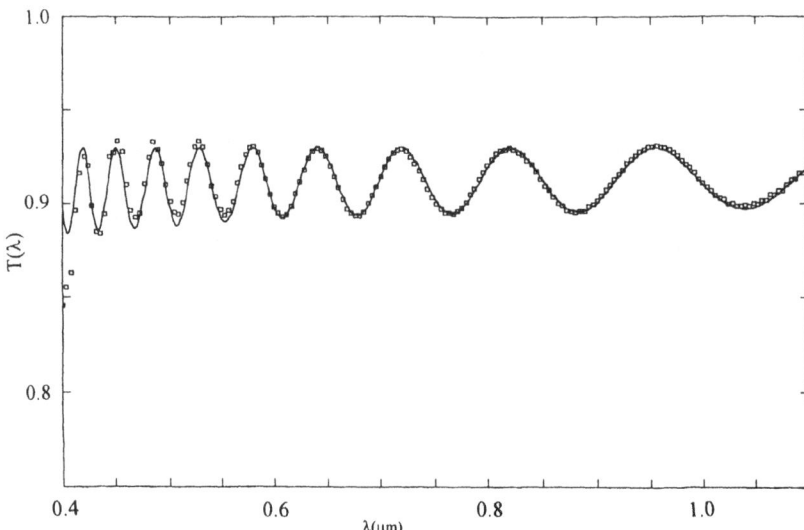

Fig.1. Transmittance $T(\lambda)$ versus wavelength λ of a PAMS film spun over a fused silica substrate ($c=0.4$ gr/ml, $v=1400$ rpm). The measured thickness and refractive index are: $h=1.9$ μm and $n_f=1.60$ at $\lambda=0.630$ μm.

$$h = \frac{\Delta\phi}{\Delta\Phi} \frac{\lambda}{n_f - 1} \quad (5)$$

In Figs. 2 and 3 are reported the thicknesses measured by the interferential microscope versus the spin speed for different concentrations, respectively for the cases of PAMS and PPA. As expected, the film thickness is an increasing function of the concentration, while it is exponentially decaying as the spinning speed is increased. In Fig. 3, the points indicated *PPA Type II* were obtained for a polymer which was synthesized at a lower temperature; since this polymer was more viscous, for the same concentration it gave rise to a thicker film.

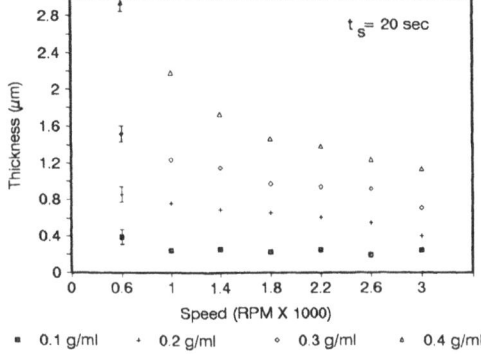

Fig.2. Measured thickness of PAMS spun films over pyrex substrates versus the spinning speed. The different symbols correspond to the following solutions' concentrations: (□) 0.1 gr/ml, (+) 0.2 gr/ml, (◊) 0.3 gr/ml, (Δ) 0.4 gr/ml.

Optical Properties of Polymer Waveguides

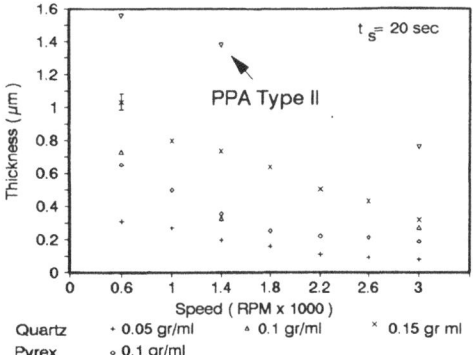

Fig.3. Measured thickness of PPA spun films over pyrex and fused silica substrates versus the spinning speed. The different symbols correspond to the following substrates and solutions' concentrations: (+) Fused silica 0.05 gr/ml, (Δ) Fused silica 0.1 gr/ml, (×) Fused silica 0.15 gr/ml, (◊) Pyrex 0.1 gr/ml. The symbols (∇) indicate the film thickness obtained for PPA synthetized at lower temperature and spun over fused silica substrates with concentration of 0.156 gr/ml.

For both polymers, at concentrations well below the saturation value, the behaviour of the film thickness h with respect to the spinning speed v and the solution concentration c can be fitted to the following expression:

$$h = c\left[Ae^{-\frac{v}{Bc}} + D\right], \quad (6)$$

where A,B and D are constants. As an example, in Fig. 4 are reported the same data as in Fig. 2 for PAMS and the curves (a,b,c) obtained using the Eq 6, respectively for solution concentrations c=0.1 gr/ml, c=0.2 gr/ml and c=0.3 gr/ml. For this case we found that A=4 10^{-3} m^4/Kg, B=7 10^{-3} m^4/(Kg s) and D=1.25 10^{-3} m^4/Kg. For concentrations close to the saturation

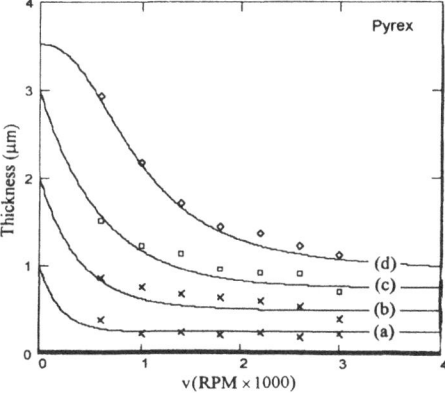

Fig.4. Comparison between measured film thickness for PAMS at several concentrations and the curves [(a) 0.1 gr/ml, (b) 0.2 gr/ml, (c) 0.3 gr/ml] obtained using the expression (4.6). Curve (d) [0.4 gr/ml] was obtained by means of a lorentzian fit.

value the behaviour changes and can be approximated by a Lorentzian shape, as shown in Fig. 4 (curve d) for the solution concentration c=0.4 gr/ml.

4. Waveguides Characterization

Once we had characterized the deposition process, we used the tarature curves of Figs. 2 and 3 to design single mode planar waveguides at $\lambda=1.064$ µm, of both PAMS and PPA, over substrates with grating couplers. We used fused silica substrates on whose surface had been etched two grating couplers. The subtrates were first coated with a photoresist, then exposed to two interfering light beams giving rise to a grating pattern and finally ion milled, yielding a periodic corrugation of the glass surface[7]. They were cleaned by the same procedure as described above before the spin coating procedure. Two gratings were fabricated on each substrate; one of them was used to couple light from a laser beam inside the film and the other to couple it out.

The theory of distributed grating couplers in the linear regime has been well understood for many years[8-10]. In the last few years in particular a great effort has been made to understand their behaviour in the nonlinear regime[11-14].

Let us consider the configuration of Fig. 5. A perturbation with period $\Lambda=0.55$ µm is present at the interface film-substrate. A laser beam is incident, from the substrate side, on the grating at an angle ϕ with respect to the normal to the plane of the waveguide. The input and guided electric fields may be expressed respectively as[14]:

$$E_{in}(x,y,t) = \frac{1}{2}a_{in}(x,y,t)e^{j(\omega t - k_0 x \sin\phi)} + c.c.,$$

$$E_{gw}^m(x,y,z,t) = \frac{1}{2}C^m a_{gw}^m(x,y,t)f^m(z)e^{j(\omega t - \text{Re}(\beta_0^m)x)} + c.c.$$

where $k_0 = 2\pi/\lambda$, λ is the vacuum wavelength, $f^m(z)$ is the field distribution of the mth mode, β_0^m is the guided wavevector of the mth mode, ϕ is the input coupling angle and C^m is a normalizing coefficient such that $\left|a_{gw}^m\right|^2$ represents the guided wave power. The interaction of the two fields, driven by the grating which diffracts the impinging light adding $K = \pm 2\pi/\Lambda$ in the x direction, can be described in the frame of the coupled-mode theory by the equation:

$$\frac{d}{dx}a_{gw}^m(x,y,t) = \\ = \gamma^m a_{in}(x,y,t) + j\left[\beta_0^m - k_0 \sin\phi \pm K + \Delta\beta^m(x,y,t) + j\alpha_1^m + j\alpha_g^m\right]a_{gw}^m(x,y,t) \qquad (7)$$

where γ^m is the coupling coefficient, α_g^m and α_1^m are the grating leakage parameter and the loss coefficient for the mth mode and $\Delta\beta^m$ has been introduced to describe a nonlinear change in the mth guided wavevector[14]. The description of the $\Delta\beta^m$ term in relation to the specific nonlinear optical mechanisms has been given elsewhere[14] and will not be discussed here.

In the case of gratings which have been designed to couple the incident beam into only one diffracted order, which is the case if $(\beta_0^m/k_0 + n_a) < \lambda/\Lambda < (\beta_0^m/k_0 + n_s)$, the grating

Optical Properties of Polymer Waveguides

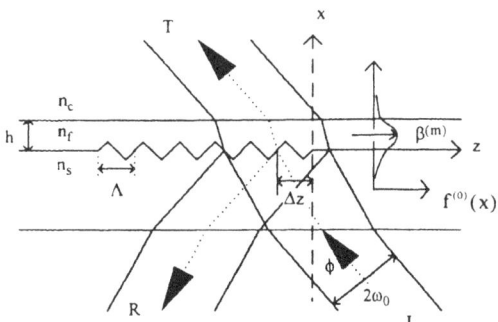

Fig.5. Geometry of the grating coupler in the case of backward coupling. The parameters are defined in the text.

leakage parameter α_g^m is connected to the coupling coefficient γ^m by symmetry[9]; $\alpha_g^m = |\gamma^m|^2/2$. In this situation the best linear coupling efficiency η (defined as the ratio of the guided power to the input one) for an input gaussian beam can be obtained by tuning the incident angle to the value f^m in such a way that the term $\beta_0^m - k_0 \sin\phi^m \pm K$ in Eq. 7 vanishes, choosing the beam waist of the impinging beam for a grating much longer than $1/\alpha_g^m$ so that $\alpha_g^m \omega_0 \cos\phi^{(m)} = 1.36$, and shifting the beam center from the grating edge by a quantity[9] $\Delta x = -1/2\alpha_g^m$. The coefficient α_g^m is the key parameter which it is necessary to know in order to optimize the coupling efficiency. It is very difficult to evaluate theoretically, starting from the physical properties of the coupler. This is the reason why it is very important to have an experimental measurement of its value.

In order to measure the propagation loss coefficient α_l^m and the grating leakage parameter α_g^m, we coupled the light from a laser beam inside the waveguides by means of one grating coupler, and detected the out of plane scattered light along the propagation direction x in the zone between the two couplers and in that above the output coupler. Since the intensity of the scattered light I(x) is proportional to the power in the waveguide P(x) and is an exponential function, as a solution of Eq. 7, from the measurement of the slopes of the graph of log(I(x)) versus x we obtain the measurement of α_l^m and α_g^m.

The detection of the scattered light was performed using a video camera in the direction perpendicular to the plane of the waveguide.

As an example, in Fig. 6, the behaviour of the intensity of the scattered light I(x) as a function of the displacement along the propagation direction is shown at $\lambda=1.152$ μm for a PAMS waveguide with 2 TE and 2 TM modes, when light is coupled into the TE_0 mode. The intensity profile of the input spot is visible at z=0. The exponential profile between the two gratings (0cm<z<1cm), which is not evidenced in this picture to emphasize the second exponential decay but which can be clearly seen in Fig.7, is due to the scattered light intensity distribution whose decay constant gives the measurement of the total loss $\alpha_l^0=0.7$dB/cm=0.16 cm^{-1}. The exponential decay for z>1cm is due to the decoupling process and permits us to measure the grating leakage parameter $\alpha_g^0=7.12$dB/cm=1.64 cm^{-1}. The value of the loss coefficient α_l^0 is bigger than those usually reported in literature of 0.1 dB/cm[15]. Nevertheless, we should point out that it has been measured for films which have not been protected with a

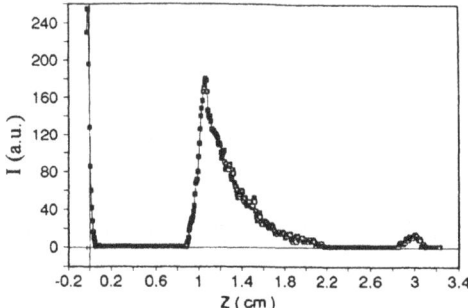

Fig.6. Intensity distribution I(a.u.) along the propagation direction z(cm), in the case of a PAMS waveguide excited at λ=1.152 μm in the TE$_0$ mode.

cladding layer after production. The measurement of α_1^0 for "as produced" films gives values which are lower than the resolution of the setup (α_1^0<0.4 dB/cm).

In Fig. 7 the same results are shown for a PPA waveguide with 2 TE and 2 TM modes at λ=0.6328 μm. In this case, log(I/I$_0$) is plotted versus the propagation distance. The light is coupled in the TE$_0$ mode; from this curve we estimated the total loss α_t^0=15.6dB/cm=3.59 cm^{-1}. In this case the grating has more than one diffracted order and it is no more possible to measure the grating leakage parameter α_g^0. For the same waveguide, which is single moded at λ=1.152 μm, we obtained for the TE$_0$ mode α_1^0=1.4dB/cm=0.32 cm^{-1} and α_g^0=38.5dB/cm=8.6 cm^{-1}.

5. Nonlinear Grating Coupling

The simplified setup for the grating coupling experiment is shown in Fig. 8. A Nd:YAG CW-modelocked laser produces 70 ps long IR pulses at λ=1.064 μm at the repetition rate f$_{rep}$=76 MHz. The laser beam has been expanded (ω$_0$=1.5 mm) in order to satisfy the condition

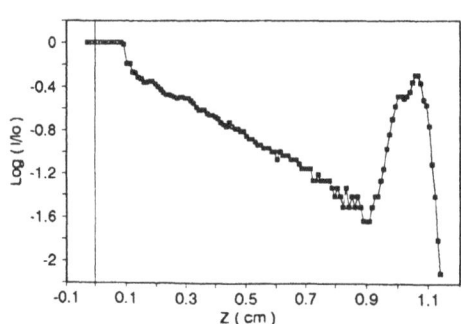

Fig.7. Logaritm of the intensity distribution log(I), normalized to the first value, along the propagation direction z(cm), in the case of a PPA waveguide excited at λ=0.6328 μm in the TE$_0$ mode.

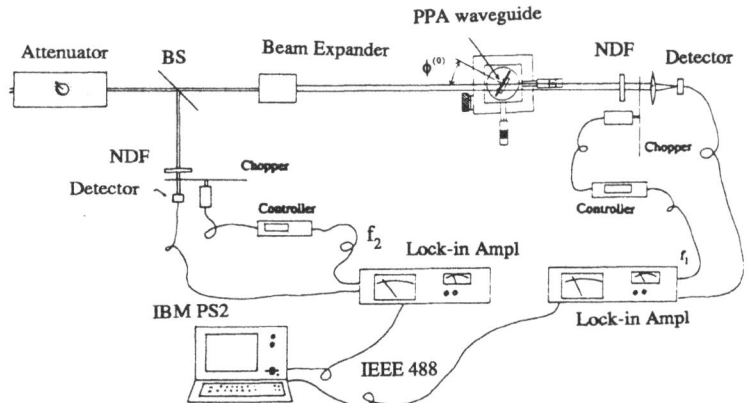

Fig.8. Simplified setup for the nonlinear grating coupling experiments with PPA polymer films.

$\alpha_g^m \omega_0 / \cos\phi^{(0)} = 1.36$ and to optimize the coupling efficiency. The coupling angle for the TE_0 mode has been optimized ($\phi^{(0)}$=-20.582 deg) at low input average power ($\overline{P} = 10$ mW).

We measured the transmitted power $\overline{P}_T$ as a function of the coupling angle ϕ^0 and of the input average power $\overline{P}$. This signal is complementary to the power coupled in the waveguide and so to the coupling efficiency η.

In Fig. 9 measurements of the transmitted power versus the coupling angle for two different input average powers are presented. We indicate with boxes the transmittance of the system as a function of the input coupling angle at low input average power ($\overline{P}$=10 mW). As expected, the transmittance reaches a minimum value corresponding to the resonant condition and to the maximum of the coupling efficiency η. The shape of the resonance curve is not gaussian, showing two shoulders (pointed out by the arrows in the figure), due to a distortion of the transverse shape of the input beam which causes a low value of the input coupling efficiency (η=0.12). On the other hand the resonance is very narrow, indicating very good quality of the polymer film, which is a necessary condition to measure a nonlinear shift of the coupling angle.

Increasing the input average power to the value $\overline{P} = 420$mW and repeating the same measurement as in the preceding case we obtained the curve reported in Fig. 9 with diamonds. The resonant coupling angle is shifted by $\Delta\phi^0 = -0.024$ deg, the minimum of the curve is less pronounced, indicating a decrease of the coupling efficiency, and the curve is no longer symmetrical. These results are in agreement with those previously obtained theoretically by Carter[11] and Assanto[14] in the case of a negative change of the refractive index of the polymer film.

From the change of the input best coupling angle, using the resonant condition for backward coupling $\beta_0^0 - k_0 \sin\theta^0 + K$, we can evaluate the change of refractive index of the film Δn_f = -3.9 10^{-4} for the high average input power case. Writing the change Δn_f in the usual Kerr-like expression:

$$\Delta n_f = n_2^{eff} I_{wg}$$

where n_2^{eff} is an effective nonlinear refractive index and I_{wg} is the intensity of the guided mode, we obtain $n_2^{eff} = -7.8 \cdot 10^{-11}$ cm^2/W.

The n_2^{eff} value is too large with respect to that reported in the literature[3], to reflect an electronic nonlinear response of the PPA film. We interpreted it as a thermally induced nonlinear refractive index, which can be written as[16]:

$$n_2^{eff} \approx 0.6 \cdot \frac{\partial n_f}{\partial T} \cdot \frac{\pi^{1/2} \alpha_0 \tau}{\rho C}$$

where $\partial n_f / \partial T$ is the optothermal coefficient, α_0 is the linear absorption coefficient, τ is the thermal relaxation time, ρ and C are the PPA density and specific heat.

Due to the high repetition rate of the laser pulses, the temperature of the film does not relax between each pulse and its successor, giving rise to an average thermal decrease of the refractive index, in agreement with the literature[17].

In order to confirm this hypothesis we performed a coupling experiment in which, using the same PPA waveguide, a weak CW probe beam at $\lambda=0.6328$ μm was coupled in the TE$_0$ mode and the coupling zone was externally heated by means of an IR incoherent source. The transmitted power was measured as a function of the coupling angle with and without the heat source. The results of the measurement are shown in Fig. 10. Curve (a) was obtained by measuring the transmitted power with respect to the coupling angle of the weak He-Ne laser beam. The curve shifts to smaller values of the coupling angle; due to the operating wavelength, the coupling is now of the forward type and a decrease of the coupling angle again

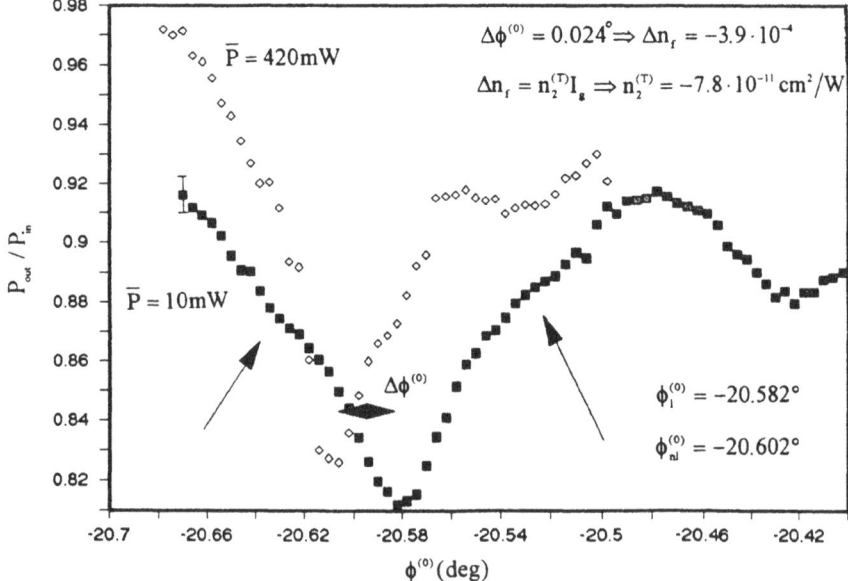

Fig.9. Transmittance P_{out}/P_{in} of the substrate-grating-film system as a function of the input coupling angle $\phi^{(0)}$ for two different input average powers: (□) 10 mW and (◊) 420 mW. The laser source is a CW-Mode-locked Nd:YAG at $\lambda=1.064$ μm, with 70 ps pulses at a repetition rate of $F_{rep}=76$ MHz.

indicates a negative change of the film refractive index and confirms the thermal origin of the nonlinearity.

In contrast with Fig. 9, the curve does not distort or change the value at the minimum. This last observation leads to the conclusion that in the experiment of Fig. 9 the nonlinear effect is really due to the local power density in the waveguide.

In the case of the IR pulsed excitation we performed some experiments in which the transmitted power was measured as a function of the input average power, while the coupling angle was kept fixed at the low power resonant value ϕ^0_{10mW}.

In Fig. 11 the measured values of the transmitted power, normalized to the linear transmittance, are plotted versus the input average power P_{in}. The transmitted power increases superlinearly as a funtion of the input power, showing that the coupling efficiency decreases, and shows hysteresis when returning to low input powers. The hysteresis is related to the resonant coupling mechanism in the coupler[13] which can give rise to bistability and optical switching.

7. Conclusions

In this chapter we have reported the characterization of the spin-coating deposition technique for the production of Poly-α-Methyl-Styrene and Poly-Phenyl-Acetylene thin films. We obtained low loss organic dielectric waveguides (< 0.7dB/cm) and achieved grating coupling of laser beams in the waveguides with good coupling efficiency (η>12%). Performing a nonlinear grating coupling experiment on a Poly-Phenyl-Acetylene thin film, we measured a thermal change of the refractive index of the film which can be represented by an effective nonlinear refractive index $n_2^{eff} = -7.8 \times 10^{-11}$ cm^2/W.

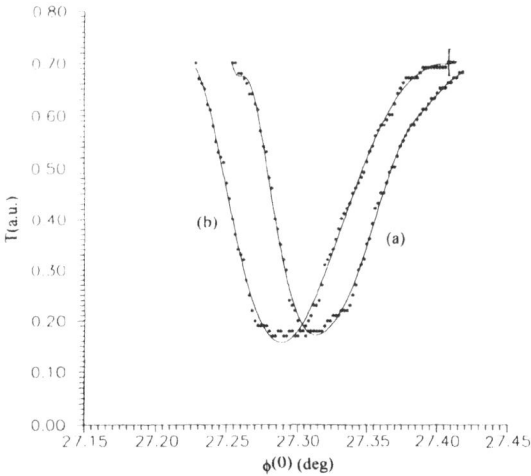

Fig.10. Transmittance T(a.u.) of the substrate-grating-film system as a function of the input coupling angle $\phi^{(0)}$ for two different conditions: (□) room temperature and (◊) under IR lamp irradiation. The laser source is a CW He-Ne at λ=0.6328 µm.

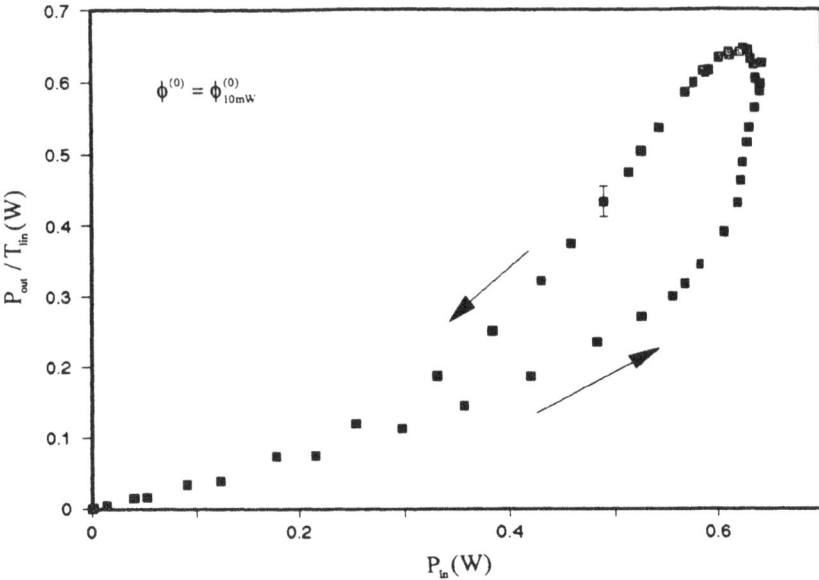

Fig.11. Transmitted power, normalized at the linear transmittance, P_{out}/T_{lin} as a function of the input average power P_{in} The coupling angle $\phi^{(0)} = \phi^{(0)}_{10mW}$ has been optimized at low input power at the TE0 resonance.

References

1. G.I.Stegeman, in "Nonlinear Waves in Solid State Physics", A.D.Boardman, M.Bertolotti, T.Twardowsky, Eds., NATO ASI Series B, Vol.247, pp.463-494.
2. P.Robin, Electrooptics with polymers, Proceedings EMRS School on Organic Materials for Photonics, G.Zerbi Ed., Obereggen (I), June 24-28, (1991).
3. D.Neher, A.Wolf, C.Bubeck and G.Wegner, Chem.Phys.Lett., 163 (2-3), 116-122 (1989).
4. A.Furlani, C.Napoletano, M.V.Russo and W.J.Feast, Polymer Bulletin, Vol.16, p.311 (1986).
5. M.Born and E.Wolf, *Principles of Optics*, Pergamon, New York (1980).
6. M.Pluta, "Advanced light microscopy", Vol.2, Chap. 7.1.3, PWN - Elsevier, Amsterdam (1989).
7. Xu Mai, R.Moshrefzadeh, U.J.Gibson, G.I.Stegeman and C.T.Seaton, Appl. Opt., 24, 19, .3155 (1985).
8. R.Ulrich, J.Opt.Soc.of Am., 63 (11), 1419-1431 (1973).
9. R.Ulrich, J.Opt.Soc. of Am., 61 (11), 1467-1476 (1971).
10. T.Tamir and S.T.Peng, Appl.Phys. 14, 235-254 (1977).
11. G.M.Carter and Y.J.Chen, Appl.Phys.Lett., 42 (8), 643-645 (1983).
12. G.M.Carter, Y.R.Chen, M.F.Rubner, D.J.Sandman, M.K.Takur and S.K.Tripaty, in "Nonlinear optical properties of organic molecules and crystals" D.S.Chemla and J.Zyss Eds., Vol.2, 85-120, Academic Press, Orlando (1987).
13. C.Liao, G.I.Stegeman, C.T.Seaton, R.L.Shoemaker, J.D.Valera and H.G.Winful, J.Opt.Soc. of Am., A2, 590-594 (1985).
14. G.Assanto, M.B.Marques and G.I.Stegeman, J.Opt.Soc.of Am., B8 (3), 553-561 (1991).
15. R.T.Chen, Appl.Phys.Lett., 61 (19), 2278-2280 (1992).
16. A.Agnesi, G.Gabetta and G.C.Reali, J.Appl.Phys., 71 (12), 6207-6209 (1992).
17. R.Burzynski, B.P.Singh, P.Prasad, R.Zanoni and G.I.Stegeman, Appl.Phys.Lett., 53 (21), 2011 (1988).

Chapter 11

FABRICATION AND CHARACTERIZATION OF CONJUGATED POLYMER WAVEGUIDES

S.SOTTINI

1.Introduction

The interest in organic optical waveguides dates back to the very beginning of Integrated Optics. For example, films of epoxy resin were extensively used by us in the seventies to demonstrate new components, in particular geodesic lenses [1]. The epoxy films, applied by dipping or spin coating, were doped with a suitable dye, usually rhodamine B, to make the light path visible via fluorescence. Fluorescence also allowed determination of the film losses[2], which were only 1 dB/cm at λ = 6328 Å . More recently, nonlinear prism coupling was demonstrated in spun film guides of MV757 epoxy resin doped with rhodamine B. From such tests, the real component of the χ^3 film non linearity was estimated[3] to be 1.5×10^{-12} e.s.u. at 616 nm . PMMA films have also been widely used, mostly to permit writing narrow structures such as efficient gratings directly in the film waveguide [4].

At present, besides plastic fibers, we have available a complete organic Integrated Optics technology that is, a variety of devices, including amplifiers, modulators and switches, have been realized by introducing functional groups into polymer structures [5].

Organic compounds are made up of molecules (polymers) whose internal cohesion arises from intense electromagnetic forces. On the contrary, the forces between molecules arise from Van der Vaals (sometimes hydrogen) bonds which are much weaker than intramolecular bonds. Because of this weak intermolecular bonding, molecules in a solid retain their individual characteristics; this has two important consequences: 1) The allowed energies for free carriers are low, about the size of the thermal or vibration energies. This results in a strong interaction between charge carriers and lattice vibrations, leading to local deformations. 2) The molecular nature of organic compounds allows one to choose molecules with certain particular properties and arrange them in a solid structure either to enhance or, to quench optical and electronic effects (the process of "molecular engineering") [6].

The ability to engineer organic materials is perhaps their most attractive feature. Other advantages are expected as well, such as low materials cost, mass production technologies, large optical boards (up to 30x30 cm) and easy patterning. On the other hand, the main disadvantage of polymers is that, with few exceptions, they have high losses in the IR region 0.8-1.6 µm, due to C-H bond vibrational absorption. To reduce this absorption, H has to be

---

S. Sottini - IROE-CNR, Via Panciatichi 64, 50127 Firenze (Italy)

replaced by heavy atoms such as deuterium or a halogen. Other problems are the thermal and time stability, and the softness of the material, which usually makes light coupling difficult.

In this Chapter the fabrication and characterization of film guides of nonlinear conjugated polymers is described. The nonlinear properties depend on the electric moments of the organic compound; as a consequence, saturated molecules, characterized by strong and localized σ bonds, have negligible nonlinear effects. On the contrary, conjugated polymers show highly delocalized π electron clouds which give rise to large third order nonlinear susceptibility, even far from the primary material resonance and with an ultrafast response time (typically ≈ 1 ps). In the tests reported here a particular polydiacetylene, known as poly-3BCMU, was used.

Conjugated polymers are attractive candidates for the realization of future nonlinear waveguide devices. Here the fabrication and characterization of polymer film guides are described, referring in particular to poly-3BCMU, a promising polydiacetylene. Two fabrication techniques have been tested: Langmuir-Blodgett and spin coating. The films were characterized by visible, IR and Raman spectroscopy. The results demonstrated the superior quality of LB oriented films with respect to electron delocalization and related nonlinear properties. On the other hand, the fabrication of spun films is much easier and their scattering losses are much smaller when they are used as waveguides. A four layer tapered guide was studied and tested to characterize polymer waveguides. In the future, this type of guide is expected to facilitate the insertion of polymer films technology into integrated optics devices.

2. Conjugated Polymers for Non-linear Optics

Currently, the most thoroughly studied semiconducting conjugated polymers are polydiacetylenes (PDA) and polythiophenes. Their typical configurations, depicted in Figs. 1a and c, are characterized by a one dimensional backbone. In Fig. 1b the side group of poly-3BCMU is shown.

The simplest description of one dimensional conjugated polymers is the free electron model [7,8]. The molecule is treated as a one dimensional box of length 2L, L being the so-called conjugation length. For a molecule with j multiple bonds of length l

$$L = l(j+\delta) \tag{1}$$

where $l\delta$ describes an additional length contribution from the end groups. δ describes the dependence of the conjugation on the polarizability of the end groups and is usually of order unity. Within this box are 2N degenerate electrons, where

$$N = (j+1) \tag{2}$$

assuming an additional contribution of one conjugated electron to the backbone by each of the end groups.

In an attempt to account for the alternating bond structure of a conjugated molecule, a periodic potential function,

$$V(\zeta) = (-1)^j V_0 \cos(\pi\zeta/L_0) \tag{3}$$

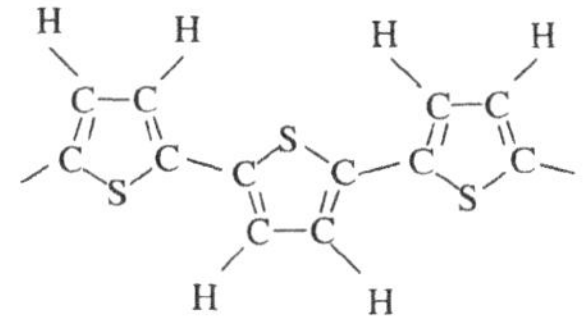

Fig.1. Bond representation of : a) polydiacetylene and c) polythiophene; b) side group formula of poly-3BCMU, a particular polydiacetylene.

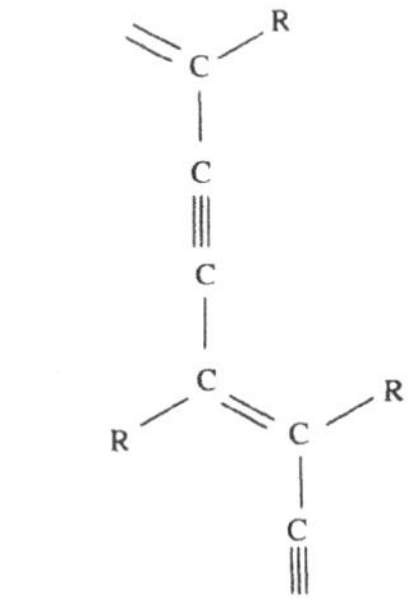

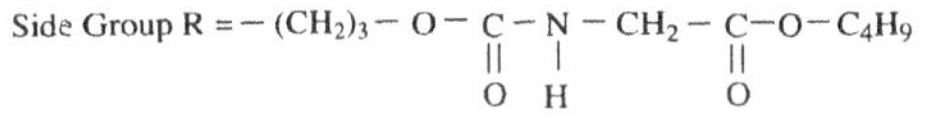

may be superimposed on the square well potential, where V_0 is the amplitude, ζ is the coordinate along the chain and L_0 is the average length of one conjugation (one multiple plus one single bond). In this case, the optical band gap is approximately given by:

$$E = V_0 + \left(\frac{h^2}{4mL_0^2} - \frac{V_0}{4} \right)(1/N + 0.5) \qquad (4)$$

where h is Planck's constant and m is the electron mass. This model therefore predicts a systematic decrease in the band separation with increasing conjugation.

The electronic transitions are localized due to lattice coupling and give rise to solitonic type lattice deformations, as shown in Fig. 2 in the case of polyacetylene. The spatial extent of such lattice deformations (~15-20 units,that is ~ 70 Å in the case of poly-3BCMU [8]) limits the electron delocalization, that is the effective conjugation length. In the case of polydiacetylenes, the first optical transition is excitonic in nature (electron-hole). Unlike phonons, polarons, etc., excitons are fast so that the relaxation time can be < 1 ps.

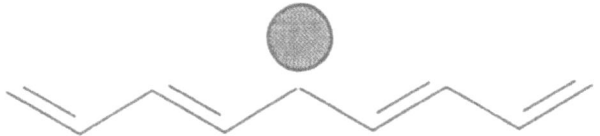

Fig.2. Solitonic type deformation in the trans polyacetylene chain.

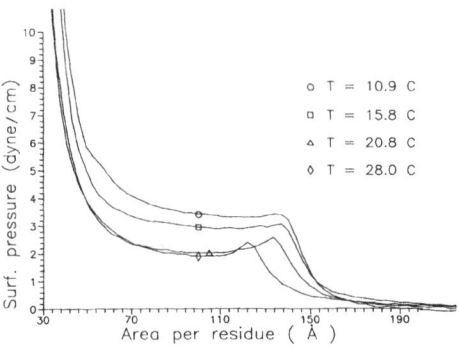

Fig.3. Surface pressure vs. area per residue isotherms at the indicated temperature for monolayers of poly-3BCMU.

The third order susceptibility χ^3 is directly related to the corresponding electronic polarizability γ [7,9]:

$$\chi^3 = \gamma L_3 N \tag{5}$$

where L_i are local field factors which, in Lorentz approximation, are given by:

Table I. Langmuir-Blodgett films.

\multicolumn{6}{c}{LANGMUIR-BLODGETT FILMS}					
sample[1]	t.s.p.[2] dyne/cm	solution	number of layers	measured transfers	spectroscopic investigations
L1	2.2	pure	3	2.6	visible
L2	2.2	pure	5	3.3	visible
L3	2.2	pure	21	11.6	Raman,vis.
L4	2.7	pure	799	307	visible
L5	2.7	pure	799	307	visible
L6	3.3	1:1[3]	7	2.4	Raman,vis.
L7	4.2	1:1[3]	183	66	Raman,vis.
L8	6	pure	6	2.2	Raman,vis.
L9	6	pure	7	1.2	Raman,vis.
\multicolumn{6}{c}{SPUN FILMS}					
			T	thickness	
S1		pure	20 °C	0.1 μ	Raman,vis.
S2		pure	20 °C	0.1 μ	Raman,vis.
ST		pure	20 °C	2.0 μ	Raman,vis.

[1] L refers to Langmuir-Blodgett films, S to spun films.
[2] t.s.p. = transfer surface pressure.
[3] Solutions in $CHCl_3$ containing 1:1 poly-3BCMU - decane.

$$L_i = (n^2 + 2/3)^{i+1} . \qquad (6)$$

and N is the density of molecules. On the other hand, γ depends on the conjugation length L, that is on the delocalization of the electronic cloud:

$$\gamma \alpha L^{10} / N^3 \qquad (7)$$

For a one dimensional system $N \alpha L$ and $\gamma \alpha L^7$. For a two dimensional system, $N \alpha L^2$ and $\gamma \alpha L^4$. The dependence on L is probably overestimated in Eq. 7: for example, according with Agrawal et al.[10], for polydiacetylenes $\gamma \alpha L^6$. In any case, it is evident that the one dimensional backbone has a strong effect on the polarizability of molecules.

3. Langmuir Blodgett and Spin Coated Poly-3BCMU Films

Future polymer waveguides for nonlinear optical devices require the fabrication of homogeneous and, hopefully, oriented thin films. The latter characteristic is necessary for second order nonlinearity but it also notably influences the third order nonlinearity due to the tensorial nature of the related susceptibility.

Oriented films of poly-3BCMU were obtained by the Langmuir-Blodgett technique (LB) [11,12]. The material, polymerized by γ irradiation (30 Mrad), had an average molecular weight of 375000 g/mol. Monolayers of this polymer were investigated by plotting the π/A isotherms (Fig. 3). Our results basically confirm those reported in Ref.13. In particular, a phase transition from a disordered coil yellow form to an ordered blue rod form occurs on monolayer compression, as indicated by the horizontal plateau which characterizes the isotherms.

The analysis of the visible spectra of the monolayers shows that the yellow form exhibits a peak at 475 nm, which persists even at higher pressure, while the blue form presents an additional peak at 620 nm, which corresponds to the maximum conjugation length. This fact is explained by relating the new peak to the ordered structure of the polymer, characterized by H bonds between the N and O atoms.

Visco-elastic properties strongly influence film transfer onto a solid substrate[11,12]; therefore, hysteresis tests were performed which proved that the phase transition is irreversible and that the films show noticeable viscosity, at least in the blue form. Fluidified monolayers were also tested, obtained by adding decan to the spreading solvent $CHCl_3$. A valuable improvement was obtained, but it was still far from being satisfactory. However, the hysteresis is much smaller below the plateau.

Several films (listed in Table I) were transferred onto hydrophilic glasses in both the expanded (yellow) and condensed (blue) states. In addition, some spun films were fabricated for comparison. These samples were investigated by visible, IR and resonance Raman (RRS) spectra. In particular, in the case of blue films a peak at 632 nm is predominant. On the contrary, spin coated films and yellow LB films show a main peak at 558 nm, related to chain distortions, responsible for more localized excitons. In the latter case, referring for example to L_1 and L_5 (Fig. 4), the spectra shift towards the red as the number of the transferred layers increases and they eventually closely resemble those of the blue form. This result confirms that the aggregation of polymer chains stimulates the color change, even if it is clearly related to an intramolecular conformational transition. This assertion is proved by the IR spectra (Fig. 5) which show an increase of the H bonds with the number of transferred layers.

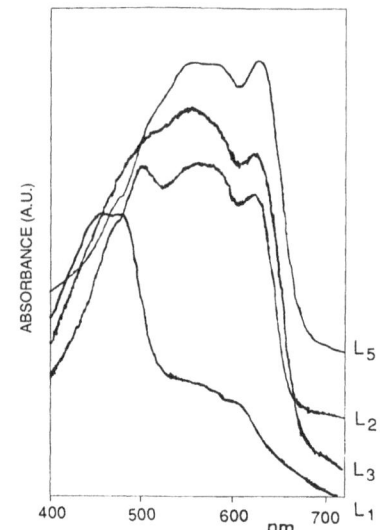

Fig.4. Visible absorption spectra of four LB films of poly-3BCMU transferred in the expanded region.

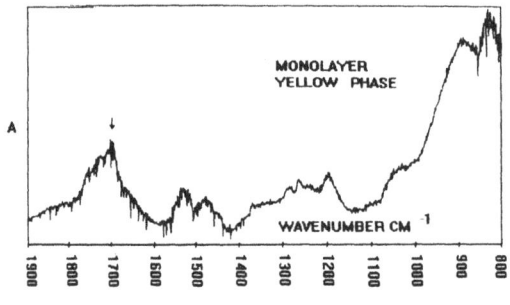

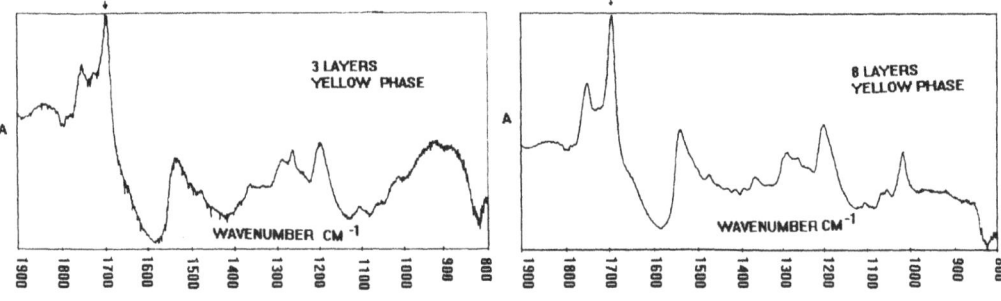

Fig.5. IR spectra of LB films transferred in the expanded region. The arrows indicate the peak due to the H bond.

Polymer Waveguides

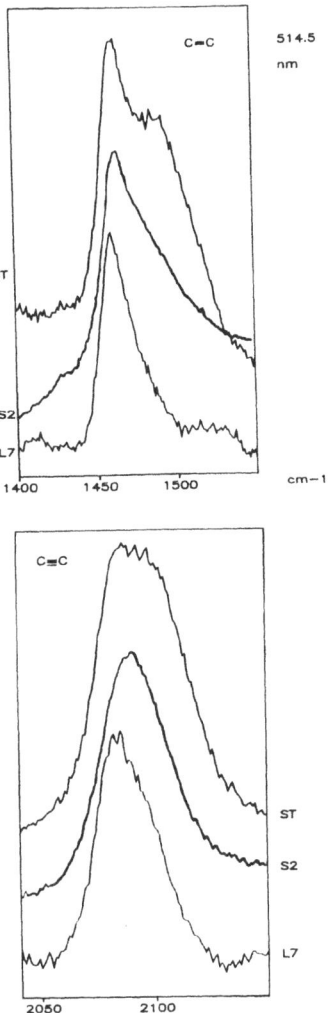

Fig.6. The RRS of a 183 layer LB film (L7) is compared with that of two spun samples (S2 and ST).

The investigation of RRS was limited to the C=C and C≡C stretching modes. The most relevant effect observed in all the samples was the shift of the Raman bands toward lower frequencies as the visible peak at 632 nm increases. This shift is usually related to a more ordered arrangement of the chains which produces increasing conjugation. For example, Fig. 6 shows the spectra (excited at 514.5 nm) of an LB film, L_7, and of two spun samples. It is worth noting the strong phonon dispersion toward lower frequencies exhibited by the spin coated films. This suggests the coexistence of different conjugation lengths.

4. Waveguide Tests Utilizing LB and Spun Films

In the future, nonlinear optical devices are expected to be implemented by using optical waveguides both to maximize the efficiency and to facilitate integration with optical fiber systems. In principle, LB films, which present an oriented structure and accurate control both of the thickness and of the refractive index, are attractive candidates for the implementation of nonlinear waveguides. Unfortunately, they also show some particular drawbacks: the propagation losses are high and increase with thickness, the minimum number of layers necessary to get waveguiding is usually too large to preserve film quality, and the butt-coupling of the organic guides with fibers or sources is made difficult by the softness of the films.

A four-layer tapered structure was investigated to alleviate these problems. It is an implementation of the 4-layer guide described in Ref. 14. As shown in Fig. 7, the tapered structure consists of an organic layer grown on top of an ion exchanged glass guide (GIW) whose depth is chosen in order to maximize the power density in the film. This region is coupled to the input and output regions by smooth tapering. Different configurations were analyzed theoretically by the Marcuse's Step Transition Method [15]. For example, the configuration depicted in Fig. 8a assures a transmitted power in the range 65-77% (Fig. 8b) as the film thickness d increases. The transmitted power decays quickly for d smaller than 0.75 μm because the propagation is near cut-off and the scattering losses become high. The tapered graded index glass guide (GIW) can be realized by a partial dipping in a KNO_3 solution which gives rise to the K+- Na+ exchange. The dipping is controlled by a motorized positioner with steps of 10 μm. Tests utilizing polystyrene proved the working principle of the device even though the efficiency was quite poor, because the organic films had no tapered edge. Then this device was used to test waveguiding both of LB and spun films of poly-3BCMU, at $\lambda = 0.850$, 1.064 and 1.31 μm.

The configuration used in the case of a 799 layer LB film of poly-3BCMU is sketched in Fig. 9. The film losses were estimated from the four layer guide, following the procedure described in Ref.14. The total scattering distribution is shown in Fig.10. In this case, the film edges were smooth, thanks both to the small thickness of the film and to the fabrication procedure. The film losses were very high, as expected because of its inhomogeneity, even though their contribution to the total losses (LB+GIW) was only $\cong 2dB$ (0.5 dB x 4 cm), since $\cong 2\%$ of the guided energy actually travelled in the film.

As shown in Fig. 10, the scattering intensity decreases in the central region. This can be explained by noticing that the polymer film was actually thinner than the expected value of 0.26 μm derived from the measured transfer ratio. Interferometric measurements gave 0.2 μm thickness, likely due to adhesion problems[16]. As a consequence, the propagation in the central region of the device was near cut-off and the energy density in the film was a bit smaller than the expected 2%.

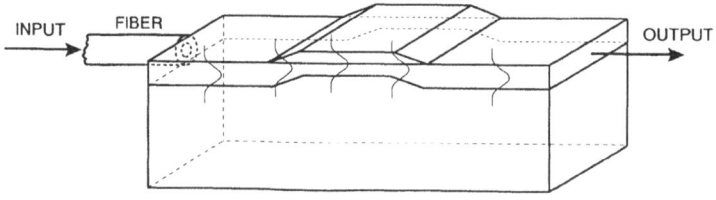

Fig.7. Four layer tapered guide.

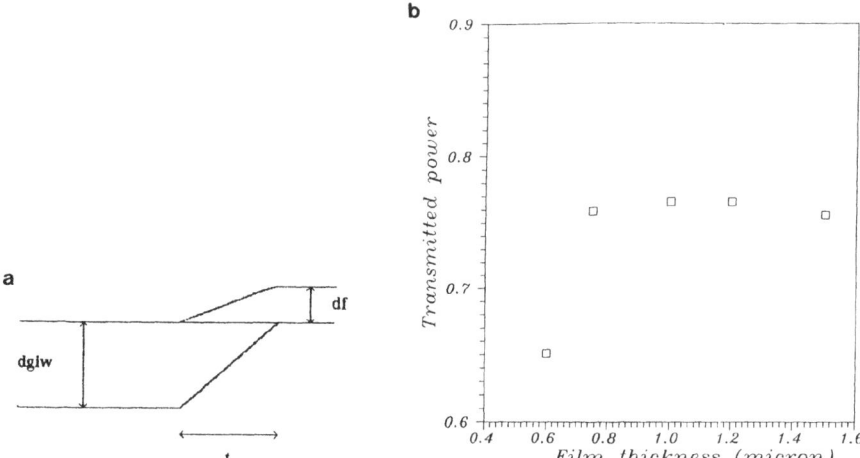

Fig.8. Theoretical behaviour of a four layer tapered guide whose configuration is sketched in a). The thickness of the graded index guide is $d_{giw} = 6$ μm and the length of the tapered zone is t = 500 μm. b) Transmitted power at λ = 1.31 μm versus the thickness of the polymer guide d_f. For $d_f < 0.75$ μm the transmitted power decays quickly because the propagation is near cutoff and the scattering losses increase.

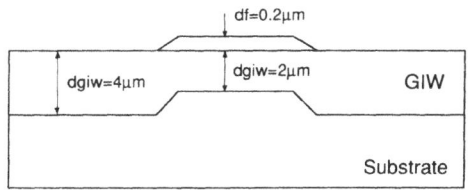

Fig.9. Configuration of the 4-layer tapered guide in the case of a 799-layer LB film.

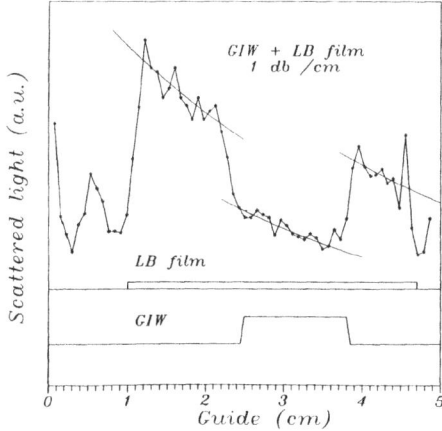

Fig.10. Scattering distribution, at λ = 1.06 μm, of the 4-layer guide of Fig.9.

Table II. Characterization of poly-3BCMU film guides.

λ (μm)	n_{TE}	n_{TM}	α (dB/cm)
0.6328	/	/	/
0.849	1.673	1.489	/
1.064	1.622	/	3.5 - 5
1.310	1.610	/	8

$n_{TE/TM}$ = refractive index of the film guide for TE or TM propagation

α = waveguide losses

Unlike LB films, spun films show an isotropic behaviour in the film plane, as shown by polarized microscopy. Their fabrication is relatively easy up to 2-3 μm thickness. In practice, the quality of the film waveguides critically depends on the homogeneity and the thickness uniformity.

Our spun films were obtained starting from a $CHCl_3$ solution (20-40 gr/l) and spinning at 600-2000 rpm per 20-30 sec. Several samples, including both conventional and four layer guides, were characterized at different wavelenghts in the linear regime.

The results are summarized in Table II. Losses were evaluated by the guide scattering. The m lines due to TM modes were particularly difficult to find because of their low contrast and the small effective refractive index of these modes. Four layer guides were of valuable help in this investigation, resulting in good agreement between theory and tests. The characterization of TM modes confirmed the anisotropy of spun films in the direction normal to the film plane[17]. This anisotropy, suggests that the spinning process results in polymer chains partially aligned in the film plane, while the aliphatic groups, characterized by small refractive index (1.41-1.43), are generally positioned normal to the plane.

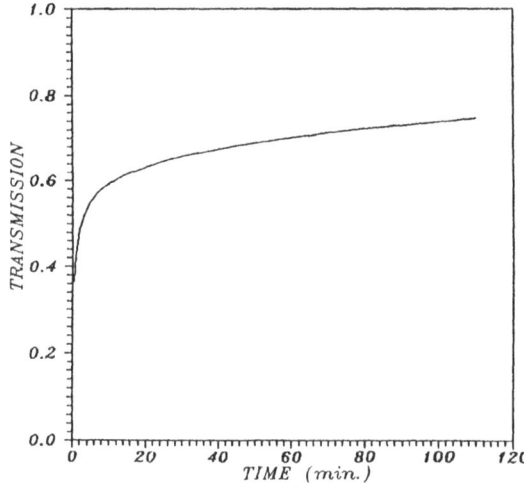

Fig.11. Bleaching of a 0.35 μm thick film utilizing a power density of 3.5 W/cm^2 from an argon laser.

The utilization of conjugated polymers in waveguide devices requests that methods be available for patterning the material to obtain channel waveguides[18]. For this reason, the photoinduced bleaching of poly-3BCMU was tested on spin-coated films, utilizing laser beams of 488 and 512 nm from a CW argon laser[16]. Fig. 11 shows a typical result. The power density of the bleaching beam was about 3.5 W/cm^2 and the film thickness was 0.35 µm The transmission of the film was monitored as it was being bleached. As shown in Fig.11, the transmission reached half of its saturation value within a minute. This behaviour is quite similar to that of poly-4BCMU[18-19]. During the bleaching, the refractive index decreases gradually from 1.674 to 1.51.

The transmission increase of bleached films is caused by the change in their absorption spectra; therefore, the spectra were measured for different bleaching exposures to monitor the process. The results are shown in Fig.12a and b, where the spectra in the UV region are reported in an expanded view: small peaks become visible in the curves corresponding to a partial bleaching of the film. Analogous peaks are present also in the monomer spectrum (Fig.12c). At the end of the bleaching process, the UV peaks disappear almost completely, indicating that also the conjugate diacetylene bonds of the monomer were broken. This allows the so called development of the bleached film; that is, the bleached filmt can be completely removed by a proper solvent, while the unbleached film remains unaffected.

5. Conclusions

LB and spin coated films of poly-3BCMU have been studied in view of their utilization in nonlinear guided optics. Poly-3BCMU is not very suitable for realizing homogeneous LB "thick" films (0.6 µm) because of monolayer viscosity. Nevertheless, several films, up to 799

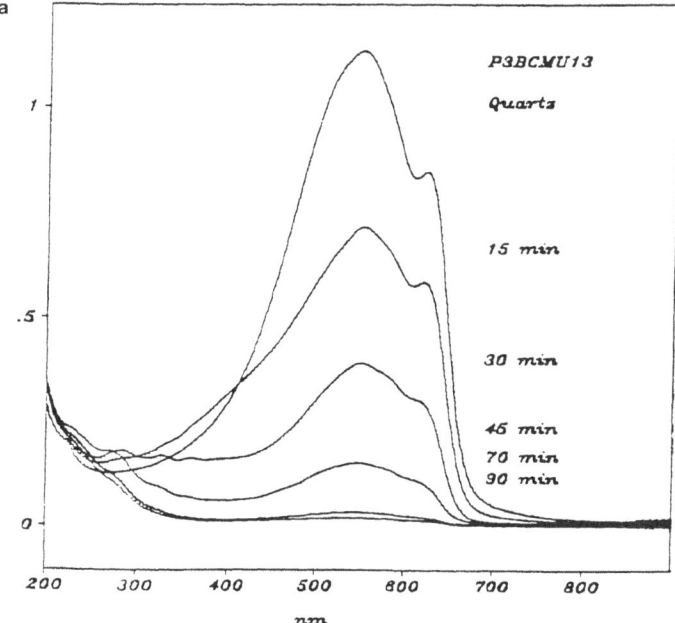

Fig.12a. Spectra of a poly-3BCMU film during the bleaching.

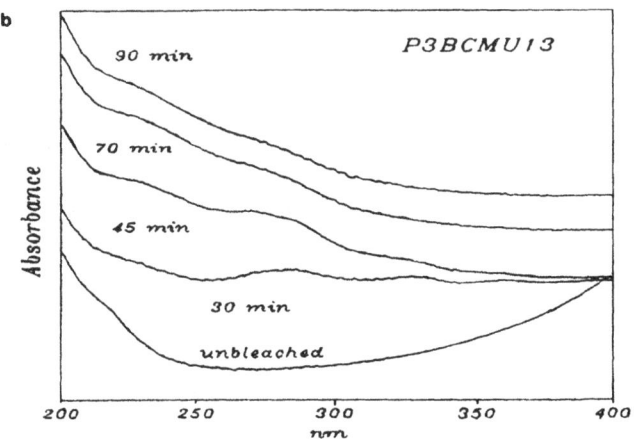

Fig.12b. Expanded view of the UV region.

layers, were transferred in both expanded and condensed forms. Their spectral analysis showed that a maximum conjugation length or chain order exists, characteristic of the polymer, which corresponds to the 632 nm peak in the absorption spectra and to the C=C stretching vibration at 1460 cm^{-1} in the RRS. These bands are present even in the spun films, but in this case the 632 nm peak is much weaker, while a strong phonon dispersion toward higher frequencies is present in the RRS. This confirms the superior quality of the LB films as to electron delocalization and related nonlinear optical properties.

In the case of spin coated films, the expected loss in nonlinear response is compensated by the relatively easy fabrication of films up to 2-3 µm thick, and by the smaller scattering losses when these films work as optical waveguides.

Poly-3BCMU films were characterized as waveguides utilizing a four layer structure

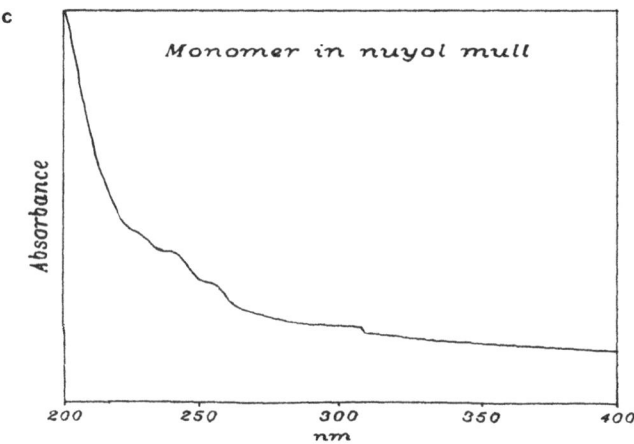

Fig. 12c. Spectrum of the 3BCMU monomer.

which includes a tapered ion exchanged glass guide. In this way, very thin films (for example, an 0.2 μm LB film) and films with high scattering losses could be tested. As expected, spun films were found to be isotropic in the film plane; however, they exhibit a strong anisotropy in the normal direction, as revealed by the propagation constants of the TM modes. The scattering losses of spin-coated films were found to be 3,5-8 dB/cm, depending on the wavelength; much higher values were found in the case of LB films, most likely due to inhomogeneities, especially on the film surface.

The four layer guide was also used to enhance the Raman signals. In the future, this structure is expected to facilitate significantly the insertion of polymer films into integrated optics devices. In particular, this guide can assure a good coupling with optical fibers.

Photoinduced bleaching of poly-3BCMU is an interesting method for patterning the material to fabricate channel waveguides: δn values > 0.1 can be achieved. It is worth noting that the refractive index of the bleached material is almost equal to that found for TM modes; this is in agreement with the presumably uniform breaking of the dyacetylene bonds of the monomer, at the end of the bleaching process.

ACKNOWLEDGEMENTS. The author wishes to thank L . Costa and M. Sparpaglione, Istituto G. Donegani, Novara, for supplying the poly-3BCMU. This research was partially supported by the Italian projects P.F.M.S.T.A. and P.F. Telecomunicazioni and by the RACE 1020 and 2012 European programs.

References

1. G.C. Righini, V.Russo, S. Sottini and G. Toraldo di Francia, Appl.Opt.,12,1477-81 (1973)
2. G.C. Righini, V. Russo, S. Sottini and G. Toraldo di Francia:"Geodesic Lenses for Integrated Optics",Proc.1973 Europ.Microwave Conference,G.Hoffman ed.,p.13.5.5,Gent, (1973)
3. B. Rossi, H.J. Byrne, W. Blau, G. Pratesi and S. Sottini, JOSA B,8,2449-2452, (1991)
4. G.C. Righini, V. Russo and S. Sottini, Opt. and Quantum Electronics, 7, 447-450, (1975)
5. Proc.Plastic Optical Fibres and Applications Conference, Paris, June 22-23, (1992)
6. D. Chemla, J.L. Oudar and J. Zyss, L'echo des Recherches, 47-59, (1981)
7. K.C. Rustagi and J. Ducuing, Opt.Comm., 10, 258-261, (1974)
8. H.J. Byrne:"On the origin and nature of the nonlinear optical properties of organic conjugated polymers", PHD thesis, Trinity College, Dublin, (1989)
9. F. Kajzar and J. Messier:"Cubic effects in polydiacetylene solutions and films",in "Nonlinear optical properties of organic molecules and crystals",v.2,D.S.Chemla and J.Zyss ed.,Academic Press, (1987)
10. G.P. Agrawal, G.P. Cojan and C. Flytzanis, Phys.Rev.B,17,776, (1978)
11. S. Sottini, D. Grando, E. Giorgetti, S. Trigari, G. Ventura, G. Gabrielli, M. Nocentini and G. Sbrana: "Langmuir-Blodgett films for waveguide nonlinear optics", in "Materials for photonic devices", D'Andrea, Lapiccirella, Marletta and Viticoli ed., World Scientific, Singapore, (1991)
12. G. Gabrielli, M. Nocentini, G. Sbrana, D. Grando, S. Sottini and G. Ventura, Thin Solid Films, 210/211, 551-554, (1992)
13. J.E. Biegajski, C.A. Cadenhead and P.N. Prasad:"Monolayer and Langmuir-Blodgett multilayer surface and spectral studies of poly-3BCMU", Langmuir, vol.4, p.689-693, (1988)
14. S. Sottini, G. Pratesi, E. Giorgetti, S. Trigari and A. Federighi: "Characterization of a Langmuir-Blodgett film waveguide by a four-layer structure", Proc.ECIO89, SPIE vol. 1141, p.37, (1989)
15. S. Sottini, D. Grando, L. Palchetti and E. Giorgetti:"Optical fiber-polymer film coupling by a tapered graded index glass guide", GRIN 92 Techn. Digest, Santiago de Compostela, (1992)

16. P.D. Townsend, G.L. Baker, N.E. Schlotter, C.F. Klausner and S. Etemad, Appl.Phys.Lett., vol. 53, p.1782, (1988)
17. S. Sottini, D. Grando, E. Giorgetti, L. Palchetti, Q. Li, G. Gabrielli, M. Nocentini and G. Sbrana, Molec. Cryst. and Liquid Cryst.,vol. 235, p. 191-200, (1993)
18. K.B. Rochford, R. Zanoni, Q. Gong and G.I. Stegeman, Appl. Phys. Lett., vol. 55, p.1161, (1989)
19. S. Aramaki, G. Assanto and G.I. Stegeman, Electron. Lett.26, 1300, (1990)

Chapter 12

LINEARIZED OPTICAL MODULATORS FOR HIGH PERFORMANCE ANALOG LINKS

G. L. TANGONAN, J. F. LAM and J. H. SCHAFFNER

1. Introduction

One role for fiber optics and integrated optics that has continued to grow in significance is analog signal transmission. In cable television distribution, for instance, "conventional" fiber-based systems deliver > 50 channels of AM video with > 50 dB S/N to the user. Given the existing analog television receiver base, one can predict that analog transmission will coexist with digital channels for some time to come. Other applications of high performance analog links fall into the broader category of antenna remoting. It is indeed possible today to design and build an antenna remoting subsystem with several GHz bandwidth, 4-5 dB noise figure, and high spur free dynamic range (SFDR) > 110 dB/Hz$^{2/3}$. Analog antenna remoting subsystems can be integrated into a variety of analog systems such as radar, cellular systems, and RF or microwave networks. Two factors that limit the dynamic range in fiber optic links are the system noise, predominantly from the laser relative intensity noise and the photodetector shot noise, and the non-linearity of the modulation process. The receiver noise level is reduced significantly using high-power solid state or DFB lasers (with > 20 mW output power and RIN < 165 dB/Hz). With noise levels approaching quantum limited performance, attention has recently focused on improving the linearity of the modulation process.

This Chapter focuses on linearized electrooptic modulators, new designs aimed at increasing the useful depth of modulation over which a modulator can be used with little distortion. We describe theoretical calculations that lead to linearized directional couplers. Predictions were made of the improvement in dynamic range possible using this new design. In addition, we present results on the experimental demonstration of linearized directional coupler modulators in LiNbO$_3$. An 8.5-11 dB improvement in the dynamic range was found in the links containing the linearized modulators.

2. Linearized Directional Coupler Modulators

In contrast with the Mach-Zehnder modulator system, directional couplers always exhibit a range of parameters that lead to a null in the third order coefficient.[1,2] The dual

---

G. L. Tangonan, J. F. Lam and J. H. Schaffner - Hughes Research Laboratories, 3011 Malibu Canyon Road, Malibu, California 90265 USA

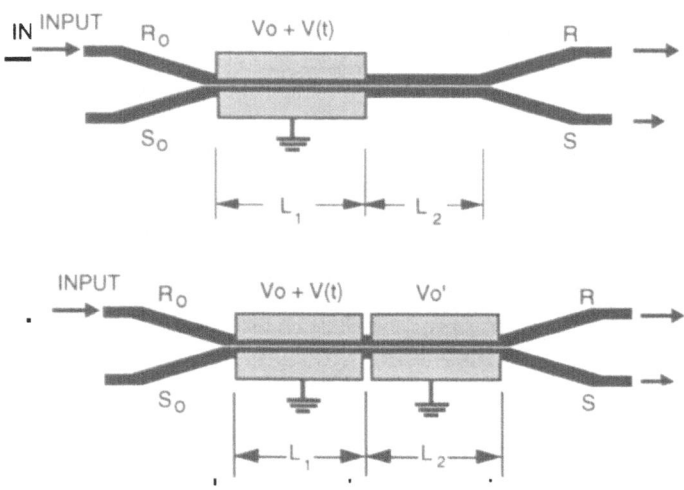

Fig. 1. Linearized directional coupler with single active (top) and two active sections (bottom).

directional coupler system consists of two coupling sections in series, as shown in Fig. 1(a). The first, or active, section has electrodes and is driven by an external voltage V_0 and a RF generator with voltage $V(t)$. It has a length L_1 and a coupling constant k_1, and is described by the parameters $\Delta\beta_1 L_1$ (determined by V_0) and $k_1 L_1$. The second, or passive, section has a length L_2 and coupling constant k_2. A more general system, depicted in Figure 1b, will contain electrodes in the second directional coupler which are biased by an external voltage V_0'. In this case, the second section is described by the parameters $\Delta\beta_2 L_2$ (determined by V_0) and $k_2 L_2$.

The principle behind the enhanced linearity of this tandem directional coupler design is based on the intuitive, notion that the second directional coupler, with an appropriate value of $k_2 L_2$ and/or $\Delta\beta_2 L_2$ will introduce an appropriate phase shift such that the derivative of the curvature in the optical transfer function is null for a specific set of parameters. The vanishing of the derivative of the curvature corresponds to a complete cancellation of the third order coefficient in the series expansion. The analysis involves the assumption of a two-frequency input signal, $V(t) = V_1 (\cos \omega_1 t + \cos \omega_2 t)$ with $V_1 \ll V_0$ to ensure operation in a small signal regime. The small signal regime assumption describes the scenario under which the system is to be implemented in most fiber optic links. The optical transfer function relates the output and input optical powers, and is obtained by multiplying the transfer matrices of the directional coupler for each.

The optical transfer function relates the output and input optical powers, and is obtained by multiplying the transfer matrices for each of the directional couplers. For our case, we chose to explore the behavior of the optical power at the output of the upper arm of the directional coupler system for a given optical power at the input of the lower arm. The result is then used to find a Taylor series expansion given by

$$\frac{P_2(\text{out})}{P_1(\text{in})} = \sum_0^\infty f_n z^n \qquad (1)$$

where $P_2(\text{out})$ and $P_1(\text{in})$ denote the output port and input port of the upper and lower arms of the system, respectively. z is given by $z = \Delta\beta_1 L_1 (V_1 / V_0)(\cos \omega_1 t + \cos \omega_2 t)$. The Taylor

Linearized Optical Modulators

series expansion is valid provided that $V_1 \ll V_0$. The coefficients f_n have simple physical interpretations. f_0 is a measure of the degree of partitioning of the static (unmodulated) optical power in the arms of the system. f_1 is the magnitude of the RF modulation at the RF frequencies ω_1 and ω_2. f_2 describes the effects of second harmonic, sum and difference frequency generation. f_3 gives rise to third order IMD and third harmonics of the RF signals, and so on.

We used the symbolic mathematical software package MATHEMATICA in a MacIntosh FX computer to obtain the analytical expressions for each coefficient in the series expansion. We then searched for null values of the third order IMD for all values of $k_1 L_1$, $k_2 L_2$, $\Delta\beta_1 L_1$, and $\Delta\beta_2 L_2$. This was done through an exhaustive procedure of obtaining 3-D plots of the third order coefficient as a function of all possible combination of pairs of parameters $k_2 L_2$, $k_2 L_2$, $\Delta\beta_1 L_1$, and $\Delta\beta_2 L_2$. After finding the set of nulls of the third order coefficient, we chose the parameter space which can be implemented in a laboratory setting, and studied the behavior of the zeroth, first, second, third, fourth and fifth order coefficients as a function of the phase shift parameters $[\Delta\beta_1 L_1, \Delta\beta_2 L_2]$. These are the only parameters that can be altered by changing the DC bias voltages. This procedure was repeated for each directional modulator coupler system.

2.1. Dual Directional Coupler System (One Active Section)

This dual directional coupler with one active section is depicted in Fig. 1(a). A single active region is followed by a second passive region. The optical transfer function is given by

$$T = \sin^2(k_2 L_2) + \frac{k_1^2}{K^2}\sin^2(KL_1)\cos(2k_2 L_2) + \frac{k_1}{2K}\sin(2k_2 L_2)\sin(2KL_1) \quad (2)$$

where k is the coupling constant between the two parallel waveguides, L is the length of the electrodes and $K = (k^2 + (\Delta\beta/2)^2)^{1/2}$. $\Delta\beta$ is the phase shift introduced by the applied voltage given by $\Delta\beta = (\pi/\lambda) n_o^3 r_o (V_o/d)$ with r_o being the electrooptic coefficient, λ is the wavelength of light, n_o is the host index of refraction, and d is the distance between the opposite ends of the electrodes.

The results of the series expansion are shown in Fig. 2. A set of nulls in f_3 can be found for a specific set of parameters. The results of Fig. 2 correspond to the following values: $k_1 L_1 = 9\pi/2$, $k_2 L_2 = 1.605$ and the null of the third order coefficient is found at $\Delta\beta_1 L_1 = 6.79$.

2.2. Dual Directional Coupler System (Two Active Sections)

The optical transfer function of the directional coupler with two active sections is given by

$$T = \sin^2(k_1 L_1)\sin^2(k_2 L_2)\left\{\left(\frac{k_1}{K_1}\right)\left(\frac{\Delta\beta_2}{2K_2}\right) - \left(\frac{k_2}{K_2}\right)\left(\frac{\Delta\beta_1}{2K_1}\right)\right\}^2 \quad (3)$$
$$+ \left(\frac{k_1}{K_1}\right)^2 \sin^2(K_1 L_1)\cos^2(K_2 L_2) + \left(\frac{k_2}{K_2}\right)^2 \sin^2(K_2 L_2)\cos^2(K_1 L_1)$$
$$+ \frac{k_1 k_2}{2K_1 K_2}\sin(2K_1 L_1)\sin(2K_2 L_2)$$

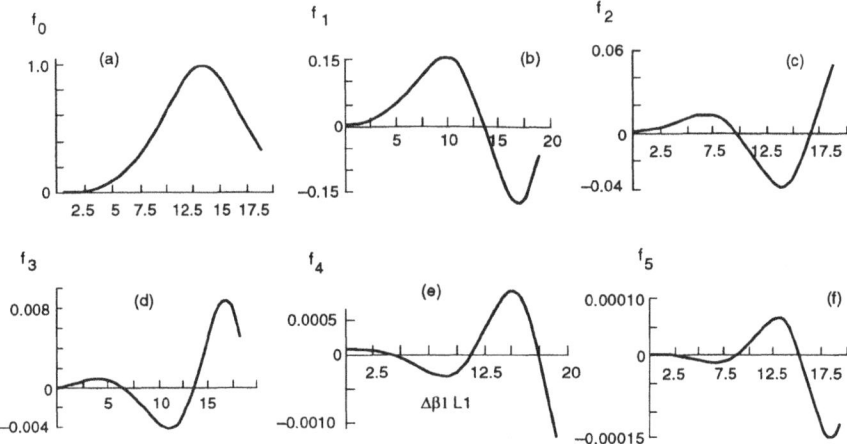

Fig. 2. Coefficients for extended directional coupler [Fig. 1(a)] with $k_1L_1 = 9\pi/2$.

Carrying out the same procedures as described above, we obtained the coefficients for the series expansion of the optical transfer function. These are shown in Fig. 3 for the case where the parameter space is defined by $k_1L_1 = \pi/2$, $k_2L_2 = \pi/2$ and $\Delta\beta_2L_2 = 1.299\,\pi$. The nulls of the third order coefficient are found to be $\Delta\beta_1L_1 = 2.577$ and 2.597.

Significantly improved linearity in the neighborhood of these two nulls is observed in Fig. 3c and 3d. From the behavior of the f_3 (as shown in Fig. 3d), one can draw the conclusion that this specific system is insensitive to changes in the operating parameters..

3. Calculation of the Spur Free Dynamic Range

The SFDR is defined as the difference of the drive electrical power between the contributions from f_1 and (f_2, f_4, f_5). We shall be interested in obtaining the values of the

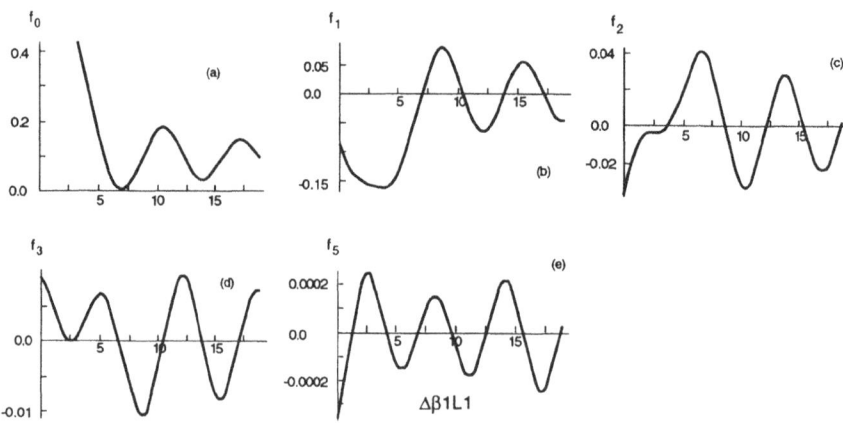

Fig. 3. Coefficients for dual-electrode directional coupler [Fig. 1(b)].

SFDR for the third order IMD term which arises from the fifth order coefficient in the Taylor series expansion of the optical transfer function. The calculation of the SFDR proceeds as follows. The contribution to the detected electrical power at the receiver end due to the term $f_n z^n$ is given by

$$P_E^{(n)} = 0.5 \, (f_n \, z_{ab}^n \, P_{opt} \, r)^2 \, R \tag{4}$$

where z_{ab} is defined by $z = z_{ab} \cos(a\omega_1 - b\omega_2)t$, a and b being integers. P_{opt} is the optical power incident on the detector, r is the detector responsivity, and R is the load resistance of the detector circuit. For example, the detected electrical power at frequency w_1 corresponds to n=1, a=1 and b=0, while the detected electrical power at the IMD frequency $2\omega_1 - \omega_2$ corresponds to n=5, a=2 and b=1.

We calculated $P_E^{(1)}$ and $P_E^{(5)}$ for the directional coupler modulator systems described above. In order to correlate with future experimental data, we converted the units of Watts for the electrical power to units of dBm. A straightforward calculation gives the following expression for $P_E^{(1)}$ and $P_E^{(5)}$ in units of dBm.

$$P_E^{(1)} = X_1 - X_0 + 10 \log_{10} \left\{ \frac{1}{2} f_1^2 (\Delta\beta L)^2 (P_{opt} r)^2 R \right\} \tag{5}$$

$$P_E^{(5)} = 5X_1 - 5X_0 + 10 \log_{10} \left\{ \frac{1}{2} f_5^2 (\Delta\beta L)^{10} (P_{opt} r)^2 R \right\}$$

where $X_1 = V_1^2/R$ is the RF electrical power, and $X_0 = V_0^2/R$ is the DC biased electrical power.

The SDFR is obtained by solving for X_1 in both expressions, setting $P_E^{(1)} = P_E^{(5)} = NF$ (noise floor), and subtracting X_1 for the first order contribution from that of the fifth order contribution. The final expression is given by

$$SSFDR = -\frac{4}{5} NF + \log_{10} \left\{ \left(\frac{1}{2} \right)^8 \frac{f_1^{20}}{m^4 f_5^4} S^8 \right\} \tag{6}$$

where $S = (P_{opt} r)^2 R$ is the detected electrical power arising from the optical beam, and m is the degeneracy factor for the fifth order term. For a concrete example we assumed the following parameters for the calculation of SFDR:

<u>Dual Directional Coupler</u>   SFDR = 117 dB $k_1 L_1 = \pi/2$ $k_2 L_2 = 1.60$
(one-active region)
<u>Dual Directional Coupler</u>   SFDR = 118 dB $k_1 L_1 = \pi/2$ $k_2 L_2 = \pi/2$
(two-active region) $\Delta\beta_2 L_2 = 1.299 \, \pi$

These results show that the directional coupler modulator system has a significantly improved SFDR, as a direct result of incorporating intermodulation cancellation into the design.

4. Experimental Studies on Linearized Directional Coupler Modulators

We investigated the two linearized directional coupler configurations shown in Fig. 1. Both configurations consist of a long integrated optic directional coupler with κL=205°, and an active electrode upon which the RF signal and DC bias are impressed. These DC bias sections allow the modulators' transfer curves to be tuned for linearity. Operation at frequencies > 500 MHz requires traveling wave electrodes to reduce the RF/optical signal velocity mismatch. Each modulator was fabricated on z-cut $LiNbO_3$, first forming a Ti- indiffused integrated optical directional coupler 2.265 cm long, and then sputtering on an SiO_2 buffer layer that was 1.4 µm thick before evaporating the 2 µm thick gold asymmetric coplanar strip electrodes. The traveling wave electrode strip was 8 µm wide and the gap between the electrode and the ground plane was 6 µm. The effective microwave index and characteristic impedance of the electrode were calculated to be 2.32 and 63 Ω, respectively. The traveling wave electrode was 1.132 cm long and the input and output ends were bent and flared to accommodate coplanar waveguide probes. Two passive bias electrodes, each 0.520 cm long, followed each active section. The ultra-thick buffer layer led to a switching voltage of 19.3 V at a 1.3 µm wavelength. The other link components consisted of a Nd:YAG laser source which delivered approximately 7 mW of optical power into the modulator and a single mode fiber pigtailed high speed PIN photodetector which collected the light exiting the modulator.

The link response with a linearized directional coupler modulator with two passive sections is shown in Fig. 4. The measured output signal, third-order intermodulation distortion, and shot noise power in a one Hz bandwidth are shown as a function of the input power per channel when the active section was biased to 3.2 V for maximum signal. Linear fits were made to the signal and intermodulation distortion powers. The SFDR was found by subtracting the signal level from the noise level at the input power where the extrapolated intermodulation distortion equaled the noise level. The dynamic range was maximized at 111.2 dB in one Hz with V_1 = 4.0 V, V_2 = 1.4 V, and V_3 = 6.2 V. The linear fit to the third-order intermodulation distortion is approximately 6 dB/dB which suggests that, for small signals (the maximum depth of modulation was 5% per channel), this transfer curve is performing third-order intermodulation distortion cancellation up to the sixth power. This is consistent with analysis of these devices[3] which showed that below saturation the intermodulation curve is steeper for high linearity modulators than the 3 dB/dB slope expected for simple directional coupler or Mach-Zehnder modulators and, in fact, the slope of the curve becomes increasingly steep as the input power decreases until a third-order intermodulation null is reached (complete cancellation). The modulator was not biased to purposely minimize the second harmonic, but at 2 % depth of modulation the second harmonic was measured as -58.5 dBc.

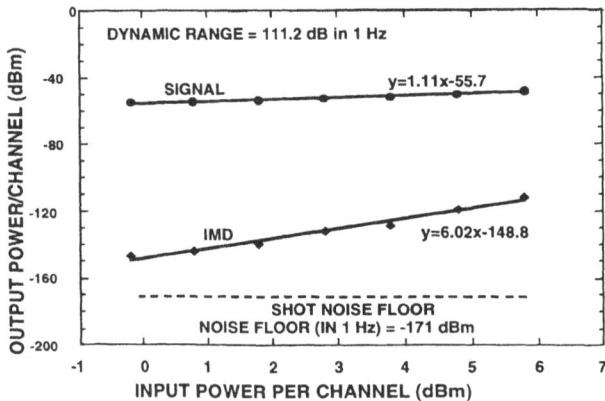

Fig. 4. Signal power and third-order intermodulation power at 500 MHz versus input power for modulator with two passive bias sections.

Table 1. Comparative results for different modulations.

TYPE	FREQ. (MHz)	DYNAMIC RANGE (dB)	LINK INSERTION LOSS (dB)	SIGNAL SLOPE (dB/dB)	IMD SLOPE (dB/dB)I
SDC	500	102.6	50.4	0.98	2.77
LDC, 2P	500	111.2	55.7	1.11	6.02
LDC, 1P	500	111.7	53.8	1.19	6.14
SDC	1000	98.6	52.8	1.10	3.00
LDC, 2P	1000	109.6	58.8	0.89	5.54

A detailed comparison was carried of the SFDR for the simple directional coupler (SDC), linearized modulator with one passive section (LDC, 1P) and linearized directional coupler with two active sections (LDC, 2P). In general we have verified that 9 - 11 dB improvement in the SFDR is attainable with this new design. Table 1 summarizes the results obtained in the experimental study.

5. Summary

In this chapter we have focused on the development of a new class of integrated optical devices - linearized directional couplers. These new components will have a significant impact on subsystems relying on analog transmission of RF and microwave signals. Theoretical analysis has defined the parameters such that third order distortions can be nulled ($f_3 = 0$) for two directional coupler configurations. Experimental studies have shown that significant improvements are indeed possible. The characterization of the linearized directional coupler modulator indicates that in excess of 111 dB dynamic range is achieved, 11 dB higher than the simple directional coupler modulator.

Reference

1. J. F. Lam and G. L. Tangonan, , IEEE Photon. Tech. Lett., Vol. 3, No. 12, December 1991, pp. 1102-1104.
2. M. L. Farwell, Z. Q. Lin, E. Wooten and W. S. C. Chang, IEEE Photonics Tech. Lett., Vol. 3, No. 9, September 1991, pp. 792-795.
3. J. Schaffner, W. Bridges, C. Gaeta, R. Hayes, G. Tangonan, R. Joyce and J. Lewis, High Fidelity Microwave Remoting, Final Report to the USAF, Rome Laboratory, Contract No. F30602-91-C-0104, November 1992.

Chapter 13

AN EXAMPLE OF TI:LINBO3 DEVICE FABRICATION: THE MACH-ZEHNDER ELECTROOPTICAL MODULATOR

P.CUSUMANO and G.LULLO

1. Introduction

Integrated optics on $LiNbO_3$ has already reached a stage of maturity. Several manufacturers are producing standard and custom devices on $LiNbO_3$ such as high speed (up to 20 GHz) phase and intensity modulators, switching matrices, hybrid optical gyroscopes, etc.[1]. Two techniques are commonly used to fabricate these devices: titanium indiffusion for 1.3 and 1.5 µm wavelength operation and annealed proton exchange (APE) at 0.8 µm, due to its higher power handling capacity.

In this Chapter we report our experience with $Ti:LiNbO_3$ technology. The design and fabrication of an electrooptical Mach-Zehnder modulator are described and its performance is evaluated.

2. Operating Principles

The electrooptical modulator described here is based on a guided wave Mach-Zehnder interferometer (Fig. 1). The optical signal, launched into the input monomode channel waveguide, is split by a symmetrical Y-junction, which operates as a 50% power divider feeding the interferometer arms. The distance between the arms must be sufficient to assure their decoupling. After a distance L the two guided waves recombine in the second symmetrical Y-junction. The resultant signal emerges from the output waveguide.

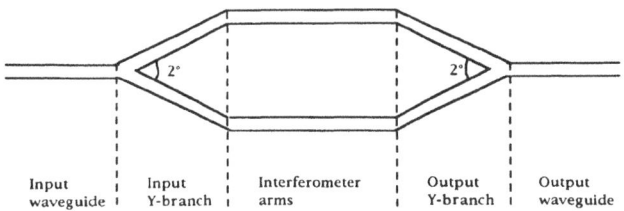

Fig.1. The guided-wave Mach-Zehnder interferometer.

P.Cusumano and G.Lullo - Dipartimento di Ingegneria Elettrica, Università di Palermo Viale delle Scienze, 90128 Palermo, Italy

The integrated Mach-Zehnder interferometer bases its operation on the interference between guided modes in the output Y-junction[2,3]. As Fig. 2 shows, a linearly tapered waveguide zone is always present in Y-junctions, even if input and output waveguides are monomodal, as in this case. By assuming that the fundamental modes propagate in the two arms of the Y-junction, in the taper zone each of the two modes can excite both the fundamental and the second order mode.

Depending on the phase relation between the two incoming modes, different situations can take place. If the two modes are in phase (Fig. 2-a), the two second order modes have opposite phases. They interfere destructively and all the power is in the fundamental mode, which can propagate in the output monomodal waveguide. On the contrary, if the two incoming modes have a half-wavelength relative phase shift (Fig. 2-b), the fundamental modes interfere destructively and all the power belongs to the second order mode. However, this mode cannot propagate in the output monomodal waveguide. It radiates through substrate modes and no power reaches the output waveguide. Of course, if the phase shift is between 0 and π, the power will be divided among substrate modes and the fundamental mode of the output waveguide.

In order to obtain an intensity modulator, it is necessary to externally induce a phase shift between the two modes. As $LiNbO_3$ is an electrooptical material, the refractive index, in the zone where one interferometer arm is placed, can be changed by an external electric field (Pockels effect)[4]. The field is created by an electrode pair with an applied voltage V, as is shown in Fig. 3-a. This causes an optical path variation between the two arms.

If Y-cut X-propagating substrates are used, as in this case, the electrodes must be placed on both sides of the channel waveguide (Fig. 3-b), in order to use the highest electrooptical coefficient r_{33} of $LiNbO_3$ for phase modulation of the TE-like mode (the mode with its electric field parallel to the z-axis which is the optical axis). In this way no buffer layer is required to avoid losses caused by a metal cladding.

The electrooptical index change is

$$\Delta n_{33} = \Delta n_e = -\frac{n_e^3}{2} r_{33} \frac{V}{G} \Gamma \tag{1}$$

where G is the gap between the electrodes and Γ is a parameter (always less than 1), that takes into account the overlapping between the electric and optical fields[4]. If the electrode length is L, the total phase shift $\Delta\Phi$ is proportional to the length and to the electrooptical index change given above.

The output optical power I_{out} in a Mach-Zehnder interferometer is a function of the phase difference and, neglecting Y-junction asymmetries and propagation losses, it is given by the following expression[5]:

$$I_{out} = I_{in} \cos^2 \frac{\Delta\Phi}{2} \tag{2}$$

Introducing the $\Delta\Phi$ expression and rearranging gives:

$$I_{out} = I_{in} \cos^2\left(\frac{\pi}{2} \frac{V}{V_\pi}\right) \tag{3}$$

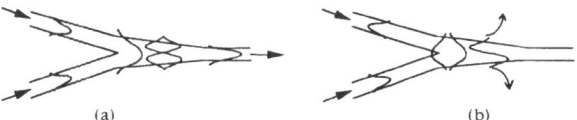

Fig.2. Recombination in the Y-junction: the two incoming modes are in phase (a) or have opposite phases (b).

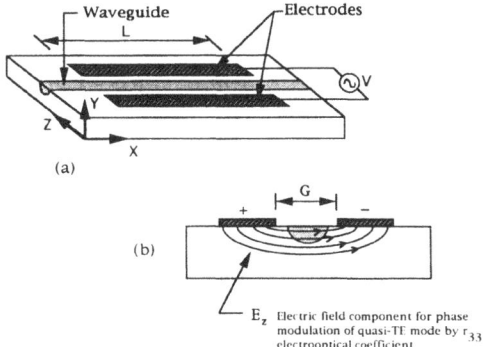

Fig.3. Ti:LiNbO$_3$ channel waveguide phase modulator (a) and transversal distribution of the applied electric field (b).

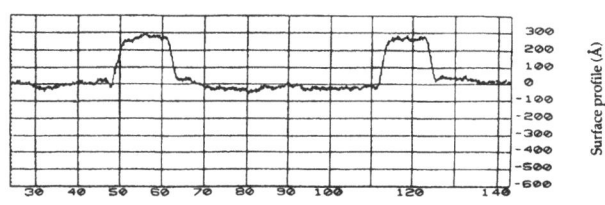

Fig.4. Transverse microprofile of the LiNbO$_3$ surface after Ti-indiffusion.

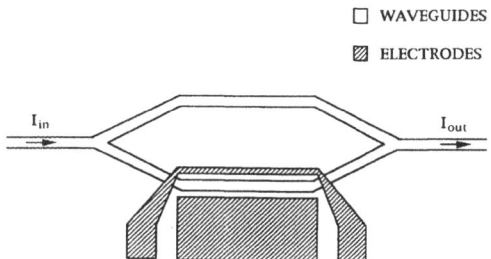

Fig.5. Sketch of the electrooptic Mach-Zehnder modulator.

where V_π is the half-wave voltage, that is the voltage at which $\Delta\Phi$ is π. With perfectly symmetrical Y-junctions, a voltage equal to V_π will ideally produce the complete extinction of the fundamental mode at the output waveguide.

3. Design and Realization of Guiding Structure and Electrodes

The modulator is fabricated by Ti-indiffusion on a Y-cut X-propagating $LiNbO_3$ substrate whose dimensions are 30x10x1 mm. This material is chosen because of its low propagation losses and high electrooptical coefficients. Monomode channel waveguides at $\lambda = 1.3$ μm are obtained by diffusion of 6 μm wide, 200 Å thick titanium stripes at 1012 °C for two hours in flowing dry oxygen. In the guiding regions where the titanium stripes have been diffused, the presence of a slight ridge above the surface, commonly known as "swelling"[6], can be observed. Fig.4 shows a microprofilometer transversal measurement of the surface near the two interferometer waveguides. The ridge height is about 300 Å, that is 1.5 times the titanium thickness. This is due to the fact that titanium indiffusion is mostly interstitial and so a volume increase is produced close to the surface. The swelling of the waveguide regions is very useful for aligning the mask for the electrodes in the subsequent lithography process. The Y branch angle is 2°; this value is a trade-off between critical angle considerations and the need to shorter the longitudinal dimension of the device. The distance between the arms is 60 μm. The interferometer arm length depends both on the half-wave voltage and the desired bandwidth. The chosen electrode geometry is the "asymmetrical" type, in which one electrode has a finite transverse dimension (W), comparable to the channel waveguide width, and the other is semi-infinite (Fig.5). For this electrode configuration, the capacitance per unit length C/L depends only on the transverse geometrical parameter G/W, where G is the gap between the electrodes. Starting with a reasonable value of 4 pF/cm for C/L, from theoretical diagrams[7] we obtain G/W= 0.6. By assuming G=12 μm to allow the 6 μm channel waveguide to be placed in the middle of the gap, we calculate W= 20 μm. The length of the interaction zone is chosen equal to 15 mm, corresponding to a half-wave voltage lower than 5 V at $\lambda = 1.3$ μm and a bandwidth greater than 5 GHz for travelling wave operation.

4. Microlithography

Two masks were used in the fabrication of the device, the first to define the guiding structure and the second for the electrodes. Both of them were produced by the authors using direct laser writing on photoresist coated chromium masks, then developing the exposed photoresist and wet etching the chromium layer in the opened windows . Fig.6 shows the scheme of the direct laser writing system[8]: the beam of a 442 nm He-Cd laser is focused over the mask surface which is moved by a four-axis step motor positioning system that provides 1 μm positioning accuracy. The spot size depends on the numerical aperture of the objective used: its minimum value is around 1 μm. A scanning unit allows the spot to be moved for raster writing on small areas (1 mm^2). The whole system is controlled by a personal computer. Pattern transfer on the substrate is carried out by mask contact UV exposure of a positive mask. The waveguide geometries are created by lift-off of a titanium thin film deposited by RF sputtering. The same technique is used to define the electrodes, which are made of a Cr-Au layer deposited by electron-gun evaporation.

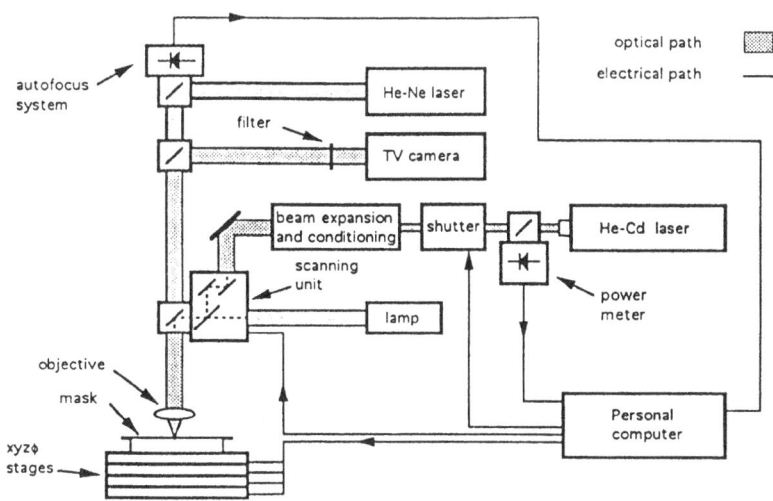

Fig.6. Block diagram of the direct laser writing system.

5. Experimental Results and Conclusions

For the input and the output of the optical signal to the modulator, we use end-fire coupling with monomode optical fibres; thus it is necessary to polish[9] both input and output sides of the device at optical grade and with sharp edges.

In Fig.7 the low-frequency transfer function of the modulator at $\lambda = 633$ nm is shown. The half-wave voltage is about 3.5 V, with a corresponding modulation depth better than 95%. This value shows the precise power division and recombination at the Y-junctions and

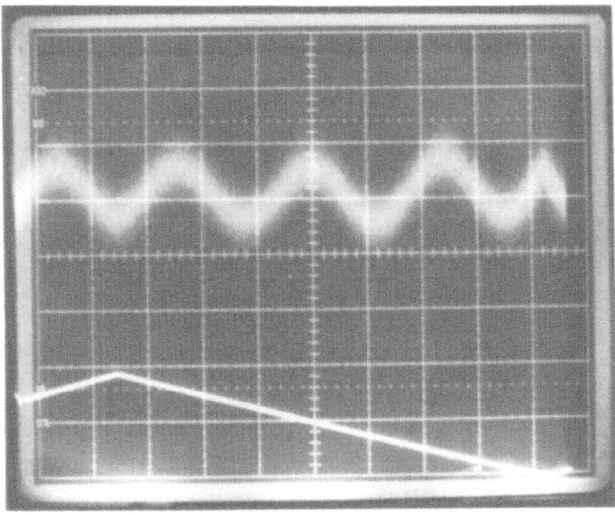

Fig.7. Low frequency response of the moulator. Below, driving voltage varies between +10 and -10 V; above, the output intensity of the optical signal.

therefore the high quality of the photolithographic process used. Also, it is in good agreement with theoretical expectations.

We have not been able to measure the modulation bandwidth above 100 MHz, because instrumentation for higher frequencies was not available in our laboratory.

ACKNOWLEDGEMENTS. The authors wish to thank CRES (Centro per la Ricerca Elettronica in Sicilia - Monreale) for allowing its microlithography facilities to be used in this work, and the Italian CNR (Consiglio Nazionale delle Ricerche) for the partial financial support through its "Progetto Finalizzato Tecnologie Elettroottiche".

References

1. F. J. Leonberger, "Status and applications of commercial integrated optics", in: "Proceedings of ECIO '93", p. 10-5, Neuchatel (1993)
2. P. Cusumano, G. Lullo, F. Trapani and C. Arnone, "Progetto e realizzazione di un modulatore ottico integrato su LiNbO$_3$", in: "Proceedings of FOTONICA '91", Sirmione (1991)
3. M. Izutsu, Y. Nakai and T. Sueta, Opt. Letts., vol.7, No.3, pp. 136-138 (1982)
4. R. C. Alferness, IEEE Trans. on Micr. Th. and Tech., vol. MTT-30, No. 8, pp. 1121-1137 (1982)
5. F. Auracher and R. Keil, Wave Electr., No. 4, pp. 129-140 (1980)
6. N. J. Parsons, "Integrated optics", in: "Principles of modern optical systems", Artech House (1989)
7. R. C. Alferness, "Titanium-diffused lithium niobate waveguide devices", in: "Guided-wave optoelectronics", T.Tamir, ed., Springer Verlag, Berlin (1988)
8. C. Arnone, Microelectronic Engineering, vol. 17, pp. 483-486, (1992)
9. G. W. Fynn and W. J. A. Powell, "Cutting and polishing optical and electronic materials", Adam Hilger, Bristol and Philadelphia (1988)

Chapter 14

ALL-OPTICAL SWITCHING IN AlGaAs SEMICONDUCTOR WAVEGUIDE DEVICES

J.S. AITCHISON

1. Introduction

Nonlinear optics, has for a long time been considered an attractive method of realising high speed, all-optical, switching devices[1,2]. To date there have been two areas of interest namely, resonant and nonresonant nonlinearities. Nonresonant nonlinearities are generally orders of magnitude smaller than their resonant counterparts, however, they respond on a very much faster time scale, typically 10^{-14} s. Combining the very fast response time of a nonresonant nonlinearity with a waveguide geometry, where the high optical intensities required can be maintained over long propagation distances can potentially result in efficient ultrafast switching devices. The devices proposed use a pipeline processing architecture, where more than one optical pulse can be present in the device, though in a different time slot.

One of the major problems, which has had to be overcome is that of a suitable material. The high optical intensities required in such devices lead to other, unwanted nonlinear effects, such as two and three photon absorption. These effects must be minimised while the magnitude of the nonlinear refraction coefficient, n_2 should be optimised. The material must also be compatible with a waveguide fabrication technique.

Semiconductors have been receiving an increasing amount of attention as possible nonlinear materials, due to their large nonlinear coefficient and the possibility of integration with other components. To date most work has concentrated on band-gap resonant nonlinearities. All-optical switching has been observed in GaAs/AlGaAs multiple-quantum well directional couplers[3]. However, due to the resonant nature of the underlying mechanism the recovery time of such devices was limited to a few nanoseconds or 100's of picoseconds with carrier sweep out using an electric field[4]. Faster recovery times have been observed in devices based on the optical Stark effect[5]. However, in both of the above cases the device through-put was limited to less than 10%.

Attention has now focused on nonresonant nonlinearities in the spectral region corresponding to half the fundamental band-gap. Theoretical predictions[6,7] have indicated the presence of an enhancement in the nonresonant, nonlinear, contribution to the refractive index in this spectral region, which has been attributed to the presence of the two-photon band. Other investigations have shown that the large Two Photon Absorption (TPA) coefficient of semiconductors can severely limit the usefulness of ultrafast switching devices[8,9]. Taken

---

J. S. Aitchison - Department of Electronics and Electrical Engineering, Rankine Building, University of Glasgow, Glasgow G12 8LT, UK

together the above suggest that the optimum point of operation for a semiconductor based device is close to half the band-gap, but at a wavelength where TPA does not limit performance[10,11].

Additionally, to be of practical interest such a nonlinear switching device must operate at a wavelength corresponding to one of the low loss telecommunications window, either 1.3 µm or 1.55 µm. In this Chapter the use of AlGaAs as a material for performing all-optical switching functions at 1.55 µm will be considered. The discussion will commence with a review of the theory which has predicted the magnitude and dispersion of the nonlinearity. Several different nonlinear switching devices will be considered in detail, in particular, the nonlinear directional coupler, asymmetric Mach-Zehnder and nonlinear X-junction. The principles of operation of each will be outlined and nonlinear switching results considered. In the final Section a novel approach to all-optical switching using cascaded second order nonlinearities will be discussed. This latter effect is currently receiving a great deal of attention and may soon result in low power all-optical switching devices.

2. Nonresonant Nonlinearities in Semiconductors

It is well known that the linear refractive index and absorption of a material are related through the Kramers-Krönig (KK) relationships. When a perturbation ξ, is introduced into the system the change in refractive index Δn, arising from changes in the absorption coefficient $\Delta \alpha$ can be written as:

$$\Delta n(\omega;\xi) = \frac{c}{\pi} \int_0^\infty \frac{\Delta \alpha(\omega';\xi)}{\omega'^2 - \omega^2} d\omega' \qquad (1)$$

The KK relationships imply that a change in the absorption of a material causes a change in the refractive index. When the absorption change is a consequence of the optical intensity the result is an intensity dependent refractive index. In particular, TPA which has the form $\alpha = \alpha_o + \beta I$ leads to an index change of the form $n = n_o + n_2 I$. It is from this starting point that recent theoretical predictions[6,7] have been able to account for both the magnitude and dispersion of n_2. The theory includes not only TPA but absorption changes due to the Raman effect, the linear Stark effect and the quadratic Stark effect.

Two-photon absorption has been the subject of intensive investigations during the last two decades[12]. Theoretical expressions for the TPA coefficient and scaling laws for multi-photon absorption processes in semiconductors have been derived, using a second order perturbation approach[13]. Assuming a two parabolic band model, the degenerate TPA coefficient can be approximated as[7]:

$$\beta_2 = \frac{K\sqrt{E_p} F_2(2\hbar\omega/E_g)}{n_o^2 E_g^3} \qquad (2)$$

Where, K is a material independent constant with a value of K=1940 (in units such that β_2 is in cm/GW and E_g and E_p are in eV), n is the linear refractive index and $E_p = 2|p_{vc}|^2/m_o$ where p_{vc} is the Kane momentum parameter. E_p is almost material independent with a value of 21 eV. The function F_2 can be expressed as

$$F_2(x) = \frac{(2x-1)^{\frac{3}{2}}}{(2x)^5} \qquad (3)$$

The form of F_2 reflects the assumed band structure of the semiconductor. Once an expression for the TPA coefficient has been obtained, this can be included in the Kramers-Krönig relation and a dispersion function for n_2 deduced. The nonlinear refractive index, of semiconductors can thus, be shown to have the following dependency:

$$n_2 = K' \frac{40\pi\sqrt{E_p}}{cn_0^2 E_g^4} G_2(\hbar\omega/E_g) \qquad (4)$$

Where n_2 has units cm^2W^{-1} and the function $G_2(\hbar\omega/Eg)$ describes the dispersion of the nonlinear index and has the form:

$$G_2(x) = \frac{2}{\pi} \int_0^\infty \frac{F_2(x':x)dx'}{x'^2 - x^2} \qquad (5)$$

The dispersion function G_2 accounts for contributions from TPA, the Raman effect, the linear and nonlinear Stark effect. The various contributions to the dispersion function are plotted in Fig.1. The most significant term is clearly that arising from TPA, except close to the band edge where the quadratic Stark effect becomes dominant.

Now that we have expressions for the dispersion of both the nonlinear refractive index and the TPA coefficient we are in a position to consider the implications for all-optical switching. A nonlinear switching device requires the accumulation of a given phase change over a certain propagation, or interaction distance, implying a large value of n_2, for low power operation. An efficient switch will also require low propagation loss and in the case of nonresonant devices, low two-photon loss. It can be shown that for any optical switching system, one must achieve a refractive index change such that:

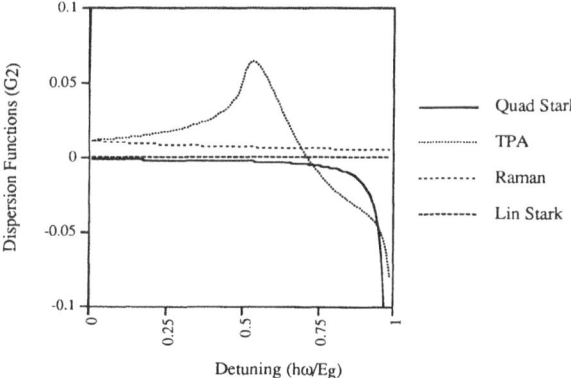

Fig. 1. The spectral dependence of the dispersion function G_2 showing the contributions from two-photon absorption, quadratic Stark effect, Raman and the linear Stark effect. (after Ref. 7)

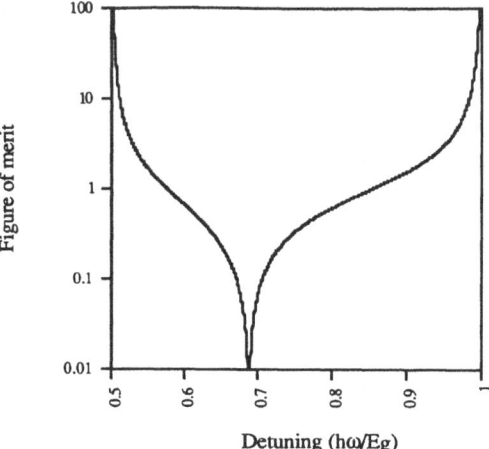

Fig. 2. Spectral dependence of the figure of merit. Large values in the figure of merit are obtained at detunings corresponding to the near resonance condition and the half band gap condition.

$$|\Delta n| = c_{sw} \alpha \lambda \tag{6}$$

Where c_{sw} is a numerical constant, of the order of unity, whose precise value depends on the particular device geometry, for example $c_{sw}=2$ for the nonlinear directional coupler. In the case of a nonresonant nonlinearity the dominant loss mechanism it that due to TPA, thus, $\alpha \approx \beta_2 I$ and the index change $\Delta n = n_2 I$, hence, the switching requirement, or figure of merit, can be written in terms of the dispersion functions F_2 and G_2 as:

$$\frac{h\omega}{E_g} \frac{|G_2(h\omega/E_g)|}{F_2(h\omega/E_g)} > 2\pi c_{sw} \tag{7}$$

The left hand side of Eq. 7 is plotted in Fig.2. From this figure it can be seen that there is a range of frequencies where the figure of merit becomes too small for all-optical switching. There are two spectral regions of interest: close to the absorption edge where the quadratic Stark effect gives rise to a large, negative value of n_2 and in the spectral region near half the band gap, where TPA results in an enhancement in n_2 and the major loss mechanism also goes to zero. It is this latter spectral region where we will concentrate most of our discussions on nonlinear switching devices.

3. Nonlinear Switching Devices

One of the fundamental requirements for the operation of any nonlinear waveguide device is that of low-propagation loss. Several groups have been involved with such studies and the following design utilises their results[14]. The waveguides used for all the half band gap measurements described were MBE grown on an n+ GaAs wafer and consisted of: a 4 μm thick $Al_{0.24}Ga_{0.76}As$ isolation layer, followed by a 1.5 μm thick $Al_{0.18}Ga_{0.82}As$ waveguide layer and finally a 1.5 μm thick $Al_{0.24}Ga_{0.76}As$ capping layer. The relatively thick isolation layer was used to reduce coupling losses into the high index GaAs substrate. The scattering losses in this

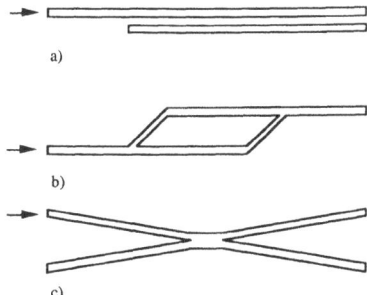

Fig. 3. Schematic representations of the devices studied: a) the nonlinear directional coupler, b) the asymmetric Mach Zehnder and c) the nonlinear X-junction.

structure were minimised by the use of small refractive index steps between the waveguide layer and the surrounding material. Two-dimensional waveguides were formed by a combination of photolithography and reactive ion etching with $SiCl_4$ to depths of between 1.2 µm and 1.45 µm. The strip loading configuration, where only a small fraction of the evanescent field comes into contact with the etched side walls, also helps to reduce scattering losses.

In the following Sections a number of different nonlinear, all-optical, switching devices will be considered, namely the nonlinear directional coupler, the asymmetric Mach-Zehnder and the nonlinear X-junction. These devices are shown schematically in Fig. 3.

The nonlinear directional coupler has been demonstrated in several different configurations and provides a reference to compare different materials. Nonlinear interferometers require a smaller phase change than the directional coupler to switch, hence they are attractive for low power operation, while the nonlinear X-junction has the interesting property of a digital like response.

3.1. The Nonlinear Directional Coupler

In this Section the principles of operation of the nonlinear directional coupler are outlined and experimental results described. The linear directional coupler consists of two waveguides in close proximity to each other. When light propagates in the input guide the evanescent tail of the mode overlaps the second waveguide, this results in a periodic exchange of power, as a function of distance. Power transfer only occurs when the two guides are phase matched, or have the same propagation constant. When the directional coupler is fabricated from a nonlinear material the propagation constants become power dependent, with the result that at high input power levels the two guides become detuned and the light remains in the input guide.

The switching characteristics of the nonlinear directional coupler were initially predicted by Jensen[15], using coupled mode theory. The variation in the output intensity of the bar state for the nonlinear directional coupler is given by the following relationship:

$$I_b(L) = I(0)\left[1 + cn\left(\frac{\pi L}{L_c}\bigg|m\right)\right]/2 \qquad (8)$$

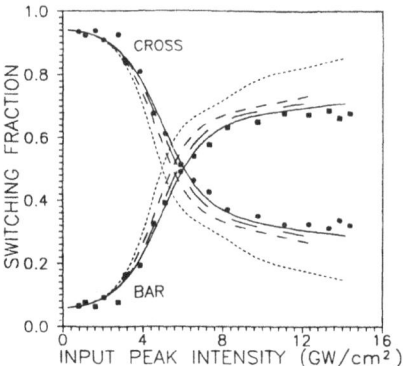

Fig. 4. Switching fraction in the AlGaAs directional coupler. The solid curve represents the results of a numerical simulation, including the effects of two and three photon absorption.(after Ref. 17)

Where L is the length of the coupling section, L_c is the coupling length, $m=[I(0)/I_c]^2$, $I_b(L)$ is the output intensity of the bar state, $I(0)$ is the input intensity, cn(u|m) is a Jacobi elliptic function and I_c is given by:

$$I_c = \frac{\lambda}{L_c n_2} \tag{9}$$

The experimentally measured transmission characteristics for a half beat length directional coupler are shown in Fig 4.

The output of a coupled cavity, mode-locked, NaCl:OH⁻ colour centre laser, producing 450 fs pulses at a wavelength of 1.545 μm was used to excite the directional coupler[16,17]. As the input intensity is increased the nonlinear phase shift induced breaks the low power phase matching condition required for coupling, with the result that the power appears in the bar state at high intensities. Complete switching of the device was impossible due to the fact that the nonlinearity had an instantaneous response time when compared to the temporal width of the pulse . The result is a break up of the output pulse, with the low intensity wings emerging from the cross channel and the high intensity centre from the bar channel. Values of the nonlinear coefficient were obtained by fitting these experimental results to theory[18], these were $n_2 = 1.1 \times 10^{-13}$ cm²W⁻¹, β_2(two-photon absorption coefficient)=0.08 cm GW⁻¹, β_3(three photon absorption coefficient)=0.05 cm³ GW⁻². These results compare well with those predicted by theory, namely $n_2 = 1.5 \times 10^{-13}$ cm² W⁻¹, $\beta_2 = 0$ and $\beta_3 = 0.08$ cm³GW⁻².

The switching performance of the nonlinear directional coupler can be improved further by increasing the length of the device. The effect of this is a reduction in the switching power and hence, a reduction in nonlinear absorption. Recent results have demonstrated all-optical switching and demultiplexing, in a 2 cm long AlGaAs directional coupler[19]. The switching power has been reduced by a factor of ~3, to a peak power of 90 W and no evidence of nonlinear absorption was observed.

Fig. 5 shows a schematic diagram of the demultiplexing experiment. Two pulse trains are simultaneously coupled into the directional coupler; a low power signal with a TM polarisation and repetition rate of 76 MHz, plus a high power, control pulse train with TE polarisation and a reduced repetition rate of 19 MHz. In the absence of the control pulse, most of the signal should appear in the

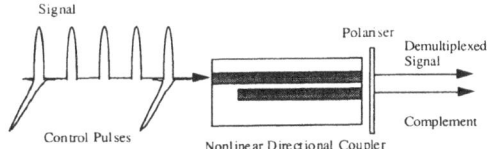

Fig. 5. Schematic representation of the all-optical demultiplexer

cross channel. When the signal and control pulse are present together the coupler will be detuned and the signal will exit from the input waveguide. A polariser at the output blocks the transmitted control light.

Experimental results from the demultiplexing experiment are shown in Fig. 6. As can be seen the presence of the control pulse causes the signal to appear in the bar channel at the output, the other channel shows the complement. This experiment demonstrated the first example of femtosecond demultiplexing in an integrated format. The short length, relative to fibre devices leads to reduced delay times and insensitivity to environmental parameters.

3.2. The Nonlinear Mach-Zehnder Interferometer

An alternative device for all-optical switching is the nonlinear Mach-Zehnder interferometer, shown schematically in Fig. 3b. The device is based on the more familiar electro-optic integrated Mach-Zehnder[20], where a phase shift is produced in one arm by the application of an electric field. In the all-optical version a relative phase change between the two arms is produced by using an asymmetric Y-branch, this results in different intensities in each arm[21,22]. The relative intensities and hence, the phase difference between channels, can be

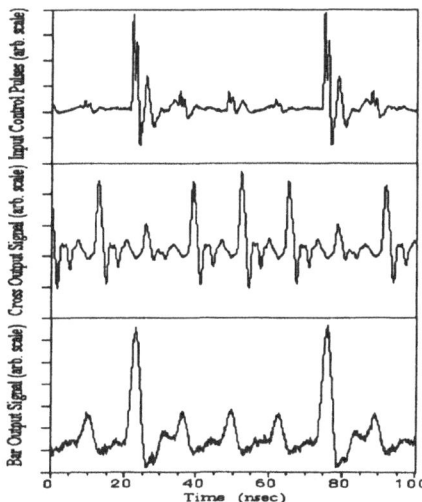

Fig. 6. Demonstration of demultiplexing. The top trace shows the 19 MHz control pulses, the centre trace shows the output of the cross state and the bottom trace shows the demultiplexed output. (after Ref.19)

controlled by varying the angle of the junction. At the output a second, identical, asymmetric Y-junction recombines the modes; when they are in phase, $\Delta\phi=0$ the fundamental mode in the output section is excited and when $\Delta\phi=\pi$ the first order mode is excited, or in the case of a single moded output waveguide radiation modes are excited. Using the scattering matrix model, the output transmission of a loss-less waveguide can be shown to have the form:

$$T = 4\delta(1-\delta)\cos^2\left(\frac{\Delta\phi}{2}+\theta\right) \tag{10}$$

Where δ is the relative power in the branching waveguide and θ accounts for any built in phase differences. The relative phase between the two arms, due to the nonlinearity has the form:

$$\Delta\phi = \frac{2\pi n_2 I_{in} l(1-2\delta)}{\lambda_0} \tag{11}$$

Where l is the length of the interferometer arm. The transmission of the asymmetric Mach-Zehnder is a $\cos^2$ function which depends on the input intensity and the branching angle.

Fig. 7 shows results taken using an interferometer with a branching angle of 3°, this gave a power split ratio $\delta:(1-\delta)$ of 0.18:0.82, the length of the two arms was 0.5 cm. The solid squares represent data taken with a coupled cavity, mode locked KCl:Tl laser operating at a wavelength of 1.52 μm and producing 330 fs pulses. As with the case of the nonlinear directional coupler the instantaneous response of the nonlinearity leads to pulse break up. The solid line in Fig. 7 shows the time average response of the nonlinear interferometer.

An alternative version of the Integrated Mach-Zehnder, where the signal is split 50:50 between the two arms and an intense control pulse is introduced into one arm has been proposed as a method of reducing the switching power requirements[23]. The Mach-Zehnder requires a phase change of π for complete switching compared with 4π for the nonlinear directional coupler. The on-off switching behaviour could be converted into a signal routing

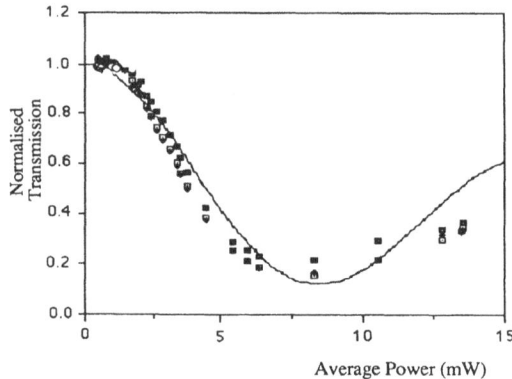

Fig. 7. Normalised transmission of a nonlinear asymmetric Mach-Zehnder interferometer, as a function of average input power. (after Ref. 22)

operation with the incorporation of a asymmetrical Y-junction designed to act as a mode sorter, with the higher order mode being routed along a waveguide instead of being converted to radiation modes.

3.3. The Nonlinear X-Junction

Another intriguing device is the X-junction[24,25,26], show in Fig. 3c, the nonlinear X-junction shows the potentially useful feature of a digital like response. The operation can be understood as follows. Light coupled into one of the input ports, on the left, propagates in a single mode waveguides until the channels are sufficiently close together that coupling between them occurs. At low powers, equal proportions of the lowest order modes, TE_{00} and TE_{01} are excited. The fraction of the power appearing in one of the output guides depends on the detailed interference between the modes in the central section. At high input power levels the symmetry of the input junction is broken, with the result that predominately the TE_{00} mode is excited in the central structure leading to a change in the output state.

X-junctions were fabricated[27], consisting of 4 µm wide, single moded input guides, an 8 µm, double moded central section and junction angles of 0.1 and 0.2°. These very small angles are required so that the junction acts as a mode sorter and not as a power divider. Typical switching results are shown in Fig. 8, Clearly there is a power dependent output from this devices. The switching response was modelled using a coupled mode analysis; similar to the case of the directional coupler, but with a changing coupling constant at each step. The results from this calculation, which includes the effects of two and three photon absorption are also shown in Fig. 8. The nonlinear X-junction requires more phase shift that the directional coupler, hence, the influence of nonlinear absorption is greater.

4. Spatial Solitons

The AlGaAs material system has proved useful not only for the demonstration of all-optical switching but also for the observation of other nonlinear guided wave phenomena, for example the study of spatial optical solitons[28]. Only a very brief discussion will be presented

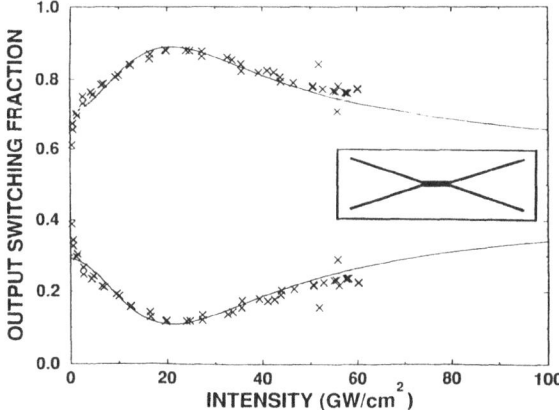

Fig. 8. Normalised transmission for a nonlinear X-junction with a 0.1° angle.

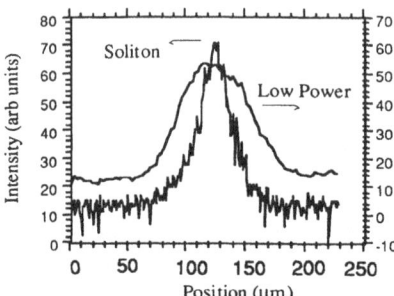

Fig. 9. Time averaged output mode profiles for the low power diffracted beam and the non-diffracting channel at 1.1 kW. (after Ref. 28)

here, since spatial solitons form the subject of Chapter 17 of this book contributed by Y. Silberberg.

Spatial solitons arise as a consequence of the competition between diffraction and self-focusing; at low input intensities diffraction dominates while at high intensities the positive nonlinearity causes the wavefronts at the centre of the beam to be slowed with respect to those on the wings, the result is a self-focusing of the beam. In three dimensions this situation is unstable and leads to a catastrophic self-focusing and material damage. However, in two dimensions, as in a slab waveguide, the situation becomes stable and allows the observation of a self-guided beam.

Initial experiments have been reported in a 5 mm long AlGaAs slab waveguide, these results are shown in Fig. 9. At low power the input beam, with a $1/e^2$ radius of 14 µm, diffracted by a factor of three as it propagated across the sample. At an input of 1.1 kW the output mode had narrowed to form a spatial soliton. The AlGaAs semiconductor system is extremely flexible and should allow the fabrication of a range of novel devices which utilise spatial solitons.

5. Nonlinearities in Active Semiconductors

Refering back to Figs. 1 and 2, there is a second spectral region where it should be possible to operate an all-optical switch, namley close to the band edge. The nonresonant nonlinear refractive index of AlGaAs has been reported in the 810 nm to 850 nm spectral range[29], where values of up to 3×10^{-12} cm^{2}W^{-1} have been observed. However, the largest measured values occured at a wavelength where the linear absorption was $\alpha=16$ cm^{-1}.

More recent investigations have reported on the gain and refractive index dynamics in bulk AlGaAs laser diodes[30]. Experiments were performed at wavelengths corresponding to the above and below band gap of the active region, the nonlinearities were attributed to spectral hole burning and delayed carrier heating in the above band gap case. In the below band gap case the results suggest that heating by free carrier absorption is important and that there is a delay of ~120 fs in the onset of this heating. In order to operate in the low loss, 1.55 µm spectral window an alternative material, from AlGaAs must be used. The refractive nonlinearities of InGaAsP optical amplifiers have recently been reported[31], instantaneous nonlinearities estimated at $n_2 \sim -3.2 \times 10^{-12}$ cm^{2}W^{-1} were observed. Active semiconductor nonlinearities hold a considerable amount of promise for future all optical devices.

6. Cascaded Second Order Nonlinearities

So far only devices which operate via an intensity dependent refractive index, n_2, resulting from the third order nonlinearity have been considered. In this Section the use of a cascaded second order nonlinearity will be described and its application to all-optical switching discussed. Several recent experiments[32,33] have shown that the nonlinear phase shift experienced by a fundamental beam, to close to a phase matched second harmonic generation (SHG), occurs in multiples of $\pi/2$. The depletion of the fundamental beam in SHG behaves in a similar manner to a third order absorptive nonlinearity. We have already seen that this leads an absorptive nonlinearity has an associated refractive nonlinearity. The physical mechanism underlying this nonlinearity is the following; the down conversion of the second harmonic to the fundamental, described by $\chi^{(2)}(2\omega,-\omega)$, occurs out of phase with the original fundamental.

The variation of the electric fields of the fundamental E_1 and the second harmonic E_2 can be described by two coupled partial differential equations, of the form:

$$\frac{\partial E_1}{\partial z} = \frac{i\omega}{4n_1 c} \chi^{(2)}(-\omega, 2\omega) E_1^* E_2 \exp(-i\Delta k z) \qquad (12)$$

$$\frac{\partial E_2}{\partial z} = \frac{i\omega}{4n_2 c} \chi^{(2)}(\omega, \omega) E_1^2 \exp(i\Delta k z)$$

Where n_1 and n_2 are the refractive indices at the fundamental, ω, and second harmonic 2ω respectively, c is the speed of light in vacuum and Δk is the phase mismatch and is given by:

$$\Delta k = 2k_\omega - k_{2\omega} + \frac{2\pi}{\Lambda} \qquad (13)$$

One intriguing property of this type of nonlinearity is the fact that by varying the detuning, phase changes with different signs can be realised. Fig. 10 shows a novel, all-optical switch which makes use of this effect[34,35]. The switch is based on a Mach-Zehnder interferometer; the two arms have different phase matching gratings. By operating on either side of perfect phase matching it is possible to have phase shifts of equal magnitude but of opposite sign. The "push-pull" switch in Fig. 10 will thus switch with a nonlinear phase shift of $\pm\pi/2$ in each arm.

Simulations carried out for this device indicate that switching powers of less that 10 W should be readily achievable. The nonlinear transmission, phase change and pump depletion for a typical device are shown in Fig. 11.

The data in Fig. 11, were calculated assuming an initial detuning of $\Delta kL=\pm 2\pi$, this results in complete recovery of the fundamental power at the switching point. In many applications it is useful to have a control pulse switching a signal pulse train. The cascaded second order

Fig. 10. Schematic of the cascaded second order, push-pull switch. Each arm has a different phase matching grating, designed to give an opposite phase mismatch and hence opposite phase shifts.

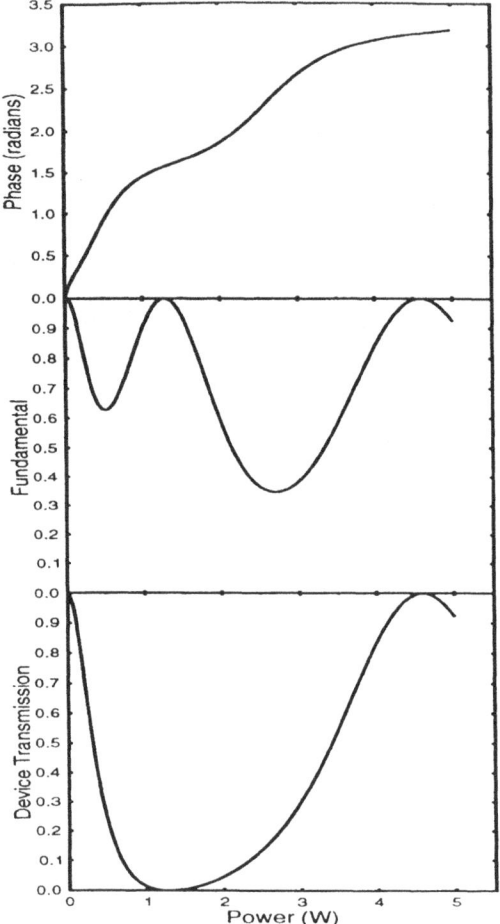

Fig. 11. The top trace shows the magnitude of the nonlinear phase shift in one arm of the interferometer, the middle trace shows the magnitude of the fundamental and the bottom trace the normalised switching fraction, as a function of input power.

nonlinearity can also perform this function by using inputs at two different frequencies[36], ω_1 and ω_2. Instead of second harmonic generation, either difference or sum frequency generation is used to provide a nonlinear phase change, which can result in all-optical switching.

7. Conclusions

In this Chapter we have considered experimental results on ultrafast all-optical switching, in semiconductor waveguide devices operating at half the band gap. In this spectral region there is an enhancement in n_2, due to the two-photon band edge. Working close to the maximum in n_2, allows an efficient, Kerr like, nonlinearity to be exploited. The resulting

Table 1. Summary of results from other, semiconductor based, nonlinear directional couplers.

Material	Mechanism	I_{sw} (GW/cm^2)	τ (ps)	P_{out}/P_{in} %	λ (μm)	Ref
AlGaAs	bound elec.	2	<0.4	32	1.55	16,17
GaAs	Optical Stark	>3	0.5	~2	0.865	5
GaAs	Carriers	0.3	1500	~1	0.855	4
AlGaAs	$\chi^{(2)}:\chi^{(2)}$	~0.1				34,36

devices have both an ultrafast response time, switching and recovering on a time scale shorter than the 450 fs pulses used and a high throughput, 32%. Working at half the band gap allows the problem associated with linear and nonlinear absorption to be minimised.

In Table 1 we summarise results from other, semiconductor based, nonlinear directional couplers. To date experiments in semiconductors have been confined to a spectral region close to the band gap, where relatively large values for n_2 can be achieved[4]. However, the results show that the through put of these devices is limited due to linear absorption. Devices based on the AC Stark[5] effect have faster switching times than their resonant counterparts, though, because they operate within the TPA band, their throughput is limited. Predictions for the switching power requirements of devices based on cascaded second order nonlinearities in AlGaAs based waveguides indicate low switching power requirements. The nonresonant nature of these nonlinearities also suggest a fast response time and high device through put. The latter depending on correct design so that all the fundamental power is recovered at the switching power.

For a nonlinear switching device to become a practical reality, the switching power levels must be reduced to a level where operation with a semiconductor laser become possible. Recent experimental results with mode-locked semiconductor lasers and travelling wave amplifiers have shown that ~160 W of peak power can be generated in the 0.8 µm spectral range, using an all semiconductor system[37]. The powers available from semiconductor sources in the 1.55 µm spectral region are lower than those obtainable around 0.8 µm. However, combining gain switched semiconductor lasers and Er doped fibre amplifiers with the decreasing power requirements for all-optical switching devices, indicates that practical systems should be realised in the near future.

References

1. G.I. Stegeman, E.M. Wright, N. Finlayson, R. Zanoni and C.T. Seaton, J. Lightwave Tech. 6:953 (1988).
2. G.I. Stegeman and E.M. Wright, Opt. and Quant. Electron. 22:95 (1990).
3. P. Li Kam Wa, J.E. Stich, N.J. Mason, J.S. Roberts and P.N. Robson, Electron. Lett. 21:26 (1985).
4. P. Li Kam Wa, A. Miller, J.S. Roberts and P.N. Robson, Appl. Phys. Lett. 58:2055 (1991).
5. R. Lin, J.P. Sokoloff, P.A. Harten, C.L. Chuang, C.G. Lee, M. Warren, H.M. Gibbs, N. Payghambarian, J.N. Polky and G.A. Pubanz, Appl. Phys. Lett. 56:993 (1990).
6. D.C. Hutchings, M. Sheik-Bahae, D.J. Hagan and E.W. Van Stryland, Opt. and Quant. Electron. 24:1 (1992).
7. M. Sheik-Bahae, D.C. Hutchings, D.J. Hagan and E. W. Van Stryland, IEEE J. Quant. Electron. 27:1296 (1991).
8. J.S. Aitchison, M.K. Oliver, E. Colas and P.W.E. Smith, Appl. Phys. Lett. 56:1305 (1990).
9. K.W. DeLong and G.I. Stegeman, Appl. Phys. Lett. 57:2063 (1991).

10. H.K. Tsang, R.S. Grant, R.V. Penty, I.H. White, J.B.D. Soole and E.Colas, Electron. Lett. 27:1993 (1991).
11. M.N. Islam, C.E. Soccolich, R.E. Slusher, A.F.J. Levi, W.S. Hobson and M.G. Young, J. Appl. Phys. 71:1927 (1992).
12. E.W. Van Stryland, M.A. Woodall, H. Vanherzeele and M.J. Soileau, Opt. Lett. 10:490 (1985).
13. B.S. Wherrett, J. Opt. Soc. Am 1B:67 (1984).
14. R.J. Deri and E. Kapon, IEEE J. Quant. Electron. 27:626 (1991).
15. S.M. Jensen, IEEE J. Quant. Electron. 18:1580 (1982).
16. J.S. Aitchison, A.H. Kean, C.N. Ironside, A. Villeneuve and G.I. Stegeman, Electron. Lett. 27:1709 (1991).
17. A. Villeneuve, C.C. Yang, P.G.J. Wigley, G.I. Stegeman, J.S. Aitchison and C.N. Ironside, Appl. Phys. Lett. 61:147 (1992).
18. C.C. Yang, A. Villeneuve, G.I. Stegeman and J.S. Aitchison, Opt. Lett. 17:710 (1992).
19. A. Villeneuve, K. Al-Hemyari, J.U. Kang, C.N. Ironside, J.S. Aitchison and G.I. Stegeman, Electron. Lett. 29:721 (1993).
20. J.E. Zucker, I. Bar-Joseph, B.I. Miller and U. Koren, IEEE Photon. Technol. Lett. 2:32 (1990).
21. K. Al-Hemyari, C.N. Ironside and J.S. Aitchison, IEEE J. Quant. Electron. 28,:2051 (1992).
22. K. Al-Hemyari, J.S. Aitchison, C.N. Ironside, G.T. Kennedy, R.S. Grant and W. Sibbett, Electron. Lett. 28:1090 (1992).
23. A. Lattes, H.A. Haus, F.J. Leonberger and E.P. Ippen, IEEE J. Quant. Electron. 19:1718 (1983).
24. Y. Silberberg and B.G. Sfez, Opt. Lett. 13:1132 (1988).
25. J.P. Sabini, N.Finlayson and G.I. Stegeman, Appl. Phys. Lett. 55:1176 (1989).
26. H. Fouckhardt and Y. Silberberg, 7:803 (1990).
27. J.S. Aitchison, A. Villeneuve, and G.I. Stegeman, Opt. Lett. 18:1153 (1993).
28. J.S. Aitchison, K. Al-Hemyari, C.N. Ironside, R.S. Grant and W. Sibbett, Electron. Lett. 28:1879 (1992).
29. M.J. LaGasse, K.K. Anderson, C.A. Wang, H.A. Haus and J.G. Fujimoto, Appl. Phys. Lett. 56:417 (1990).
30. C.T. Hultgren, D.J. Dougherty and E.P. Ippen, Appl. Phys. Lett. 61:2767 (1992).
31. K.L. Hall, A.M. Darwish, E.P. Ippen, U. Koren and G. Raybon, Appl. Phys. Lett. 62:1320 (1993).
32. N.R. Belashenkov. S.V. Gagarskii and M.V. Inochkin, Opt. Spectrosc. 66:1383 (1989).
33. R. DeSalvo, D.J. Hagan, M. Sheik-Bahae, G.I. Stegeman and E.W. Van Stryland, Opt. Lett. 18:13 (1992).
34. C.N. Ironside, J.S. Aitchison and J.M. Arnold, to appear in IEEE J. Quant. Electron. (1993).
35. G. Assanto, G.I. Stegeman, M. Sheik-Bahae and E.W. Van-Stryland, Appl. Phys. Lett. 62:1323 (1993).
36. D.C. Hutchings, J.S. Aitchison and C.N. Ironside, Opt. Lett. 18:793 (1993).
37. P.J. Delfyett, L.T. Florez, N. Stoffel, T. Gimitter, N.C. Andreadakis, Y. Silberberg, J.P. Heritage and G. Alphonse, IEEE J. Quant. Electron. 28:2203 (1992).

Chapter 15

INTEGRATED OPTICS SENSORS

O. PARRIAUX

1. Introduction

The very first papers on Integrated Optics and on Fibre Optics date back to about the same time, at the end of the sixties. However, the subsequent commercial developments in these two fields have not been at all comparable.

Fibre optics is now a well established technology based on a single material, which has led to internationally standardized products and related components. This is not the case with Integrated Optics, which is not a single technology. Depending on the desired functions, the waveguide circuit is fabricated on a planar chip of glass or silicon, $LiNbO_3$, or a III-V semiconductor substrate, and this is likely to remain the case for a long time. There is no single substrate performing all the needed optical functions sufficiently well, with sufficient yield and cost effectiveness.

This lack of unity in underlying materials and fabrication technologies is one of the reasons for the slower industrial development of Integrated Optics. Another reason is that signal processing can often be performed without resorting to opto-electronic conversion: optical signal processing is *not a passage obligé*. Even when optical processing is desired, all-fibre components often perform sufficiently well, at reasonable cost, and are easy to insert into fibre systems. This last statement naturally reveals one of the major difficulties faced by integrated optic technologies: the access problem. Once an integrated optic technology has become mature, as is now the case on glass [1] and $LiNbO_3$ [2], access is the remaining central technical problem, and it represents by far the major component in the cost structure of devices.

In spite of the intrinsic technical difficulties of this field, integrated optics has advanced steadily under a constant technology push, and also under the pull of some expected areas of application such as $LiNbO_3$ modulators and switching matrices and passive power splitters that can be realized in glass[1] and silicon based technologies [3]. The two sessions of the *6th European Conference on Integrated Optics*, ECIO'93, devoted to *Waveguide Technologies* and *Signal Processing on LiNbO₃* illustrated very well the technical maturity stage that has now been reached; and the session on *Hybridization Technologies* described the status of this key topic, which presently limits the practical applicability of Integrated Optics (IO hereafter).

O. Parriaux - CSEM Swiss Center for Electronics and Microtechnilogy, Maladière 71, CH 2000 Neuchâtel

The potential of IO extends well beyond the field of communications. The first idea that comes to mind is the use of IO for sensors, analogously to the case of fibre optics, from which emerged the field of Optical Fibre Sensors. The analogy is too restrictive, however. The planar configuration of an IO circuit allows far more flexibility. Planarity makes IO naturally compatible and possibly associable with other planar technologies such as silicon based micromechanics, optoelectronics integration on a semiconductor substrate, surface acousto-optic interactions, silicon motherboard based hybridization and, last but not least, the possibility of planarizing the very access from and to the chip in the form of coupling gratings[4]. One should therefore speak of microsystem applications rather than of sensors. The IO circuit would just be the flattened optical part of a system, allowing the latter to be more compact, more stable, more easily mounted and packaged and also to involve novel optical functions that cannot be performed in volume optics.

Optics has long been a technology for reference metrology and instruments. Now that coherent light has become cheaply available with long life, low consumption and small size semiconductor lasers, one can expect optical technology to emerge in a much wider market. The best example is the remarkable development of micro-optical heads for the compact disc market. In spite of the impressive compactness and low cost of the optical assemblies of the present generation of CD consumer electronics, a new generation of smaller, more integrated optical heads is arriving with the new mini-disk [5], and integrated optics is bound to be a serious candidate technology for future flat, extremely low weight, ultra-stable optical circuits to be incorporated in competitive microtechnological modules and microsystems. This was anticipated and demonstrated as early as 1985 when H. Nishihara presented a whole family of monolithic devices for CD reading, magneto-optic Read/Write and distance interferometry [6]. This beautifully illustrated the capability of the combination of a waveguide with a coupling grating to "flatten" an entire optical system, including input/output access, with a substantial gain in space, stability and especially cost, since the optical circuit can be entirely made using planar production techniques. With integrated optics, optical technology moves from the workshop of the craftsman to the microelectronics foundry, where costs may tend towards zero.

2. The Trapped Optical Field

Before discussing examples of sensors, let us list the few features of the waveguide modal field which are relevant for IO sensing structures.

* The size of the modal field determines the accuracy with which a second waveguide must be located in front of the first one when they are joined[7], and the requirement on the stability of the immobilized joint. The transverse positioning accuracy is of the order of a few tenths of a micrometer in the case of fibre-compatible single mode waveguides of approximately 5 µm cross-section diameter. It is even more stringent in the case of semiconductor laser coupling. The tolerance can be of the order of one micrometer in the case of very multimoded waveguides where the requirements are within reach for high standard conventional mechanics.

* Even when formed in a isotropic substrate, an optical waveguide is often strongly birefringent because of its non-symmetrical shape and stress [8]. This implies, for instance, that a waveguide does not in general maintain a given input polarization state except in the case that the polarization is linear and parallel or orthogonal to the chip surface.

* In an "open waveguide" the modal field, though guided, extends into the cladding regions. This has a number of interesting and also detrimental consequences: 1. The evanescent field can be designed so as to perform an efficient sensing of the cladding material if the index or the optical absorption of the latter is made sensitive to a physical or chemical parameter to be measured[9]; 2. Since the modal field "sees" the waveguide surface, a periodic corrugation of the latter or of a thin film overlay can act as a light coupling element[10]; 3. Since the field is guided by total internal reflection, bending the waveguide will increase the incidence angle at the outer wall of the bend resulting in frustration of the total reflection. Guided light will leak out[11]. The leakage rate is very much dependent on the field size: the more confined the field, the weaker the bending loss[12]. This is what limits the integration density of an integrated optical circuit. In fibre compatible waveguides the tolerated bending radius is about 20 mm. In waveguides with stronger guiding (e.g., CVD of SiON, epitaxial growth on III-V substrates), much sharper bends can be accommodated: ring resonators of 50 micrometer size are possible[13]. A higher integration density can also be achieved by resorting to intrawaveguide mirrors such as vertical trenches or corrugated reflection gratings.

2.1. Integrated Signal Processing Functions

The first contribution of integrated optics in the field of optical sensing will be in performing a set of optical preprocessing functions in a monolithic fashion.

The solutions that integrated optics brings may have various degrees of complexity. Compared with the numerous existing multimode fiber sensors, deep and high numerical aperture channel waveguides can be very valuable as they allow elementary functions such as routing, distribution and even wavelength (de)multiplexing[14] to be performed monolithically without too great difficulties in fiber-chip assembling.

However, it is in single mode fiber sensing that the potential of integrated optics is the greatest, and this will be the focus of the present discussion. With single-mode IO the optical signal to be processed has a uniquely defined phase and state of polarization. This allows the most important metrology functions such as interference or polarimetry to be performed monolithically without any concern about the field shape and wave front, since a single-mode waveguide is an almost perfect spatial filter.

The very nature of the electromagnetic eigenfield of a trapped and confined wave determines the configuration of the waveguide elements performing prescribed functions. In Fig. 1 some basic waveguide functional elements are illustrated, described and compared with their better known free space optics counterparts.

Fig. 1 showed examples of simple, single optical functions. Integrated optics can do much more than this; it can combine several such single elements and, moreover, find its own way to combine several functions in single elements. Consider the case of an optical module performing the non-degenerate interferometric detection of the phase shift between two optical fields. Whereas free space optics necessitates mirrors, a polarization retarder and a beam recombiner, a single IO element of a few millimeters length, stable and polarization independent, can do the job [21,22] as illustrated with two variations in Fig. 2a, b.

Basically, instead of using the two available orthogonal polarization states in free space, one generates at will the desired orthogonal spatial modes by a smart combination of waveguide elements, and makes them interfere with the desired phase shift so as to deliver + & - sine and + & -cosine of the input optical phase shift [23] or signals having a 120 degree phase difference if preferred [24].

The two variants of Fig. 2 rest on two very different principles. The first one is a three waveguide proximity coupler. The two incoming fields excite all three eigenmodes of the three

Free space element	Integrated element

Beam splitting

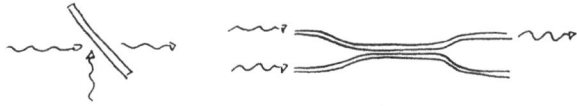

A beam splitting plate or cube is replaced by a Y shaped waveguide separation [15].

Beam superposition

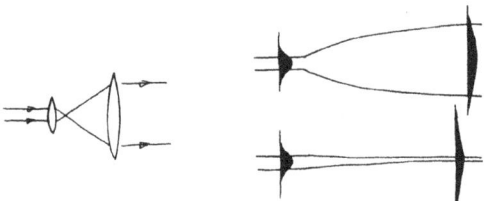

The recombination or interference of two beams can be performed by a single port Y junction or by a 2 x 2 port directional coupler. Depending on the optical phase shift between the two input ports, the optical power will be distributed between the two output ports accordingly [16]. This phase demodulator however is of limited use since it does not allow for the non-ambiguous determination of the input phase. More sophisticated waveguide elements as illustrated in figure 2 are able to deliver, for instance, sine and cosine functions of the phase shift.

Beam expansion

The set of two lenses is replaced by a waveguide horn of either increasing or decreasing cross-section [17].

Phase retardation

A phase plate is replaced by a local widening or narrowing of the waveguide cross-section.

Polarizer

An analyser can be replaced by a metal layer on top of the waveguide which filters out the TM polarization by plasmon wave coupling [18]. Very efficient polarizing action can be achieved by using a self polarizing waveguide such as a proton exchanged LiNbO$_3$ guide [19].

Multilayer filtering

A multidielectric mirror is replaced by a corrugation grating at the waveguide surface [20].

Electrooptic modulation

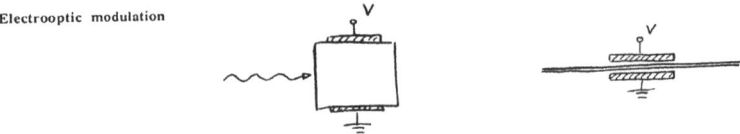

A millimeter size high modulation voltage crystal block is replaced by a low voltage, low capacitance waveguide modulator on LiNbO$_3$ [16]

Fig. 1. Some basic waveguide functional elements illustrated, described and compared with their better known free space optics counterparts.

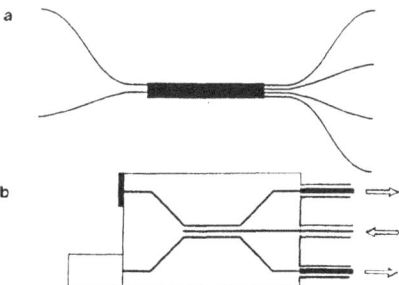

Fig. 2. Two examples of waveguide elements performing non-degenerate phase detection: a) The three-waveguide coupler; b) The self-multiple imaging multimode waveguide section.

waveguide section with a relative amplitude which depends on the optical phase between the input waves. After propagation, these three eigenmodes will in turn excite the modes of the three separating output waveguides. The resulting three output power signals are sinusoidal functions of the input phase. An appropriate choice of the coupling length leads to three power signals of near 100% modulation depth that are phase shifted by 120^0, allowing therefore excellent and stable detection conditions and unambiguous interpolation.

The second variant uses the property of multiple self imaging of a step index waveguide. This property, which was first described by R. Ulrich in 1975[25], has recently been the subject of a sudden and strong interest since semiconductor waveguide technologies have been able to define good quality step index waveguides with sharp lateral edges.

Self imaging along a step index, laterally multimode waveguide originates from the sequence of propagation constants of its modes, except from a minority of its high order ones. After propagating down the waveguide for a certain length, the phase of all modes relative to the first one has rotated an integer multiple of 2π. At this point of collective constructive interference there is created an image of the upstream field distribution considered as the object. A number n of n-fold images are produced at $1/n$ propagation distance by n families of modes of different symmetry. When $n=4$ as illustrated in Fig. 2b, this property of multiple imaging gives the possibility of superimposing the field of two input waveguides onto 4 output subimages where they interfere, delivering there power signals that can be picked up by four output waveguides. The phase relationships between the contributing mode families are such that the resulting power signals are in quadrature. This approach was used with ion exchange glass technology to define the key component of a wavelength stabilized distance interferometer [26].

More complex polarization functions such as polarization splitting and polarization conversion are not so easy to perform because of the planar symmetry of an IO circuit. In one dimension IO (slab waveguide geometry) it is possible to perform such functions by using the polarization properties of non-collinear intra-waveguide grating couplers[55]. In fully confined IO, wide band polarization splitting can be performed in directional couplers in which the coupled waveguide branch is metal-loaded so as to suppress TM mode guiding while the coupler remains a 100% coupler for the TE polarization[56]. Polarization conversion can be performed in a channel waveguide by tilting the birefringence axes periodically every beat length. This can be achieved by tilting the waveguide strip itself[56]. In diffused waveguides it is possible to tilt the birefringence axes by exerting a non-symmetrical stress on one side of the channel.

3. Examples of Integrated Optic Sensors

Some examples of sensors will now be described. It is impossible to consider them all. Only directions which the author thinks are close to applications will be considered.

3.1. Integrating Fibre Sensor Concepts

As compared with fibre communication links, optical fibre sensors involve a higher number of transition and termination components per fibre length and have to withstand difficult environmental conditions. This translates into higher costs and problematic competitiveness against established alternative sensing technologies. In some applications it may be attractive to integrate the needed componentry and to take advantage of batch technologies to implement the whole optical circuit on a single wafer. This is particularly the case as regards rotation sensors[27], Faraday current sensors[28] and possibly photoelastic force/weight sensors. The main problem faced by these middle propagation length structures are losses which limit the sensing sensitivity. Nothing can be made as cheap and low loss as fibre-like high purity CVD technologies applied to flat surfaces[29]. However these losses will always remain much higher than those of the fibres made by the very same technology, because of the surface roughness due to the necessary photolithographic transfer and subsequent etching/deposition process, and the impossibility of benefitting from the smoothing effect of high temperature fibre pulling. Nevertheless, loss figures of the order of 1 dB/m are sufficient for low sensitivity gyros and for a current sensor.

Another problem faced by IO sensor circuits is that of their yield, which is linked with the issues of technological uniformity and reproducibility. It is not certain yet that the potentially best waveguide technologies such as CVD of doped silica are suitable for obtaining a high yield of critical functional elements. Where branching elements are concerned, the problem of yield is dictated by simple geometrical symmetry conditions which are not difficult to achieve. However, as soon as critical elements are concerned, such as phase demodulators, directional couplers or wavelength (de)multiplexers (that is to say, spatially resonant structures), the uniformity across a wafer and the reproducibility from wafer to wafer of two decisive quantities becomes of paramount importance. These decisive quantities are film (or waveguide) thickness and, more importantly, waveguide refractive index. In this regard, it is interesting to point out that a simple waveguide technology such as ion exchange in glass is much closer to meeting uniformity and reproducibility criteria than more sophisticated waveguide technologies.

One should add to this the fact that IO technologies in general lead to birefringent waveguides, due to form and stress birefringence. On one hand this can be an advantage, since the waveguide will naturally be polarization preserving. However, this can also be a very severe drawback if the inherent birefringence is temperature dependent as it is in all silicon based IO circuits.

There are two obvious applications where large propagation length waveguide circuits may be of interest:

The first is rotation sensors using the Sagnac effect. In such a sensor, a fibre loop can be integrated in form of a ring resonator[27] or in form of a non-resonant spiral circuit, which necessitates waveguide cross-over. Low sensitivities of 100-1000°/hour should be obtainable in a ring resonator configuration with Si-based technology[30]. Waveguide scattering and nonlinear effects are the limiting factors in the resonant approach, although the Fibre Brillouin laser concept[31] may well be considered in a planar configuration.

Integrated Optics Sensors 233

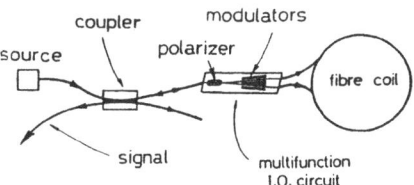

Fig. 3. The termination of a gyro fiber coil performing power splitting, polarization filtering and phase modulation on LiNbO$_3$ (Photonetics document [33]).

By the way, we should mention here the significant system and technological efforts that have been carried out during the past 10 years in order to achieve an integrated optical circuit which performs the signal processing for a fibre gyro. Passive 3-waveguide couplers can do the job for low sensitivity sensors[24], providing non-zero sensitivity around zero rotation rate. However, all-fibre couplers can do this as well, and it is significant that all gyros that are now produced for automotive applications are all-fibre systems[32].

More sensitive gyros make use of a LiNbO$_3$ monolithic circuit which performs the functions of beam splitting, beam interference, polarization and phase modulation[33]. Quite a few companies are in a position to supply such components, which can therefore be considered as an integrated optic product. Fig. 3 illustrates one of these modules.

The second application for large propagation length waveguide circuits is for current sensors using the Faraday effect. One of the problems of Fibre Faraday sensors is the handling and gripping of the fibre coil and tip. These problems would be considerably relieved by a purely monolithic technology of low loss waveguides. Glass waveguide technologies at their present industrial stage of development produce waveguide channels which can have an extremely low birefringence. These guides would allow a maximum Faraday rotation.

The Faraday effect can be detected in usual highly birefringent IO waveguides by the introduction of a periodical magnetic field screening to restore the synchronism between the TE and TM modes[28].

However, the most annoying problems encountered by current sensors are the perturbations exerted by acoustic noise and thermal drift. These are likely to affect IO versions of the sensor as well. The best approach may therefore be the development of glass compositions exhibiting high Verdet constant and low photoelastic effect.

3.2. Displacement Sensors

The second class of sensors is displacement measurement systems, with such variations as vibration and proximity sensors. There is a real need in the very next generation of machine tools for accurate translation systems of submicron accuracy over displacement ranges from a few centimeters to one meter. With the availability of CD laser sources, IO technology is a serious contender for taking optical interferometry from the reference industrial laboratory to the factory floor. Two technologies are attempting to penetrate this market. The first one is a result of work at LETI in silicon based technology[34]. This sensor, developed by CSO, implements a Michelson configuration using a planar waveguide concept[3].

Glass based technology is presently providing two systems using fully confined waveguiding in ion exchanged channels. The first one is CSEM's monolithic dual-interferometer with feedback on the laser frequency for wavelength stabilization in the

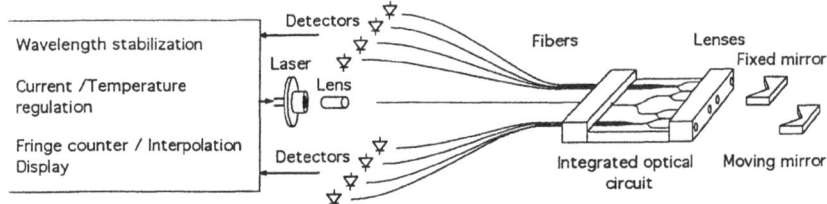

Fig. 4. CSEM's wavelength referenced displacement interferometer [26].

measurement path[26], as illustrated in Fig. 4. An absolute accuracy of 10^{-6} has been reported. It uses a monolithic array of 6 diffractive lenses as a waveguide - free space interface[35]. The second one is a single circuit achieved by means of IOT's advance ion exchange technology[36] dedicated to optical communication passive components. It uses a low polarization dependent buried three-waveguide phase demodulator of excellent reproducibility[22].

All three examples given above are passive waveguide circuits, although a similar system based on slow thermo-optic modulation has been demonstrated[37].

An active waveguide circuit can lead to better performance systems such as the one proposed by Paderborn University[38] and illustrated in Fig. 5. This demonstrator is a wonderful example of a multifunction LiNbO$_3$ monolithic circuit, performing acousto-optic frequency shifting, mode coupling, polarization splitting, electrooptic polarization coupling, control phase modulation, field superposition and laser isolation. This circuit performs heterodyne proximity detection.

Although the systems described above are very promising, their near future fate very much depends on the availability of low feedback sensitive laser sources. In the meantime, various tricks such as intentional angular misalignment and cavity coupling are being tried; they lead, however, to lower useable power and to higher system cost.

For these reasons, the better known and very safe displacement and rotation measurement principle of the "moving grating"[39] is still expected to have a successful industrial future. Here too, at the level of the read out head, Integrated Optics can already bring interesting practical solutions which bypass the laser stability problem.

3.3. *Opto-chemical Sensors*

The third class of sensor is that of evanescent wave opto-chemical sensors. This field of applications may well be the one where Integrated Optics can find its largest market in the future in terms of the number of IO probes. The aim of such sensors is to detect a chemical or physical parameter which characterizes a medium, usually a fluid or an ultra-thin film placed on

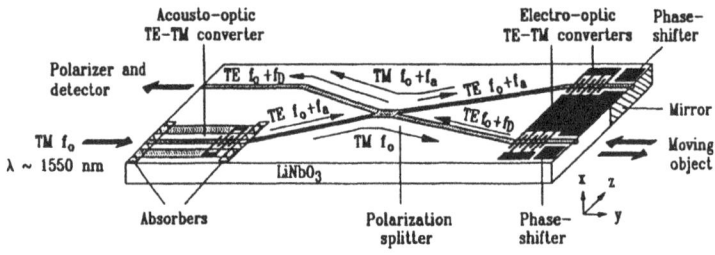

Fig. 5. LiNbO$_3$ multifuncion chip performing heterodyne vibration detection[38].

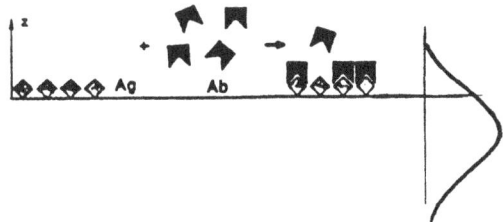

Fig. 6. Scheme of evanescent field detection of immunoassays[10].

top of a planar or channel surface waveguide. The evanescent part of the modal field which propagates in the sensed film converts the material parameter into an optically measurable quantity by means of a change of absorption, of refractive index or by fluorescence generation. It is probably the refractive index change transducing mechanism which presently attracts the greatest interest, since it best exploits the possibilities offered by guided wave optics for immunoassays in pharmacology.

As illustrated in Fig. 6, the evanescent field detects the presence and the thickness of its dielectric load by the modification of the propagation constant of the modal field. Changes in this propagation constant can be detected by the change in the synchronism condition (incidence angle or excitation wavelength) in a grating coupled planar waveguide[40] or with an interferometric configuration. It is the latter which provides the highest sensitivity, since the interaction length can be large.

There are many practical cases where the index changes to be detected are very small. Such optical transducers must therefore be very sensitive and very stable. The stability problem lies in the homogeneity and density of the waveguide film. Significant efforts have been devoted to thin film deposition techniques giving porefree layers while still offering low cost. Ion plating[41] has been considered as well as pulsed CVD[42] and ion beam assisted sputtering. Sol-gel technologies can still improve to offer very cheap but more stable layers. The sensitivity question lies in the design of the sensing waveguide. Basically the sensitivity is proportional to the normalized integral of the squared electric field in the sensing area. The latter is the entire semi-infinite superstrate in the case of homogeneous refractive index monitoring; it is an ultra-thin layer in the case of the detection of the interaction between large molecules as illustrated in Fig. 6. As a rule, the higher the guide-substrate refractive difference, the higher the sensitivity[9]. The best sensitivity achieved so far has been reached by using TiO_2 films on glass or silica[42]. It can be shown that the optogeometrical conditions for the highest sensitivity are all characterized by a single value of a normalized variable in both cases mentioned above.

Several configurations can be adopted to cancel out the undesired parasitic effects of acoustic noise and temperature differentials which may affect the optical phase measurement. One is to use an IO circuit with reference and measurement channels very close to each other[30]. The other is to use the interference between two modes propagating in the same waveguide[43]. These two modes can be the two dominant TE and TM modes or the first two spatial modes of a given polarization.

These types of waveguide probes must be very inexpensive to produce and install, since in most cases they will be single use. This places a severe requirement on the waveguide deposition technology and especially on the excitation technique: fibre pigtailing faces the problem of cost and of severe modal field mismatch. End face excitation requires very accurate

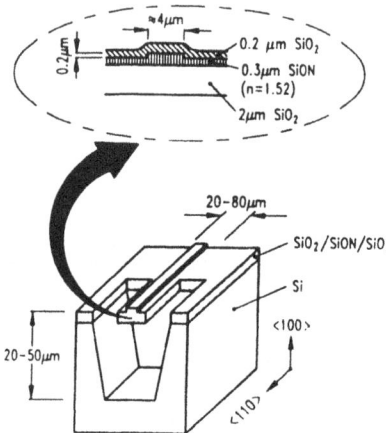

Fig. 7. SiO$_2$/SiON/SiO$_2$ cantilever resonator with optical waveguide[57]. The resonance frequency of the cantilever is about 150 kHz for a 130 micrometer cantilever length with a quality factor of about 100.

adjustment and stability. Consequently, waveguide coupling gratings are meeting here one of their most interesting and challenging opportunities. Gratings offer moreover specific features which have so far been largely unexploited. An example is a grating coupler with a spatial frequency spectrum containing the synchronism frequencies for the simultaneous coupling of the TE$_o$ and TM$_o$ modes of a planar waveguide. A single excitation beam can launch the two guided modes necessary for a common rejection scheme[43].

3.4. All-integrated Sensors

There is a whole field of practical sensors and actuators which uses single-crystal silicon for the definition of microstructures obtained by anisotropic etching[46] and possibly for electronic signal preprocessing. These are products for which assembling and packaging techniques exist. It looks attractive to use this well developed capability together with optical technologies. Microelectronic technologies on silicon offer two entry points for optics: p/n junctions for photoelectronic detection, and dielectric films for deposition/etching of waveguides. The optical read out of microstructure transducers such as pressure or accelerometer membranes and vibrating beams can be carried out by the phase variation or a birefringence variation under the photoelastic effect. Examples of achievements in this promising field can be found in the ECIO'93 review by E. Voges et al.[57]. Fig. 7 illustrates a typical optical microstructure defined on silicon.

As to the full use of Si properties inclusive of photoelectric detection and signal processing, some important work on the technological compatibility and matching of IO with CMOS technology has been carried out in a national German project[58]. Fig. 8 illustrates the interface between an optical waveguide and a detector and its electronics. These applications benefit from well established technological processes, from the necessary tools and machinery, and from suitably equipped silicon foundries, and can therefore be produced cheaply.

Here again, however, the main condition for practicability is the technological and micromechanical hybridization question which determines the cost and competitiveness of possible products versus alternative read-out technologies. Regarding the issue of cost, it is believed by the author that high efficiency coupling gratings should play an important role, since they make input/output access compatible with planar batch technologies.

This naturally leads to a family of devices which are bound for a mass market and for which the access question is of vital importance: the family of read/write devices with CD-like and magneto-optic applications. Early designs were proposed and demonstrated by H. Nishihara about ten years ago[6]. The problems of integrated optic heads with grating couplers for magneto-optic storage applications were discussed by B. Kurdi at ECIO'93[5]. The same conference saw a very different approach proposed by LETI, which uses a simple proximity input/output[55] without a focusing element.

4. Integrated Optics Hybridization

III-V semiconductor materials offer the potential of performing most optoelectronic signal processing functions on a single chip: light generation and amplification, detection, waveguide routing and electrooptic modulation. It is however not likely that the communication systems of the next generation will fully exploit this potential. Hybridized assemblies where each function is performed by the best suited material and technology will therefore remain attractive. Furthermore, every monolithic or hybrid module has to be interfaced with the outside world (free space or fibres).

This implies that hybridization techniques and technologies are a *passage obligé* for this whole field as well as for the neighbouring field of sensor applications. This is where most of the added value of future products lies. There are two different groups of tasks where both alignment and immobilization functions have to be performed accurately: *a) bringing together the elements of an opto-hybrid module; and b) interfacing the latter with the outside world.*
These requirements call for the design and development of dedicated tools and inspection systems. The relative positions of the hybridized components must be maintained within 0.2 micrometer typically over a temperature range of -40 to 85 degrees, which necessitates clever assembly designs and a suitable hierarchy of soldering processes. The list of specifications of packaged devices as well as the recommended testing procedures have been established. They can represent a concrete guide for the interested reader[44]. Points a) and b) above will be further described below.

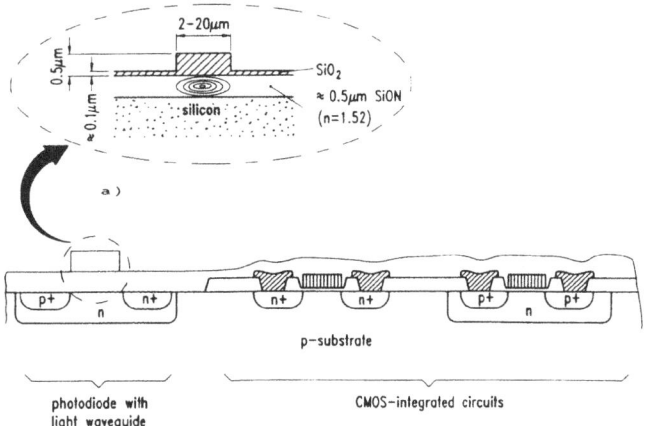

Fig. 8. Scheme of the interface of a SiON waveguide and a CMOS detection and its electronic circuit[64].

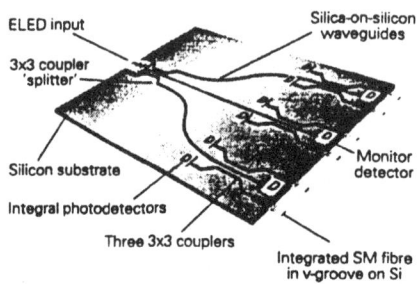

Fig. 9. Opto-hybrid module on silicon for the detection of a three-axis gyrometer[48].

a) Assembling an opto-hybrid. As far as technology is concerned, wide use will be made of single crystal "silicon mother boards" for assembling the various elements with optical and possibly electrical connections[45]. Silicon wafers represent a cheap and ideal mechanical substrate with good thermal properties and excellent surface characteristics. Furthermore, a wide range of microelectronics technologies and high reliability tools and machines is readily available for the preparation of the substrate and the fabrication of optoelectronic circuits. Microelectronic thin film deposition techniques are available for the fabrication of the optical waveguide interconnection and processing network within the mother board by CVD of oxynitride and phosphosilicate films. Anisotropic etching can be used for the definition of lead structures, of positioning features for fibres, of lenses and of reference edges[46]. Self alignment solder bump immobilization techniques (preferably fluxless) are used for the accurate placing of active elements and for electrical connections as in the case of laser chips[47]. Fig. 9 shows a recent achievement in the sensor field. This is the opto-hybrid module on a silicon motherboard for the signal processing of a 3-axis fiber gyro[48] involving integrated photodetectors, laser, electronic circuitry and fibres.

b) Interfacing an opto-hybrid with the outside world. In telecommunication applications, the technologies mainly concern the attachment of fibres to the integrated optic component. Several approaches have been proposed. In silicon based IO, the substrate can also be used for the definition of V-grooves where the fibres will be glued or soldered after the grooves and fibres have been metallized[49]. Anisotropic etching of silicon is the obvious technology. However, deep trenches of precise depth with vertical walls can also be achieved in a silicon substrate[50] or in a thick dielectric waveguide multilayer[29] by dry etching. This relieves the orientation constraints imposed by wet etching and can be very valuable for IO devices of increasing complexity performing fan-in/fan-out of a large number of channels[51]. In the case of waveguide materials where the definition of deep and accurate V-grooves is difficult (glasses,

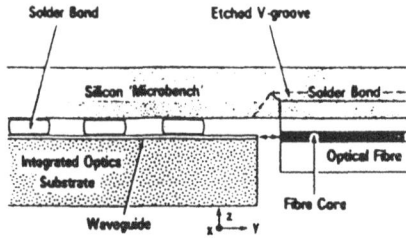

Fig. 10 Flip-chip solder bump assembly of an IO chip and a fiber loaded Si mother board[52].

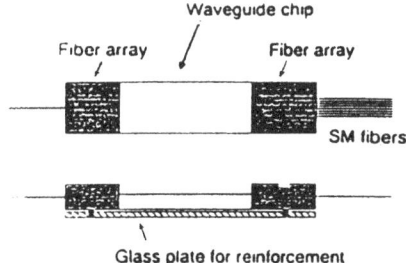

Fig. 11. Schematic illustration of a pigtailed multiport splitter[53].

lithium niobate as well as in III-V materials) one resorts to two principal approaches: 1) A silicon chip, comprising an array of fibre positioning V-grooves and an array of solder bumps, is mounted upside down onto the IO chip in a flip-chip configuration[52] (Fig. 10); and, 2) An end-face polished multifibre ferrule involving Si V-grooves or accurately machined grooves is adjusted and glued to the polished end-face of the IO circuit, their front faces being in mechanical contact[53] (Fig. 11). In sensor and microsystem applications, where fibres are not necessarily involved, the technology of highly efficient coupling gratings[54] represents the most coherent access technology. An important R&D effort is going on in this direction.

5. Integrated Optic Coupling Gratings

This subject would deserve one course on its own since this is the most coherent response to the general access problem. The interested reader can refer to the invited and extended ECIO'93 papers of H. Nishihara[59] and V. Sychugov[60] on this subject. An IO coupling grating is a planar structure which is in principle compatible with the planar waveguide technologies. Such an access approach needs two conditions to be fulfilled:

1). High coupling efficiency. The efficiency does not have to be as high as in a butt coupling, but it should reach a typical value of 50 %. A review of the possible means of increasing the coupling efficiency can be found in Ref. 61.

2). Easy production by standard photolithographic techniques. The typical spatial period is 0.5 µm, which corresponds to the case of vertical incidence on a glass or silica based waveguide at 0.8 µm wavelength. Practical microsystem applications involving such coupling gratings cannot afford direct writing for reasons of cost. A suitable and dedicated mask transfer technique is needed[62] which still has to be developed in form of an industrial mask transfer machine.

Waveguide gratings are not just another way of achieving the access function to and from an optical waveguide. They have their specific properties such as abnormal reflection[63] which could be profitably used in a wide range of waveguide, microoptic and laser applications.

6. Conclusions

Integrated optics is the domain of confined optical fields where several functions can be integrated in a single monolithic circuit. Several technologies are now available for implementing this general concept. For active signal processing, $LiNbO_3$ technology is now well at hand, thanks to progress made to satisfy telecommunication needs. A number of companies are in a position to offer standard and custom made devices[2]. These are still quite expensive and will find their place in systems where there is no alternative. For passive functions such as light routing, multiplexing, and phase demodulation, glass is an interesting candidate since it leads to very reproducible waveguide elements and does not require a heavy technological infrastructure; however, silicon based IO offers by far the most numerous possibilities and the largest flexibility in defining structures and in material doping. It also offers the potential for a strong cost decrease with increased product quantities.

The intrinsic problems that most IO technologies have to solve to achieve high yield and low cost are uniformity across a wafer, reproducibility from wafer to wafer and stability versus temperature and other environmental conditions.

There are also extrinsic problems which for the time being hinder IO applications. First is the question of hybridization and of the related costs and reliability problems. Progress is being rapidly made in this direction. The second problem, which is out of the control of IO itself, is the question of the light source, its modal stability and feedback sensitivity. Some competitive applications require long coherence length sources having low feedback sensitivity without isolators, and a large tuning range without mode hops in the 0.8 micrometer wavelength window. These should ideally be DFB lasers, which are not yet available in this wavelength range.

Finally, it can be said that there is no obvious application in the sensor field where Integrated Optics is the only way to proceed, unlike the field of optical communications where more or less monolithic OEICs represent a *passage obligé*. It is therefore likely that silicon based IO will benefit from the developments needed for OEICs and will be ready to take on a large share of the possible applications as soon as large quantities of product are available. However, other waveguide technologies (glass and vacuum techniques) will by then respond to specific needs here and in the fields of displacement/rotation metrology, in chemical sensors and in microsystems, where the integration of optical functions is desirable and possible.

References

1. M. Mc Court, "Status of glass and Si-based technologies for passive components", Proc. of ECIO'93, Neuchâtel, Switzerland, p.9-1, April 18-22 1993.
2. L. Leonberger, Status and applications of commercial integrated optics, ibid. p. 10-5.
3. S. Valette, S. Renard, J. P. Jadot, P. Gidon and C. Erbeia, Sensors & Actuators, A21-A23, 1990, pp. 1087-1091.
4. T. Suhara and H. Nishihara, IEEE QE-22, pp. 845-867, 1986.
5. B.N. Kurdi, "Integrated optics for optical data storage", Proc. ECIO'93, Neuchâtel, Switzerland, , p. 12-17, April 18-22 1993.
6. T. Suhara, S. Ura, H. Nishihara and J. Koyama, "An integrated optic disc pickup device", Proc. ECOC-IOOC Venezia, pp. 117-120, Oct. 1-4, 1985.
7. R.G. Hunsperger, A. Yariv and A. Lee, Applied Optics, Vol. 16, pp. 1026-1032 ,1977.
8. G. Voirin, P. Debergh, O. Parriaux and O. Zogmal, "Integrated optic polarimetric refractometer", Micro System Technologies 90, Berlin, Springer, pp. 785-790, Sept. 10-13, 1990.
9. O. Parriaux, Chapter 4 in "Fiber Optic Chemical Sensors and Biosensors", Ed. O. Wolfbeis, CRC Press, 1991.
10. W. Lukosz, Biosensors & Bioelectronics, Vol. 6, pp. 215-225, 1991.
11. M. Heiblum and J. H. Harris, IEEE QE-11, pp. 75-83, 1975.

12. R. Baets et al., J. Opt. Soc. Am., Vol. 73, pp. 177-182, 1983.
13. T. Krauss, P. J. R. Layboum and R. M. De La Rue., "Directionally coupled semiconductor ring lasers", Proc. of ECIO'93, Neuchâtel, Switzerland, p. 7-6, April 18-22, 1993.
14. M. Seki, R. Sugawara, E. Okuda, H. Wadw, T. Yamasaki and Y. Harada, "Low-loss, guided-wave multi-/demultiplexer using embedded gradient-index ion exchange waveguides", 12th ECOC, 1986, Barcelona, Spain, p. 439.
15. M. Izutsu, Y. Nakai and T. Sueta, Optics Letters, Vol. 7, pp. 136-138, 1982.
16. R.V. Schmidt and R. C. Alferness, IEEE CAS-12, pp. 1099-1108, 1979.
17. W.K Burns, in "Guided-Wave Optoelectronics", Ed. T. Tamir, Springer, p. 89.
18. Y. Tong and Wu Yizun, IEEE QE-25, pp. 1209-1213, 1989.
19. A. Watanabe, T. Kawazoe and H. Mori, "A multi-function gyro chip based on Ti-indiffusion and proton-exchange", OFS'93, Florence, May 4-6, 1993, pp. 317-320.
20. J. Van Roey and P. E. Lagasse, Applied Optics, Vol. 20, pp. 423-429, 1981.
21. P. Roth and O. Parriaux, "Integrated optic interferometer with phase diversity", IOOC'89, Kobe, Japan, pp. 120-121, 18-21 July 1989.
22. H. Grübel, G. Nitsch and R. Fuest, "Laser Interferometrie: Flexibel durch LWL-Strahlführung", Feinwerktechnik & Messtechnik, 1992/10.
23. Th. Niemeier and R. Ulrich, Optics Letters, Vol. 11, pp. 677-679, (1986).
24. K. Dolde, "Integriert optischer (3x3) - Koppler auf $LiNbO_3$ zur passiven Signalauswertung bei faseroptischen Mach-Zehnder Interferometern", AMA-Seminar, Heidelberg, 28-29 November 1988, pp. 113-127.
25. R. Ulrich, Optical Communications, Vol. 13, pp. 259-264, 1975.
26. G. Voirin, L. Falco, O. Boillat, O. Zogmal, P. Regnault and O. Parriaux, "Monolithic double-interferometer displacement sensor with wavelength stabilization", Proc. ECIO'93, Neuchâtel, Switzerland, April 18-22, 1993, p. 12-28.
27. J. Bismuth, P. Gidon, F. Revol and S. Valette, Electronics Letters, Vol. 27, pp. 722-724, 1991.
28. P. Debergh and O. Parriaux, "Current sensing by magneto-optic coupling in a birefringent waveguide", Proc. of ECIO'93, Neuchâtel, Switzerland, p. 12-36, April 18-22, 1993.
29. N. Takato, K. Jinguji, M. Yasu, H. Toba and M. Kawachi, J. Lightwave Tech, Vol. 6, p. 1003, 1988.
30. S. Valette, "Integrated optical sensors", Proc. ECIO'93, Neuchâtel, Switzerland, p. 12-1, April 18-22 1993.
31. D. Garus, R. Hereth, F.Schlieq and H. Nolte, "Experimental investigation and theoretical modelling of a Brillouin ring laser gyroscope", OFS'93, Florence, pp. 15-17, May 4-6 1993.
32. K. Hotate, "Fiber optic gyro: technologies and applications in Japan", OFS'93, Florence, pp. 89-95, May 4-6 1993.
33. H.C. Lefèvre, P. Martin, J. Morisse, P. Simonpiétri, P. Vivenot and H.J. Arditty, SPIE Vol. 1585, pp. 42-47, 1991.
34. S. Valette, J. Modern Optics, Vol. 35, pp. 993-1005, 1988.
35. H. Buczek, SPIE Vol. 1574, 1991.
36. L. Ross, Glastechn. Ber., 62(8) 285, (1989).
37. D. Jestel, A. Baus and E. Voges, "High resolution interferometric displacement sensor using integrated optics in glass", Micro System Technologies 90, Berlin, pp.733-738, 10-13 Sept 1990.
38. F. Tian,R. Ricken and W. Sohler, "High performance integrated acousto-optical heterodyne interferometer in $LiNbO_3$", OFS'93, Florence, pp. 263-266, 4-6 May 1993.
39. A. Spies, Feinwerktechnik & Messtechnik, 98, pp. 406-410, (1990).
40. K. Tiefenthaler and W. Lukosz, J. Opt. Soc. Am. B, Vol. 6, pp 209-219, 1989.
41. H.K. Pulker and M. Reinhold, Glastech. Ber. 62 100-105, (1989).

42. M. Heming, B. Danielzik, J. Otto, V. Paquet and Ch. Fattinger, "Plasma impulse CVD deposited TiO_2 waveguide films: properties and potential applications in integrated optical sensor systems", Mat. Res. Soc. Symp. Proc. 276 Pittsburgh, PA, p. 117, 1992
43. Ch. Fattinger, M. T. Gak, B. J. Curtis, H. Schütz, M. Heming and J. Otto, "The bidiffractive grating coupler: a universal platform for optical surface probing", Proc. ECIO'93, Neuchâtel, p. 4-12, April 18-22 1993.
44. Bellcore TA 1221: Generic requirements for passive fibre optic component reliability assurance practices.
45. C.H. Henry, G. E. Blonder and R. F. Kazarinov, J. Lightwave Tech., Vol. 7, pp. 1530-1539, 1989.
46. P.P. Deimel, J, Micromech. Microeng., Vol. 1, pp. 199-222, 1991.
47. J.W. Parker, J. Lightwave Tech., Vol. 9, pp. 1764-1773, 1991.
48. I.R. Croston, "Silicon optohybrids for advanced optoelectronic multi-chip modules", Proc. ECIO'93, Neuchâtel, p. 11-1, April 18-22 1993.
49. G.-D. Khoe, H. G. Kock, D. Küppers, J. H. F. M. Poulissen and H. M. de Vrieze, J. Lightwave Tech., Vol. LT-2, pp. 217-227, 1984.
50. S. Valette, "Vers les circuits optoélectroniques hybrides", OPTO 92, pp. 205-210, (1992).
51. H. Takahashi, Y. Ohmori and M. Kawachi, Electron. Letters, Vol. 27, 1991, pp. 2131-2133, 1991
52. M.J. Goodwin et al., J. Lightwave Tech., Vol. 9, pp. 1639-1645, 1991.
53. K. Grosskopf, N. Fabricius and R. Fuest, "Performance of integrated optical single-mode multiport splitters in glass under environmental test conditions", Proc. ECIO'93, Neuchâtel, p. 9-6, April 18-22 1993.
54. I.A. Avrutsky, A. S. Svaknim, V. A. Sychugov and O. Parriaux, Optics Letters, Vol. 15, pp. 1446-1448, 1990.
55. V. Lapras, P. Labeye and P. Gidon, "Development of a reading integrated optical circuit for magneto-optical head", Proc. ECIO'93, Neuchâtel, p. 12-34, April 18-22 1993.
56. H. Heidrich, "Progress and prospects towards coherent receiver frontend OEICs - Issues, challenges and perspectives", Proc. ECIO'93, Neuchâtel, p. 2-17, April 18-22 1993.
57. E. Voges, H. Bezzaoui and M. Hoffmann, Integrated optics and microstructures on silicon, Proc. ECIO'93, Neuchâtel, p. 12-4, April 18-22 1993.
58. P. Salomon, "Integrated optics: Integration into microsystems - The German innovation support programme"Microsystem Technology", Proc. ECIO'93, Neuchâtel, p. 12-20, April 18-22 1993.
59. H. Nishihara, T. Suhara and S. Ura, Integrated-optic grating couplers, Proc. ECIO'93, Neuchâtel, p. 4-1, April 18-22 1993.
60. V.A. Sychugov, A.V. Tishchenko and A.S. Svakhin, "Peculiar aspects of theory, technology and applications of corrugated waveguides", Proc; ECIO'93, Neuchâtel, p. 4-5, April 18-22 1993.
61. O. Parriaux, Technisches Messen,Vol. 58, No 4, 1991, pp. 158-164, 1991.
62. J.-M. Verdiell, T. L. Koch, D. M. Tennant, R. P. Gnall, K. Feder, M. G. Young, B. I. Miller, U. Korer, M. A. Newkirk and B. Tell, "Single step printing of Bragg gratings using a conventional incoherent source and a phase mask: Application to a multi-wavelength DBR laser array", Proc. ECIO'93, Neuchâtel, p. 4-8, April 18-22 1993.
63. I.A. Avrutsky and V.A. Sychugov, J. of Modern Optics, Vol. 36, pp. 1527-1539, 1989.
64. "Integrierte Optik auf Silizium, Abschlussbericht des Verbundprojektes - Entwicklung eines CMOS-kompatiblen Gesamtprozesses zur Herstellung optoelektronischer Schaltungen auf Silizium", VDI/VDE-IT GmbH, 1992.
65. U. Hilleringmann, K. Knopse, C. Heite, K. Goser and K. Schumacher, "Eine monolithische Integration von Wellenleiter und VLSI-Komponenten auf Silizium - Technologie und Schaltungstechnik", Proceedings of Sensor'91, Nurnberg, pp.73-82, 13-16 Mai 1991.

Chapter 16

MULTI-QUANTUM WELL INTEGRATED STACKS FOR DETECTION IN THE MID-INFRARED

I. GRAVÉ, A. SHAKOURI, N. KUZE and A. YARIV

1. Introduction

The development and improvement of advanced epitaxial crystal growth techniques such as molecular beam epitaxy (MBE) and metal-organic chemical vapour deposition (MOCVD) during the last two decades, has opened the door for the realization of devices in the quantum size regime. Quantum size phenomena can be observed when the experimental dimensions approach the order of magnitude of the DeBroglie wavelength associated with the system under investigation. The tools needed to understand and design artificial semiconductor structures, are known under the label "bandgap engineering."

A particle in one dimension, confined between two potential barriers represents the basic model of a quantum well. The Schroedinger equation for this system is the simplest, involving only the kinetic term and boundary conditions. Quantum wells can be realized with advanced epitaxial techniques by growing alternate layers of a low bandgap material and a higher bandgap material (e.g., GaAs and $Al_xGa_{1-x}As$); the dimensions of such layers range from a few Angstroms to a few hundred Angstroms. Many material systems can be used to implement quantum well structures, beyond $GaAs/Al_xGa_{1-x}As$, which was the material system used in this work.

Bulk GaAs has a bandgap of 1.42 eV at 300 K, while in $Al_xGa_{1-x}As$ the bandgap increases monotonically with the Al percentage x. Some attributes of this material systems are well known. The $GaAs/Al_xGa_{1-x}As$ system is lattice matched over the whole alloying range. This allows the growth of heterojunctions which do not induce detrimental misfit dislocations. GaAs and $Al_xGa_{1-x}As$ (for x < 0.42) are direct bandgap materials. This allows high efficiency in light generation through the recombination of electrons and holes. The refractive index of GaAs is larger than that of $Al_xGa_{1-x}As$ at near infrared wavelengths: this fact is used for guiding the coherent radiation in quantum well lasers.

As a result of the fact that the electron effective mass in GaAs is quite small, and by using growth control to subnanometer scale, quantum wells with energy levels in the range of few tens to hundreds of millielectronvolts (meV), with respect to the GaAs bulk band edges, can be prepared. Transitions between these levels can be observed and used to the advantage of applications.

I. Gravé, A. Shakouri, N. Kuze and A. Yariv, California Institute of Technology, 1021 California Ave., Caltech, Pasadena, CA 91125, USA

The GaAs layer, sandwiched between the $Al_xGa_{1-x}As$ barriers, confines both the electrons and the holes, since at a $GaAs/Al_xGa_{1-x}As$ heterojunction about 60% of the bandgap difference is offset by the conduction band, while 40% is offset by the valence band. The carriers (electrons in the conduction band, holes in the valence band) are thus confined in one dimension, along the direction of the growth. In the additional two dimensions there is no confinement and the carriers behave like free carriers in the crystal, in the effective mass approximation. The carriers are then said to occupy energy subbands, rather than energy levels.

The allowed energies of a carrier in a GaAs quantum well, relative to the bulk GaAs band edges (the "bottom" of the well) are given, in the approximation of an infinitely deep square well by

$$E = E_n + \frac{\hbar^2}{2m^*}\left(k_x^2 + k_y^2\right) \quad (1)$$

where m^* is the effective mass and has different value for different types of carriers, and k_x, k_y are the wave vectors in the plane parallel to the layers. Denoting by d the width of the quantum well, the confinement energy is given by

$$E_n = \frac{\hbar^2}{2m^*}\left(\frac{n\pi}{d}\right)^2 \quad (2)$$

Interband transitions between the conduction band quantum well and the valence band quantum well have been the object of an extensive study. In the $GaAs/Al_xGa_{1-x}As$ system, the interband wavelengths correspond to the near infrared, around 0.8 μm. Some quantum well interband devices were found to largely benefit in performances, (mainly) due to the modified density of states arising from the confinement. These included the most important optoelectronic device to date, the semiconductor laser.

Intersubband transitions, or the transitions among different subbands within the conduction band quantum well (or the valence band quantum well) have been only more recently the subject of an intense research effort. One can probably consider the studies of inversion layers in silicon, during the seventies, a precursor of the present research on intersubband transitions in quantum well systems. In 1977 a few ideas about intersubband transitions in quantum wells and their applications were disclosed.[1] However, only in 1983, the two most important intersubband applications were proposed in details in two separate papers. Smith et al.[2] proposed the intersubband and bound-to-continuum infrared detectors, while Gurnick and DeTemple discussed the possibility of enhanced optical nonlinearities associated with these transitions.[3]

Later on, the large oscillator strength of the transition was measured,[4] and intersubband infrared detectors displaying large responsivities and good detectivities were demonstrated.[5] The generation of second harmonic generation via intersubband transitions was observed too.[6-9]

Additional topics involving intersubband transitions are nowadays being addressed by various research groups. Among them the monitoring of spontaneous emission,[10] and the fabrication of modulators for the mid-infrared.[11] A few research teams have started looking into the very challenging goal of designing an intersubband-based mid or far infrared semiconductor laser.[12-14] Such an invention would be a breakthrough and would push further the scope of integrated optics in the mid-infrared.

2. Intersubband Versus Interband Transitions

Interband transitions involve two levels in two different bands. Thus the wave functions of the two states have different Bloch functions. Intersubband transitions instead involve states within the same band, displaying same Bloch functions but different envelope functions. In the envelope function approximation,[15] we can write the wavefunction of the states in the wells as

$$\psi_{s,n,k_{\|}}(r) = \Phi_{s,n}(z) U_{s,n}(r) e^{ik_{\|} \cdot r_{\|}} \qquad (3)$$

where s is the index for the conduction or valence bands, n is the index for the subband in the quantum well, $r_{\|} = (x,y)$ is the component of the position vector in the direction parallel to the surface of the layers, $k_{\|} = (k_x, k_y)$ is the wave vector in the same parallel direction. $U_{s,n}(r)$ is the Bloch function, $\Phi_{s,n}(z)$ is the envelop function for subband n in band s. The orthogonality and normalization relations are

$$\frac{1}{d} \int_{-d/2}^{d/2} \Phi_n^*(z) \Phi_{n'}(z) dz = \delta_{n,n'} \qquad (4a)$$

$$\frac{1}{a} \int_{\text{unit cell}} U_s^*(r) U_{s'}(r) dr = \delta_{s,s'} \qquad (4b)$$

The matrix element for the interaction of the quantum well system with incident radiation is proportional to

$$\langle \psi_{s,n,k_{\|}} | A \cdot P | \psi_{s',n',k'_{\|}} \rangle \propto \frac{1}{a} \int_{\text{unit cell}} \varepsilon_{\|} dr_{\|} [U_s^* \nabla_{\|} U_{s'} + U_s^* i k'_{\|} U_{s'}] \delta_{n,n'} \delta_{k_{\|}k'_{\|}} + $$

$$+ \frac{1}{d} \varepsilon_z \int_{-d/2}^{d/2} dz [\Phi_n^* \nabla_z \Phi_{n'}] \delta_{s,s'} \delta_{k_{\|}k'_{\|}} \qquad (5)$$

A is the vector potential of the electromagnetic field and its direction is in the direction of the polarization vector ε. For interband transitions the second term vanishes since $s \neq s'$ and the first term requires for n (in the conduction band quantum well) to be equal to n' (in the valence band quantum well). These are the well known selection rules for interband transitions for radiation in the near infrared in QW semiconductor lasers, for example.

We can see that the selection rules for the intersubband transition are quite different. In particular an intersubband transition can be induced only by light polarized in the plane of the quantum well,[16] with the electric field in the direction of the growth, (the z direction in our notation), as portrayed in Fig. 1. Whenever the configuration is symmetric, e.g., in a square well with symmetric barriers, additional selection rules allow nonvanishing transitions only for subbands of the same parity, within the same quantum well.

Another important difference is shown in Fig. 2. Contrary to the large spread in energies allowed for interband QW transitions, for an intersubband transitions the energy range is very narrow, ideally a delta function, if one neglects the natural linewidth, nonparabolicity of the bands and a few additional broadening effects. In other words the joint density of states of the transition is much narrower for intersubband transitions.

3. Intersubband Infrared Detectors

One can use intersubband transitions to detect infrared light. The idea, as proposed by Smith et al.,[2] involves the excitation of carriers from a ground-state subband inside one of the quantum wells, to an excited subband located in energy just close to the top of the well, or even above the barrier level, in the continuum. Technically we call these bound-to-continuum transitions (or bound-to-extended-states transitions), as opposed to bound-to-bound excitations. An electric field, applied through electrodes on opposite side of the quantum well or multi-quantum well region, collects the carriers and a photocurrent can thus be measured. The transport is through the continuum in the bound-to-extended configuration, or involves some tunneling through the barriers, if the bound-to-bound design is used. The principle of operation of a bound-to-continuum infrared detector is shown in Fig. 3. Experiments showed that performances of detectors (in terms of internal quantum efficiencies or responsivities) are largely enhanced in the bound-to-extended configuration. An optimum is achieved when the excited subband is just very close to the top of the well at the conduction band edge of the barrier[17]. Bound-to-continuum transitions display wider bandwidth than bound-to-bound configurations. An optimized detector can display a spectral bandwitdh 10-20% of its central wavelength. Designing the quantum well so that the excited level will be well above the barrier, deep into the continuum, allows for a somewhat larger bandwidth at the expense of a reduced responsivity.[18] Optimizing the performances of devices, along with figures of merit used in detector characterization, required the understanding of basic limits in the detection mechanism,[19-20] as well as some efforts in optimizing materials, epitaxial growth, band engineering and overall design of these structures. Additional efforts have also addressed the versatility of these detectors and the improvement of many aspects for different applications. One example is the quest to expand the wavelength range where these detectors are effectively used (7-12 µm) towards longer or shorter wavelengths.[21-22] Other efforts have tried to implement the same schemes of intersubband detection in various materials, in a search for systems displaying improved performances[23-24].

4. Nonlinear Optics Via Intersubband Transitions

The transitions between the subbands of a quantum well have extremely large oscillator strengths. The dipole moment associated with these transitions is of the order of the electric charge e times the QW size d ($\sim 20\text{-}100$ Å).

The nonlinear susceptibility of order n is proportional to the product of (n+1) of these large dipole matrix elements. Although other factors do influence the relative magnitude of the susceptibility, such as the density of available carriers and the dephasing time T_2 of the transitions, it is apparent that large nonlinear susceptibilities can be expected near intersubband resonances. Enhanced nonlinear effects (over what is observed in bulk) are possible close to the resonance. Even-order nonlinearities can be further enhanced by careful design and bandgap engineering: asymmetrical well configurations can be used to enlarge the mean electronic displacement during optical transitions. Electrical fields can be used to finely tune subband separations to achieve multi-resonance situations.[25]

Experimental results on second order nonlinearities, second-harmonic generation and optical rectification, confirmed that large second-order susceptibilities could be obtained in QW structures via intersubband transitions. Enhancement over bulk value of up to three order of magnitudes were measured.

Detection in the Mid-infrared

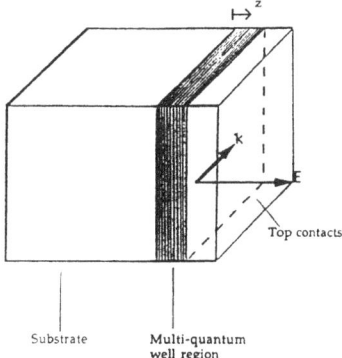

Fig. 1. Intersubband selection rules for electric dipole transitions require the incident light to be polarized in the direction of the growth and the confinement, the z direction, as in the figure.

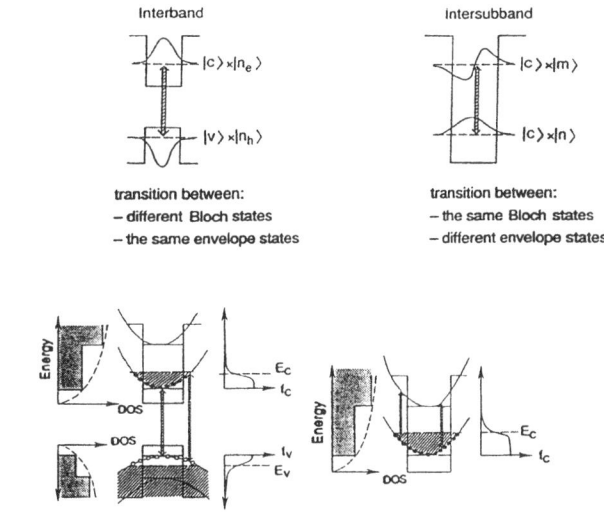

Fig. 2. The nature of interband and intersubband transitions.

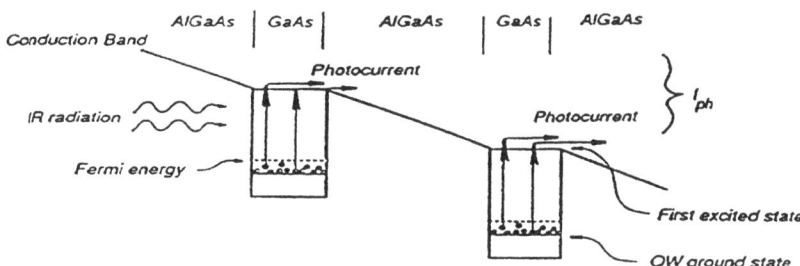

Fig. 3. Operation of a multi quantum well intersubband infrared detector under applied bias. Infrared radiation, incident on the device, must have a component propagating in the plane of the grown layers, according to the selection rules. The wells are heavily doped with electrons, which absorb the infrared photons and are excited from the ground subband to a second subband located on top of the well. The excited carriers are then swept towards the anode by the applied electric field.

Third-order nonlinear effects, were also recently observed[26-30] and large enhancements over bulk values were measured (4-6 orders of magnitudes). There is hope that third-order nonlinearities can be used for a large number of optical signal processing applications in the mid-infrared range.

5. Integrated Multi-Stack Quantum Well Infrared Detectors

Among the features of MQW infrared detectors, as mentioned above, a narrow spectral range of detection, ranging around 10-20% of the peak wavelength. A bound-to-continuum detector designed for peak responsivity at - say - 10 µm, would typically have the response larger than half the peak value between 9.5 µm and 10.5 µm. In some applications this could be considered an advantage; in many other situations one would like to achieve as broad a range of detection as possible. Attempts have been made to address this issue. For example, Levine et al. noted that by designing a detector in which the excited subband is "pushed deep in the continuum" the spectral bandwidth is enlarged to cover the important athmospheric window 8-12 µm.[31] This came, however, at the expense of responsivity and of a lower temperature of operation for acceptable performances.

A second issue which can be seen as limiting in many applications is the lack of flexibility, as for the peak wavelength and the whole spectral range, once the device has been designed and grown. During the design stage one can choose and tune the detector's spectral parameters to his/her own needs. Once grown and prepared as an operating device very little can be made to change, tune or adjust the active spectral range. Few attempts to obtain a MQW tunable detector have been made, mainly using Stark shift effects in various configurations.[32-34] However, these attempts suffered from either a low responsivity or too small a tunability range.

Last, but not least, future needs in the use of thermal imaging systems in the mid infrared, will undoubtedly include a multispectral capability for the system and/or the device. This projection can hinge on the hystory and the trends observed in the much more developed visible and near infrared range.

Following these considerations we designed a multi-color detector based on the serial integration of different quantum wells, or stacks of quantum wells, each stack designed for response at a different peak wavelength. As we will show in this chapter, the results of our measurements not only achieved, in principle, the goals outlined above, but also revealed some new basic physical effects linked to the formation, propagation and readjustment of electric field domains in superlattices.[35] There is hope that these effects could be used, in the future, not only for the design of efficient multispectra detectors, but for an even larger class of optoelectronic switching devices.

5.1. Design and Characterization of the Multi-Stack Infrared Detector

This new type of bound-to-continuum GaAs infrared detector consists of three different stacks of quantum wells arranged in series. All the wells in a given stack are identical, but each stack is designed for absorption and detection at a different wavelength, featuring distinct well widths and barrier heights.

The structure was grown by molecular beam epitaxy on a semi-insulating GaAs substrate. The multi quantum well region, clad by two n-doped contact layers, consisted of 3 stacks of 25 quantum wells each; the first 25 wells were 3.9 nm wide and were separated by

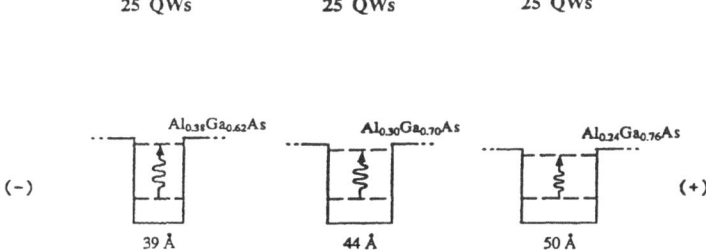

Fig. 4. The conduction band profile of the multistack detector. Starting from the semi-insulating substrate, there are three stacks of quantum wells. Each stack consists of 25 identical quantum wells. Each stack is designed to yield a peak photoresponse at a different wavelength. This is achieved by changing the width of the wells and the Al percentage in the composition of the barriers, among the different stacks. The shortest wavelengths are detected by the stack closer to the substrate and the longest ones by the stack on top of the device.

$Al_{0.38}Ga_{0.62}As$ barriers; the second stack consisted of 25 quantum wells 4.4 nm wide with $Al_{0.30}Ga_{0.70}As$ barriers; the last stack had 25 wells 5.0 nm wide and $Al_{0.24}Ga_{0.76}As$ barriers. All the barriers were 44 nm long; the wells and the contacts were uniformly doped with Si to $n=4\times10^{18}$ cm^{-3}. The schematic conduction band diagram is shown in Fig. 4.

Our design was based on the accepted values for the GaAs/AlGaAs material system, including band nonparabolicity[36], and a band offset value of 0.60. The structure was thus designed to display peaks of absorption at 1335, 1052 and 880 cm^{-1}. Each peak corresponds to absorption from one stack, starting from the one grown close to the substrate and designed for shorter wavelength detection, up to the stack grown last, and designed for the longer wavelengths.

The absorption at zero field and room temperature is shown in Fig. 5. The measurement was taken with a Fourier transform infrared spectrometer in a multipass waveguide geometry; the absorption of light polarized in compliance with the selection rules was normalized by the absorption of light polarized in the perpendicular direction, to allow for only the intersubband

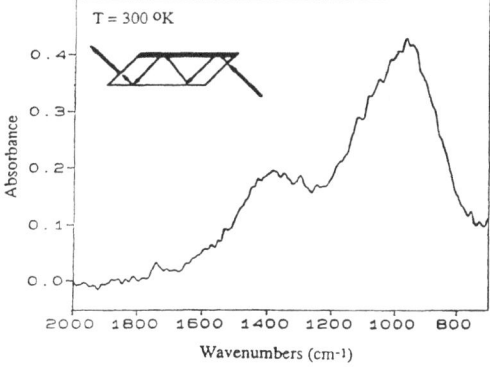

Fig. 5. Absorption spectrum at room temperature. The measurement was performed with a Fourier transform spectrometer using a waveguide geometry as shown in the inset; the spectrum is normalized to reflect the contribution of the intersubband absorption alone. An absorption coefficient $a_{45} = 600$ cm^{-1} for the peak at 1364 cm^{-1} was derived.

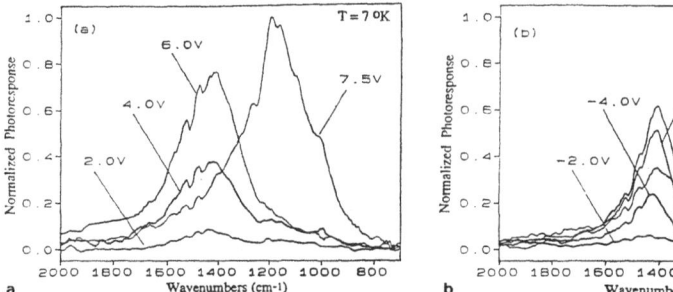

Fig. 6. (a) Spectral photoresponse for few value of applied positive voltage. Note the switching in peaks at an applied voltage around 6.5 V. The responsivity at the peak of 1140 cm and the applied voltage of 7.5 V is 0.75 A/W. (b) Spectral photoresponse for few values of applied negative voltage. Note the broadening in the spectral response below -8.0 V. For still lower voltages (around -13 V), the third peak begins to contribute (not shown in the figure). The units are the same for both (a) and (b).

contribution. The absorption peak at 1364 cm^{-1} is due to the 3.9 nm wells while the stronger absorption centered at 964 cm^{-1} is the composite contribution of the two other species of quantum well, which, individually, have absorption peaking at 1080 and 920 cm^{-1}. These results fit well with our design values. We see that in each of the 3 different types of wells, light is absorbed by electrons excited from the first subband to a second subband which is located close to the top of the well. The blue shift in the experimental values versus the calculated ones can be accounted for with the inclusion of the exchange interaction[37] in the calculated values; in these heavily doped samples, the correction supplied by many-body effects is noticeable. The existence of two absorption peaks that merge into a wide and strong peak was also experimentally verified by analyzing the absorption of a few additional MBE grown control wafers, which were designed to include, each time, only two of the stacks described above.

Devices were processed out of the grown wafer and prepared as etched mesa, 200 µm in diameter. Fig. 6(a) displays the smoothed photocurrent spectroscopy of a device at a temperature of 7 K, for different values of the applied voltage; the polarity is defined here as positive when the higher potential is applied to the cap layer on top of the mesa. It is seen that, for low applied field, the first stack of 3.9 nm wells, closer to the substrate, provides most of the photocurrent at the appropriate excitation energies around the peak of 1411 cm^{-1}. When the bias is increased above a threshold of 6.5 V, a sharp transition takes place and the responsivity peak switches to 1140 cm^{-1}; it is apparent that the second stack of quantum wells is now responsible for most of the photocurrent, while the contribuition from the first stack has sensibly decreased. The small shifts of the photocurrent peaks with regard to the absorption peaks are due to the different experimental temperatures[38] and to the applied electric field.[39] If we apply a negative bias to the detector [Fig. 6(b)], again, at low voltages, the photocurrent is due mostly to electrons excited in the first stack of wells; the responsivity increases with the applied voltage , but its magnitude is always less than that corresponding to the same forward bias; in addition one observes that the photocurrent peak around 1400 cm^{-1} is much broader in the forward bias mode. When the bias is increased to more negative values, the response extends to lower energies, showing increasing contributions from the second stack: the first stack continues to contribute a constant value to the photocurrent, in contrast to the reduction in response experienced in the opposite polarity of the applied electric field. For still more negative voltages, it is apparent from the results that the spectral domain of significant response expands to still lower energies, to include contribuitions from the third stack of wells, around 900 cm^{-1}.

From these results it is apparent that the detector can operate in one of a number of modes. At forward and low bias voltages, the response peaks at a single wavelength (~ 1400 cm^{-1}) and the device functions as a standard bound-to-continuum infrared detector. When exceeding a critical applied voltage, the detector's spectral response switches to a different peak wavelength (~1140 cm^{-1}), while the detection at the previous peak is significantly reduced. Thus the detector can operate as a two-color, voltage controlled switching device. At a reverse bias the detection is again centered on the higher energy peak (~ 1400 cm^{-1}) up to a specific applied voltage; for moderately higher voltages, two peaks yield a significant photoresponse. At still higher values of the reverse voltage a third response peak appears, which results in operation as a wide-band detector. Here the detector works as a voltage-controlled, adjustable-spectral-domain detector. And finally, at large negative bias, the detector has a wide spectral range of response.

The dark current, measured with a cold shielded window is shown at different temperatures in Fig. 7. One can note a strong asymmetry between the two polarities. A fine stucture in the plateaux of the I-V curves (not resolved in Fig. 7), corresponding to regions of negative differential resistance, was observed and is shown in Fig. 8. This measurement gives a most important clue to the origin of the switching phenomena, as will be discussed later on.

The data of the photocurrent spectral measurements were taken with a Fourier transform spectrometer, complemented by a set-up including a calibrated black-body source and a set of cooled filters at different wavelength. The noise equivalent voltage was measured directly with a spectrum analyzer in the cold, shielded window configuration.

One should also note the very low values of the dark current, which, combined with a responsivity ranging up to 0.75A/W, ultimately yield high D* for this detector. Table I displays figures of detectivities as measured at 40 K. The noise characteristics were measured directly with a spectrum analyzer.

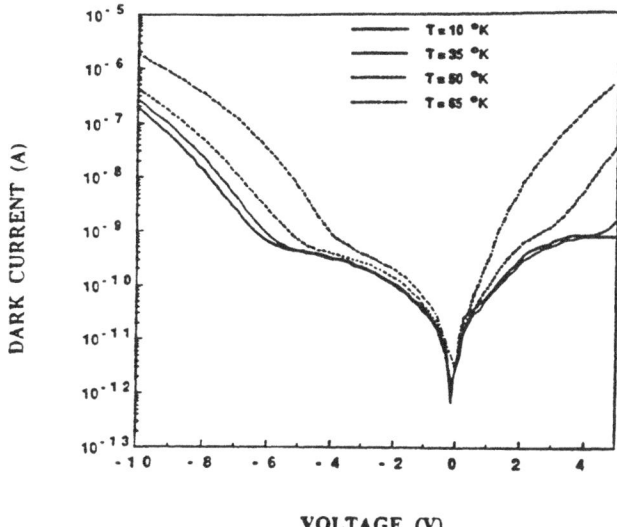

Fig. 7. Dark currents at different temperatures as measured in the shielded, cold window configuration.

Table I. Measured detectivity at 40 K for different wavelengths.

Wavelength	[μm]	7.2	8.0	9.0	9.8	11.6
Measured D^*_λ	[cmHz$^{1/2}$/W]	7.9×10^{11}	4.6×10^{11}	4.1×10^{11}	2.6×10^{11}	7.0×10^{10}

6. Electric Field Domain Formation and Expansion

Before proceeding with the analysis and the interpretation of the experimental results, we present here a short background review on some related topics, namely the formation and expansion of high-field domains.

In 1963, Gunn was able to monitor coherent microwave output generated when a dc electric field was applied across a randomly oriented, short n-type sample of GaAs or InP.[40] The electric field had to exceed a critical threshold value of several thousand Volts per centimeter. The frequency of oscillations was found to be related to the reciprocal of the transit time across the length of the device. This effects were found to be consistent with previous theories, which invoked the presence of negative differential resistance (NDR) in the sample. The specific mechanism responsible for NDR in the Gunn effect was found to be a field-induced transfer of conduction band electrons from a low energy, high mobility valley, to higher energy, low mobility satellite valleys. This transferred electron effect has also been referred to as the Ridley-Watkins-Hilsun effect, honoring those who theoretically predicted and explained such effects.[41]

It can be shown[42] that a semiconductor exhibiting bulk NDR is inherently unstable; a random fluctuation of carrier density at any point in the semiconductor tends to produce an instantaneous space charge that grows exponentially in time; this is true provided the semiconductor is biased in the NDR region of its characteristics. If the semiconductor device displays a voltage-controlled NDR (N-shaped I-V characteristics), it can be shown that high field domains, or, equivalentely, a space-charge accumulation, will form and will travel across the sample, regenerating at the first electrode when collected at the second electrode. Similarly, for current-controlled devices, whose I-V characteristics are S-shaped, high current filaments will form.

A similar, yet different scenario occurs when applying an electric field parallel to the growth direction of a superlattice. In an undoped superlattice, or when otherwise the density of carriers is low, the electric field is constant along the superlattice. The voltage drop along each period of a uniform superlattice is the same. The picture can change drammatically, when a significant number of carriers is introduced into the superlattice, either by doping, injection or photoexcitation. Under this scenario, the electric field can brake into two or more regions with different field strengths (electric field domains), separated by space-charge regions. These domains form a static configuration, contrary to the dynamical pattern observed in bulk samples for the Gunn effect. Another difference is found in the mechanism responsible for NDR, which is resonant or sequential tunneling within the bandgap-engineered conduction band of the superlattice, rather than intervalley scattering as observed in bulk transferred-electrons devices.

The first evidence of electric field domains in superlattices was provided by Esaki and Chang.[43] They observed an oscillatory behavior in the conductance versus voltage characteristics of a GaAs/AlAs superlattice, and they interpreted it as the formation and expansion of a high-field domain along the sample. Furthermore, the oscillations displayed a

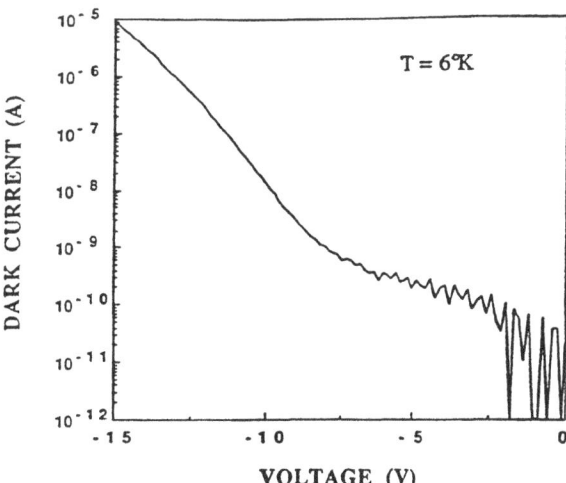

Fig. 8. The fine structure in the I-V characteristic at low temperatures reveals regions of negative differential resistance. This is an indication of the formation and expansion of a high-field domain in the sample.

period in agreement with the conduction-subband spacing of the ground state and the first excited state. These experiment was performed in the regime of miniband conduction through thin barriers (40 Å AlAs). Due to these thin barriers, the wells are strongly coupled and a large ground state bandwidth (ΔE = 5 meV) results, allowing electron transport by miniband conduction. The expanding high-field domain is thus produced by the electric-field-induced breaking of the miniband conduction.

Choi, Levine et al. demonstrated the formation of an expanding high-field domain in a sample consisting of weakly-coupled quantum wells,[44] with a much narrower ground-state bandwidth (ΔE = 0.4 meV). The transport mechanism in these sample was sequential resonant tunneling; still, negative conductance oscillations were monitored in the electrical characteristics of the device.

In the last few years, there have been additional transport experiments on different kinds of superlattices, including $In_{1-x}Ga_xAs/In_{1-x}Al_xAs$, $GaAs/Al_xGa_{1-x}As$, and $In_{1-x}Ga_xAs/InP$, that showed evidence of high-field domain formation.[45-49] In all these experiments, again, the presence of an expanding high-field domain was inferred from an oscillatory region in the current-voltage characteristics, and the period of the oscillations was in most cases related to the subband spacing in the corresponding superlattice. Fig. 9 shows the progressive stages of the formation and expansion of a high-field domain in a uniform superlattice as presented in Ref. 44. For low applied bias, the drop in potential is uniform along the superlattice. The transport is mediated by sequential resonant tunneling from the ground subband in one well to the ground subband in an adjacent well. In principle, considering the quasi 2-D nature of the electron gas in the wells, resonant tunneling should be possible only when the energy levels in each well coincide. This condition is disrupted by the presence of even a small electric field. However, Kazarinov and Suris showed[50] that acoustic phonons and impurity scattering within each well can relax conservation of energy and momentum, provided that $eV \ll \eta h/\tau_1$. Here V is the potential difference between the adjacent wells and τ_1 is the ground-state scattering time. When the applied field is large enough for eV to become of the same order of magnitude as $\eta h/\tau_1$, the continuous sequential resonant tunneling channel, through adjacent wells in the sample is disrupted and the first NDR peaks appears. Each period cannot sustain the resonance condition and the resistance becomes much larger across the sample. One of the periods takes on all the additional resistance and a high-field domain is formed. Increasing the bias will increase the voltage drop only across this high-field domain. Eventually, the ground state level rises to within $\eta h/\tau_2$ of the first excited leve of the adjacent well and a resonant channel of transport is reinstalled. The conductance is large again. Further increase in bias will cause additional wells to break off from the resonant condition in the low-field domain and to be progressively introduced to the high-field domain, while the I-V characteristic repeats the oscillatory behavior in conductance. Eventually, for large enough a bias, all the wells will fall within the HFD. The appearance of high-field domains in semiconductor superlattices can be

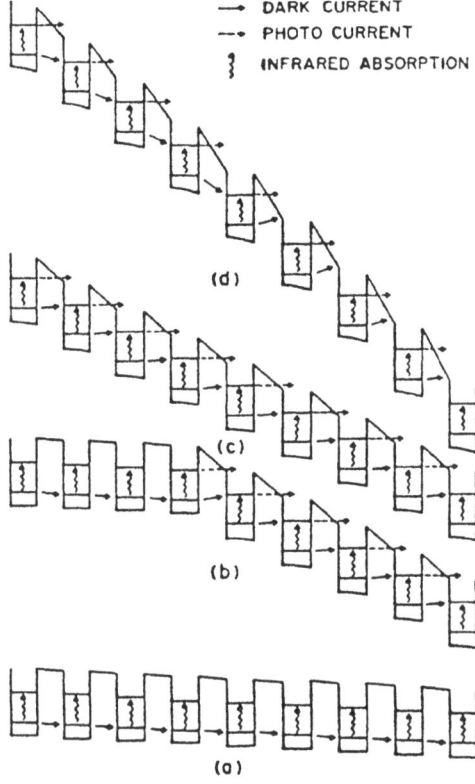

Fig. 9. Formation and expansion of a high-field domain in a uniform superlattice. (a) At very low applied voltage the voltage drops uniformaly on each period, and the dark current flows by sequential resonant tunneling between the ground subbands in adjacent wells. (b) For an intermediate applied voltage a high-field domain has formed and expanded to include 5 periods at a larger applied voltage. Transport in the high-field domain region is by sequential resonant tunneling between an excited subband and the ground subband in the adjacent well. Note that, in general, the transport is nonresonant at the boundary between the two field domains. (c) For an even larger voltage the whole superlattice is under high-field domain. (d) An increase in the applied voltage will result in more band bending; in some configurations additional excited subbands can play a role in the transport.

explained qualitatively on the basis of the continuity equation. The current density along the superlattice, at steady state, $j = env_d$, where n is the carrier density and v_d is the field-dependent drift velocity, is a constant, for a given applied voltage. For a static configuration in the presence of large carrier densities, the electric field F(z) can, in principle, be calculated solving Poisson's equation, and this is generally done by some self-consistent algorithm. One way to interpret Poisson's equation is the statement that the carrier density is proportional to the gradient of the electric field. This, together with the continuity equation for the current density, leads to the condition[51]

$$\frac{dF}{dz} v_d(F) = \text{const} \tag{6}$$

The drift velocity of a carrier is much larger for resonant configurations, i.e., when the relevant subbands in adjacent wells are aligned, than in non-resonant patterns. Accordingly, the change in the electric field has to be large in the non-resonant portions of the sample. This, in other worlds, implies the formation of high and low electric field domains along the multi quantum well region. A simplified, yet intuitive model for the formation of high-field domains can be solved for uniform superlattices.[52] A uniform superlattice has a number of identical quantum wells separated by barriers. In the simplest configuration, there are only two subbands, with energies E_1 and E_2. The drift velocity is written as:

$$v_d(F) = v_1 \delta(F - F_1) + v_2 \delta(F - F_2) \tag{7}$$

with the fields $F_1 = 0$ and $F_2 = (E_2 - E_1)/ed$, where d is the superlattice period. (The finite width of the resonance is neglected here; the condition $F_1 = 0$ also allows for a nonphysical finite drift velocity under a vanishing electric field). Substituting for $v_d(F)$ into Eq. (6) and integrating, one can obtain z(F). Inverting z(F), integrating again, and imposing boundary conditions based on the applied external voltage, one can find a solution for the field

$$F(z) = F_1 \Theta(z)\Theta(z_1 - z) + F_2 \Theta(z - z_1)\Theta(L - z) \tag{8}$$

where Θ is the step unit function and

$$z_1 = (1 - F_{appl}/F_2) L \tag{9}$$

z_1 is the domain boundary. As the applied field F_{appl} is increased from $F_1 = 0$ to F_2, this boundary moves from one electrode to the other, i.e., the high-field domain expands at the expense of the low field domain. It is also clear, from this solution, and by taking the derivative of the solution for the field, that a space charge builds up at the the domain boundary. A more intricated solution is obtained in the case of a uniform superlattice designed to include three subbands. The solution, in this case, shows that, depending on the strength of the applied electric field and to the energy spacing of the subbands, one can have different solutions involving the cohexistance of two or three domains. The drift velocities are also important parameters in this regard. The solutions indicate that the domains are always spatially ordered, i.e., the domain with the lowest field strength is near one electrode, followed by the the domain with the intermediate field strength and the high-field domain resides near the opposite electrode. As pointed out, the tool mostly recognized and used to identify the formation and expansion of high field domains in a multi quantum well structure has been the observation of oscillation in the I-V characteristic of the device. Recently, Schneider, Grahn and Von Klitzing have performed a study of the pattern of electric field domains in a superlattice using photoluminescence.[52] They used the fact that, under different electric fields, the quantum well subbands experience Stark shifts of different magnitudes. By monitoring the different energy shifts of the photoluminescence peaks, they were able to infer the pattern of field domains in their sample. This tool is particularly fit to study undoped superlattices, the carrier population being generated by the creation of photoexcited electron-hole pairs.

7. Interpretation of the Experimental Results

We now turn back to the analysis of the experimental results presented above. We can interpret the oscillations observed in the I-V characteristics (Fig. 8) and the features observed in the photocurrent spectral measurements (Fig. 6) as evidence for a rich pattern of high and low electric field domains in the sample. When the device is biased with a positive polarity, a high-field domain is created in the region close to the cathode within the first stack of quantum wells; as the applied voltage is increased, the domain spreads to include more and more of the 25 wells in the stack, while a complementary low-field domain in the same stack shrinks progressively. At the same time the remaining two stacks of quantum wells experience only low-field domains all along their extent. Thus, photons of all appropriate energies are absorbed by the correspondingly matched, highly doped quantum wells; however only electrons excited within the high-field regions are swept efficiently towards the contacts and contribute to the photocurrent. The carriers excited in the region of low field have a high probability of being recaptured by their own well, contributing only negligibly to the current. Once the applied voltage reaches a value large enough to extend a high-field domain to as many of the 25 wells in the first stack as allowed, the corresponding responsivity reaches its maximum value. When the applied voltage is increased even further, a high-field domain is formed in the region of the following 25 wells of the second stack, while at the same time quenching of the high-field domain in the first stack occurs. Response from a third peak, corresponding to the third stack, was not achieved even at the highest values of the applied forward voltage, indicating that the spatial region of the last stack is always under low field domain.

When the bias is applied with the opposite negative polarity, the situation is not symmetric. The high-field domain starts expanding, as before, in the region of the first stack of quantum wells; however, for this polarity, it means that the formation of a HFD nucleates from the anode. In addition, beyond a certain magnitude of the applied voltage (~-8.5V), different high-field domains coexist in two different stacks, and ultimately, for even more negative applied bias, a third high-field domain forms also in the third stack of quantum wells.

The pattern of high-field domain formation in the multi-stack structure displays new interesting features. First of all, for the positive bias case, one can witness the formation and expansion of a high-field domain starting from the cathode side, contrary to what expected from the screening effect of the space-charge build-up. This means that some stronger or more important effect govern the charge arrangement and the subsequent domain pattern. Second, for the same positive bias case and after the switching-peak voltage of +6.5V, we observe that the set-up (in strength) of electric field domains within the sample is not spatially ordered. This is contrary to the general belief that all possible configurations of electric field domains along a superlattice tend to be spatially ordered, in the sense that, from one electrode to the other, along the sample, the strengths of the various domains follow a monotonic ordering. This is not the case in the multi-stack structure, since, for voltages larger than 6.5V, a low-field domain in stack #1 is followed by a high-field domain in stack #2 and by another low-field domain in stack #3. Another interesting and novel feature is the quenching of high-field domains with increasing external bias. The multi-stack structure presents a pattern of electric field domains, with features and characteristics much more intricated than those observed or thought possible beforehand. One can witness not only the formation and the expansion of electric field domains in the superlattice, as a function of increasing applied voltage, but also the subsequent readjustment of these domains to a totally new configuration.

Based on this kind of analysis, one can hope to better understand and maybe control the patterns of electric field domains along a superlattice.

8. Conclusions

We have presented the design and characterization of a new multi-stack intersubband detector. From the device aspect, this design addresses a number of issues recognized as problematic or limiting in intersubband detection, such as a narrow bandwidth and the lack of tunability. This detector can be used in a number of modes, with extended versatility, to address and solve these issues. Options include a voltage-controlled spectral range of detection and a multicolor, switching peak operation. From the performance aspect, very large responsivities and detectivities have been measured, comparable to the state-of-the-art results in intersubband detectors. The complex switching behavior of the sample has been attributed to the formation, expansion amd readjustment of electric field domains along the sample. New and more involved phenomena have been observed in the interplay of these domains. We have shown that intersubband photocurrent spectroscopy is a natural tool to study the structure and the dynamics of such domains in carrier-rich superlattices. A simple model is good enough to predict the formation and expansion of electric field domains in simple, uniform superlattices, but not for the readjustment dynamics in more complex structures similar to our device. Finally it is important to note, that a fully-controlled pattern of switching high and low electric field domains along a sample, might lead to interesting applications within a larger class of optoelectronic or semiconductor devices, including, but not exclusively, infrared intersubband detectors. Also, in principle, these ideas and schemes might be employed in a wide range of material systems, beyond the AlGaAs system.

References

1. L.L. Chang, L. Esaki and G.A. Sai-Halash, IBM Techn. Discl. Bull. 20, 2019 (1977).
2. J.S. Smith, L.C. Chiu, S. Margalit, A. Yariv and A.Y. Cho, J. Vac. Sci. Technol. B1, 376 (1983).
3. M.K. Gurnick and T.A. DeTemple, IEEE J. Quantum Electron., QE-19, 791 (1983).
4. L.C. West and S.J. Englash, Appl. Phys. Lett. 46, 1156 (1985).
5. B.F. Levine, C.G. Bethea, G. Hasnain, J. Walker and R.J. Malik, Appl. Phys. Lett. 53, 296 (1988).
6. M.M. Fejer, S.J.B. Yoo, R.L. Byer, A. Harwit and J.S. Harris, Jr., Phys. Rev. Lett. 62, 1041 (1989).
7. E. Rosencher, P. Bois, J. Nagle and S. Delaitre, Electron. Lett., 25, 1063 (1989).
8. E. Rosencher, P. Bois, J. Nagle, E. Costard and S. Delaitre, Appl. Phys. Lett. 55, 1597 (1989).
9. A. Sa'ar, I. Gravé, N. Kuze and A. Yariv, Proceedings of the OSA Topical Meeting on Nonlinear Optics (NLO90), Kauai, Hawaii, July 1990.
10. A. Helm, P. England, E. Colas, F. DeRosa and S.J. Allen, Jr., Phys. Rev. Lett. 63, 74 (1989).
11. Y.J. Mii, R.P.G. Karunasuri, K.L. Wang, M. Chen and P.F. Yuh, Appl. Phys. Lett. 56, 1986 (1990).

12. P.F. Yuh and K.L. Wang, Appl. Phys. Lett. 51, 1404 (1987).
13. A. Kastalsky, V.J. Goldman and J.H. Abeles, Appl. Phys. Lett.59, 2 636 (1991).
14. Q. Hu and S. Feng, Appl. Phys. Lett. 59, 2923 (1991).
15. G. Bastard, Phys Rev B 24, 5693 (1981); Phys Rev B 25, 7594 (1982).
16. F. Stern, Phys. Rev. Lett. 33, 960 (1974).
17. D.D. Coon and P.G. Karunasuri, Appl. Phys. Lett. 45, 649 (1984); K.W. Goossen and S.A. Lyon, J. Appl. Phys. 63, 5149 (1988).
18. B.F. Levine, G. Hasnain, C.G. Bethea and N. Chand, Appl. Phys. Lett. 54, 2704 (1989).
19. M.A. Kinch and A. Yariv, Appl. Phys. Lett. 55, 2093 (1989)
20. I. Gravé and A. Yariv, "Fundamental Limits in Quantum Well Intersubband Detection," in "Intersubband Transitions in Quantum Wells", Rosencher, Vinter and Levine Eds. (Plenum, New York 1992).
21. A. Zussman, B.F. Levine, J.M. Kuo and J. de Jong, J. Appl. Phys. 70, 5101 (1991).
22. H. Schneider, F. Fuchs, B. Dischler, J.D. Ralston and P. Koidl, Appl. Phys. Lett. 58, 2234 (1991).
23. S.D. Gunapala, B.F. Levine, R.A. Logan, T. Tanbun-Ek, and D.A. Humphrey, Appl. Phys. Lett. 57, 1802 (1990).
24. S.D. Gunapala, B.F. Levine, D. Ritter, R. Hamm and M.B. Panish, Appl. Phys. Lett. 58, 2024 (1991).
25. C. Sirtori, F. Capasso, D. L. Sivco, A.L. Hutchinson and A.Y. Cho, Appl. Phys. Lett. 60, 151 (1992).
26. I. Gravé, M. Segev and A. Yariv, Appl. Phys. Lett. 60, 2717 (1992).
27. A. Sa'ar, N. Kuze, J. Feng, I. Gravé and A. Yariv, Appl. Phys. Lett. 61, 1263 (1992).
28. M. Segev, I. Gravé and A. Yariv, Appl. Phys. Lett. 61, 2403 (1992).
29. D. Walrod. S.Y. Auyang, P.A. Wolff and M. Sugimoto, Appl. Phys. Lett. 59, 2932 (1991).
30. C. Sirtori, F. Capasso, D.L. Sivco and A.Y. Cho, Phys. Rev. Lett. 68, 1010 (1992).
31. B.F. Levine, G. Hasnain, C.G. Bethea and N. Chand, Appl. Phys. Lett. 54, 2704 (1989).
32. K.K. Choi, B.F. Levine, C.G. Bethea, J. Walker and R.J. Malik, Phys. Rev. B 39, 8029 (1989).
33. S.R. Parihar, S.A. Lyon, M. Santos and M. Shayegan, Appl. Phys. Lett. 55, 2417 (1989).
34. B.F. Levine, C.G. Bethea, V.O. Shen and R.J. Malik, Appl. Phys. Lett. 57, 383 (1990).
35. I. Gravé, A. Shakouri, N. Kuze and A. Yariv, Appl. Phys. Lett.60, 2362 (1992).
36. Z.Y. Xu, V.G. Kreismanis and C.L. Tang, Appl. Phys. Lett. 43, 415 (1983).
37. J.W. Choe, Byungsung O, K.M.S.V. Bandara and D.D. Coon, Appl. Phys. Lett. 56, 1679 (1990).
38. G. Hasnain, B.F. Levine, C.G. Bethea, R.R. Abbott and S.J. Hseih, J. Appl. Phys. 67, 4361 (1990).
39. A.Harwit and J.S. Harris, Jr., Appl. Phys. Lett. 50, 685 (1987).
40. J.B. Gunn, Solid State Commun. 1, 88 (1963); IBM J. Res. Dev. 8, 141 (1964).
41. B.K. Ridley and T.B. Watkins, Proc. Phys. Soc. Lond. 78, 293 (1961); B.K. Ridley, J.Appl. Phys. 48, 754 (1977); C. Hilsum, Proc. IRE 50, 185 (1962); Solid State Electron. 21, 5 (1978).
42. S.M. Sze, "Physics of Semiconductor Devices", 2nd edition, Wiley & Sons Publishers, chapter 11.
43. L. Esaki and L.L. Chang, Phys. Rev. Lett. 33, 495 (1974).
44. K.K. Choi, B.F. Levine, R.J. Malik, J. Walker and C.G. Bethea, Phys. Rev. B 35, 4172 (1987); K.K.Choi, B.F.Levine, C.G.Bethea, J.Walker and R.J.Malik, Appl. Phys. Lett. 50, 1814 (1987).
45. Y. Kawamura, K. Wakita, H. Asahi and K. Kurumada, Jpn. J. Appl. Phys. .25, L928 (1986); Y. Kamamura, K. Wakita and K. Oe, ibid. 26, L1603 (1987).
46. H.S. Newman and S.W. Kirchoefer, J. Appl. Phys. 62, 706 (1987).
47. M. Helm, P. England, E. Colas, F. DeRosa and S.J. Allen, Jr., Phys. Rev. Lett. 63, 74 (1989).
48. T.H.H. Vuong, D.C. Tsui and W.T. Tsang, Appl. Phys. Lett. 52, 981 (1988).
49. R.E. Cavicchi, D.V. Lang, D. Gershoni, A.M. Sergent, H. Temkin and M.B. Panish, Phys. Rev. B 38, 13474 (1988).
50. R.F. Kazarinov and R.A. Suris, Sov. Phys. Semicond. 6, 120 (1972).
51. H.T. Grahn, H. Schneider and K. von Klitzing, Appl. Phys. Lett. 54, 1757 (1989).
52. H.T. Grahn, H. Schneider and K. von Klitzing, Phys. Rev. B 41, 2890 (1990).

Chapter 17

SPATIAL OPTICAL SOLITONS - EXPERIMENTS

Y. SILBERBERG

1. Introduction

In the very early days of nonlinear optics it was proposed that an intensity dependent refractive index can lead to self-trapping of optical beams. The seminal work of Chiao, Garmire and Townes[1] discussed the concept of a light beam forming its own dielectric waveguide, thereby compensating for diffraction. It was realized soon after that stable self-trapping in a Kerr medium occurs only in two dimensional propagation, where the beam diffracts along one transverse direction. In a three dimensional medium, where light diffracts along two transverse directions, self-trapping is not stable and leads to catastrophic self-focusing[2]. Self-focusing in three dimensions attracted many investigators in the following years[3]. The two-dimensional problem was revisited later by Zakharov and Shabat[4], who made the first connection between self-trapping and soliton theory. They also showed that there is a complete analogy between spatial solitons, where the nonlinear index profile balances diffraction, and temporal solitons, where nonlinear phase modulation balances dispersion.

Temporal solitons have received much attention since Hasegawa and Tappert proposed that they could be obtained in the anomalous dispersion regime in optical fibers and applied to high-speed communications[5]. In the past decade we have witnessed impressive progress in the area of temporal solitons in fibers. Spatial optical solitons, on the other hand, have been relatively little studied. Only in the recent few years there was a growing activity in experimental studies of bright and dark spatial solitons, which are reviewed in this Chapter.

While from the mathematical point of view temporal and spatial solitons are identical, being governed by the same nonlinear Schrödinger equation, there are several differences that should encourage further study of spatial solitons. Optics has a developed understanding of spatial phenomena; in particular, spatial solitons (or self-trapped beams) can be viewed as the modes of the waveguides they are creating. Linear waveguide theory can then enhance our understanding of various soliton phenomena. In addition, many interesting questions arise when we try to extend the concept of the spatial solitons, a two dimensional entity, into a three dimensional spatial problem. And finally, there is the possibility of an interplay between the spatial and temporal problem, and perhaps a coexistence of a spatio-temporal soliton, sometimes called a "light-bullet"[6].

Y. Silberberg - Bellcore, Red Bank, New Jersey 07701-7040, USA

In this Chapter, studies of spatial optical solitons are reviewed. Although our main goal is to review the experimental work in this area, we outline in Section 2 the theoretical framework for this discussion. Section 3 reviews some of the experimental demonstrations of bright soliton propagation, and Section 4 deals with experiments with dark solitons. Section 5 discuss the temporal aspects of spatial solitons, and in particular the possibility of temporal-spatial solitons. Finally, this Chapter concludes with Section 6 where we summarize and discuss possible applications of spatial optical solitons.

2. Theoretical Background

2.1. Bright Spatial Solitons

A monochromatic scalar wave of frequency ω propagating in a homogeneous nonlinear medium with intensity-dependent refractive index $n(I)$ is described by:

$$\nabla^2 E + \frac{\omega^2}{c^2} n^2(|E|^2) E = 0. \tag{1}$$

The nonlinear index of refraction can often be described by the lowest order (Kerr-like) response $n = n_0 + n_2 |E|^2$.

Soliton solutions are obtained only when propagation is limited to two spatial dimensions, assumed here to be z and x. When the propagation is primarily along the z direction, the paraxial approximation $\partial^2 A/\partial z^2 \ll k \partial A/\partial z$, where $E = A(x,z)\exp(ikz)$ and $k = n_0 \omega/c$, can be used. Eq. (1) then takes the form of a nonlinear Schrödinger (NLS) equation:

$$2ik \frac{\partial A}{\partial z} + \frac{\partial^2 A}{\partial x^2} + 2k^2 \frac{n_2}{n_0} |A|^2 A = 0. \tag{2}$$

The second term on the left describes the diffraction in the x direction, while the third term describes the self-focusing by the nonlinear index profile. We note that in the spatial version of the NLS a diffraction term replaces the dispersion term of the temporal equation. Since the sign of diffraction is not controllable, the regimes of bright and dark solitons are determined by the sign of the nonlinear term n_2.

It will be useful to rewrite the well known fundamental soliton solutions in a non-normalized way, spelling out the spatial parameter dependence. We first consider the case of self-focusing nonlinear index, i.e. $n_2 > 0$. The fundamental bright soliton describes a beam that propagates along the z axis without changing its spatial shape, and it is given by:

$$A(z,x) = A_0 \cosh^{-1}(x/a) \exp(iz/2ka^2) \tag{3}$$

with $A_0 = (n_0/k^2 a^2 n_2)^{\frac{1}{2}}$. Here a is a measure of the beam width. This solution of a self-trapped beam in two dimensions was already found by Chiao et al.[1]. The intensity and phase profiles of the fundamental soliton are shown on the left of Fig. 1. A more general fundamental soliton solution of Eq. (2) describes a self-trapped beam which propagates at a small angle θ to the z axis:

Spatial Optical Solitons

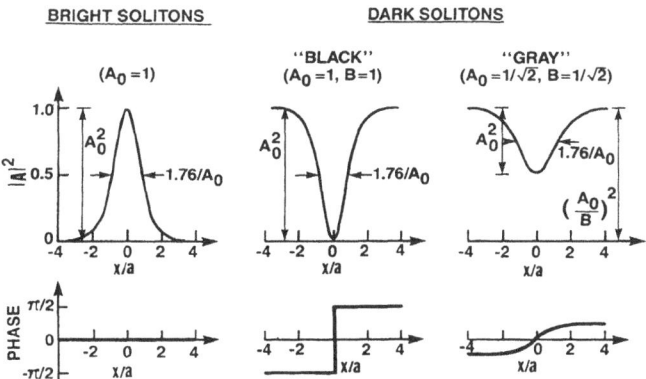

Fig. 1. Intensity and phase profiles for bright and dark solitons.

$$A(z,x) = A_0 \cosh^{-1}(\frac{x-\theta z}{a}) \exp(iz\frac{1-k^2a^2\theta^2}{2ka^2}). \tag{4}$$

The fundamental soliton is the only finite-extent beam that propagates without changing its shape. There are, however, additional solutions to Eq. (2) that describe unbounded fields with profiles that do not change while propagating. These field patterns are periodic in x, and they are equivalent to fully developed modulation instabilities in the time domain.

Higher order solitons describe optical beams that refocus periodically along the propagation direction, with a periodicity of the soliton period

$$z_0 = \frac{\pi k a^2}{2}. \tag{5}$$

Note that in the spatial case the soliton period resembles the diffraction length for a beam of width a.

2.2. Dark Spatial Solitons

When the nonlinear refractive index is of the self-defocusing type, i.e. $n_2<0$, self-trapped beams are not possible. Diffraction and self-defocusing will tend to broaden any finite size beam. However, such a medium can support *dark solitons* in the form of a depression in a field that extends to $x = \pm\infty$. It is easy to see that in such a case the nonlinear induced index variation acts again as a focusing lens that could counteract diffraction. The fundamental dark solitons[5], sometimes referred to as *black solitons*, are the simplest dark solitons:

$$A(z,x) = A_0 \tanh(x/a) \exp(-i z/ka^2). \tag{6}$$

A black soliton is characterized by an antisymmetric field distribution with a zero field at its center.

The black soliton is a special case of a broader family of dark solitons given by:

$$A(z,x) = A_0 / B[1-B^2 \cosh^{-2}(x/a)]^{\frac{1}{2}} \exp[-i\phi(x,z)], \tag{7}$$

with

$$\phi(x,z) = \tan^{-1}[\frac{B}{\sqrt{1-B^2}} \tanh(x/a)] + \frac{\sqrt{1-B^2}}{B} \frac{x}{a} - \frac{3-B^2}{B^2} \frac{z}{2ka^2}, \tag{8}$$

with $|B| \leq 1$. The black soliton is a dark soliton with $B = 1$. Dark solitons with $B < 1$, sometimes called *gray solitons*, are characterized by a depression that does not vanish. The intensity and phase profiles of black and gray solitons are shown in Fig. 1. Unlike black solitons, gray solitons have a nonuniform phase distribution along x. This means that the gray soliton feature (the depression in the field) propagates at an angle to the background field.

2.3. Spatial Solitons as Waveguide Modes

A spatial soliton can be thought of as an eigenmode of its self-generated waveguide. This picture translates the nonlinear soliton problem into an equivalent linear waveguide problem, and thereby it can help to explain in a physically intuitive way various features of spatial solitons and their interactions.

The equivalent linear problem is a particularly valuable approach when investigating problems that deviate from the ideal NLS soliton case, for example, due to different nonlinearities, or in three dimensions, as detailed in Section 2.4 below.

The equivalent linear waveguide of the bright and dark NLS solitons leads to a spatial index distribution which has a 1/cosh² profile:

$$n(x) = n_1 + \Delta n \cosh^{-2}(x/a). \tag{9}$$

The various spatial solitons must be eigenmodes of this graded index profile. The index distribution of Eq. (9) is one of a few profiles that have exact analytic solutions for their mode distribution[7]. The fundamental TE mode of this waveguide has the following amplitude distribution:

$$A(x) = A \cosh^{-s}(x/a), \tag{10}$$

where the parameter s is defined as

$$s = \frac{1}{2}(\sqrt{1+V^2} - 1) \tag{11}$$

and $V = 2ka\sqrt{2\Delta n/n_1}$. The waveguide is monomode when $s < 1$.

By comparing Eqs. (3) and (10) it is possible to identify the fundamental bright soliton with the fundamental eigenmode of the waveguide it is forming. The soliton waveguide always has $s = 1$, that is the waveguide is just on the borderline between a single-mode and a double-mode guide. The effective index of the soliton is exactly the average of the background index and the peak index value in the center of the beam.

The dark soliton solutions of Eqs. (6) and (7) are obviously radiation modes of the same type of waveguide, as their field extends to infinity. Note, however, that the radiation field of the black soliton propagates in the same direction as the waveguide, i.e., the background field has no wavevector component in the x direction. The black soliton is, therefore, the slowest radiation mode. Since the black soliton also leads to a waveguide with $s = 1$, we can conclude that the black soliton is just the lowest order asymmetric mode precisely at the cutoff point, where it becomes extended to infinity. Its effective index is then equal to the background index.

The gray solitons are true radiation modes of the nonlinear waveguide: they have an extended radiation field which is propagating at an angle to the waveguide axis. This angle increases as B decreases, as is evident from the second term on the right-hand side of Eq. (8).

This equivalent linear picture can explain in an intuitive way the coexistence of soliton pairs[8,9]. For example, consider a strong bright soliton in a focusing nonlinear medium; the waveguide it generates can then support a weak second order mode, which as explained above is a black soliton. Similarly, a strong black soliton in a defocusing medium can guide a copropagating weak bright beam.

2.4. Trapped Beams in Three Dimensions

The catastrophic self-focusing in three dimensions can be avoided if some saturation mechanism limits the increase of the refractive index. In that case, a stable self-trapped beam can exist. This was realized theoretically quite early[2], and has even been demonstrated experimentally in atomic sodium gas[10].

The realization that a spatial soliton can be translated to a linear waveguide problem has led a number of investigators to re-examine the question of self-trapping in three dimensions, looking for self-generated fibers. For example, Snyder et al.[11] consider an ideal saturating nonlinearity, i.e., a nonlinear response that is characterized by a step function. Any cylindrically symmetric beam induces a cylindrically symmetric step-index profile. Using the well known modal solutions for optical fibers, the existence of a stable self-trapped beam is proven. A similar result has been derived for more general saturable nonlinear media[12]. Such self-trapped beams are stationary solutions of the three dimensional NLS equation. They have not proven to be solitons and there are still some open questions regarding their stability properties. However, they do resemble the two dimensional solitons in many aspects. For example, two such beams can attract each other to form a double-helix orbit[13].

Recently, Chen and Snyder have suggested that even in an ideal Kerr medium stable self-trapped solutions can be found when one considers the vectorial nonlinear wave equation[14]. These are TM solutions that, at least for some parameter range, seem stable against small perturbations.

3. Experiments with Bright Solitons

Since true soliton solutions of Eq. (1) are possible only in a two dimensional space, it could be questioned whether spatial solitons are not merely a theorist's dream. Experimentalists have used one of two approaches to simulate two dimensional propagation and to demonstrate spatial solitons.

3.1. Experiments in Waveguides

When a beam of light propagates as a single mode in a planar waveguide, as shown in Fig. 2, it is free to diffract only along one spatial dimension, x. The electric field distribution in

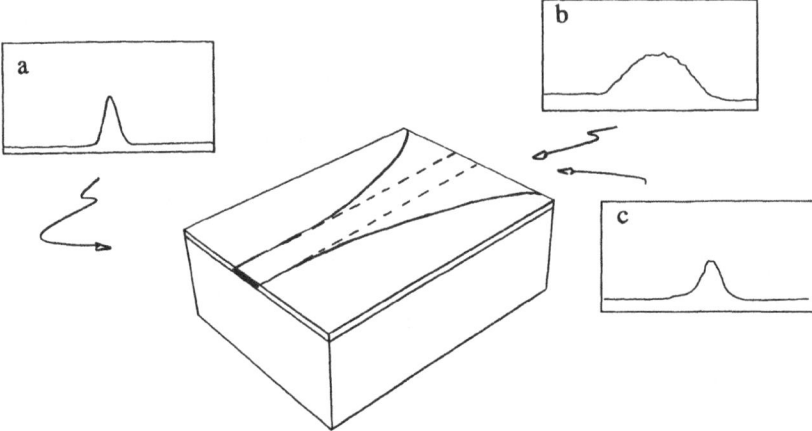

Fig. 2. Spatial soliton formation in a planar waveguide. A beam which diffracts at low power keeps a constant shape at high power. Insets show measurements of the input beam profile (a) and the output beam profile at low (b) and high (c) power. (After Ref. 18).

the direction y normal to the waveguide plane is determined by the modal field of the guiding structure, which remains constant as long as the nonlinear index changes are small compared with the index structure which defines the waveguide. The propagation of the field is determined then by the two-dimensional Eq. (2). This is, of course, the same argument that allows for a one-dimensional analysis of propagation in optical fibers. The transverse mode size w will not be affected by the nonlinearity if $w \ll a$. The total power required to launch a fundamental soliton of width a is

$$P_t = \frac{2n_0 w}{n_2 a k^2}. \tag{12}$$

As can be expected, when the nonlinear index difference required to support the spatial soliton approaches the magnitude of the index step defining the planar waveguide, the spatial soliton becomes more and more three dimensional in nature. This point was demonstrated recently by a numerical study[15].

Fundamental and high-order bright solitons were observed in experiments in a multimode CS_2 waveguide by Maneuf et al.[16,17]. Single mode propagation was obtained by careful excitation of the 10 µm wide waveguide, and square optical pulses at a wavelength of 532 nm were used to eliminate the uncertainty introduced by temporal averaging. The propagation length was chosen to be half the soliton period, where the high order solitons acquire distinct spatial features. Fig. 3 shows the observed beam shapes for three input powers that correspond to launching the three lowest order solitons, together with the theoretical shapes expected for ideal soliton beams[17]. The observed shapes clearly resemble the expected soliton shapes.

A series of experiments in single-mode glass waveguides was more recently reported by Aitchison et al.[18-20]. These experiments required high peak powers to obtain the required index change, and were therefore performed with femtosecond pulses in 5 mm long waveguides. These experiments demonstrated the formation of the fundamental bright soliton[18], and some of the observed features are shown in inserts in Fig. 2. High-order solitons could not be generated

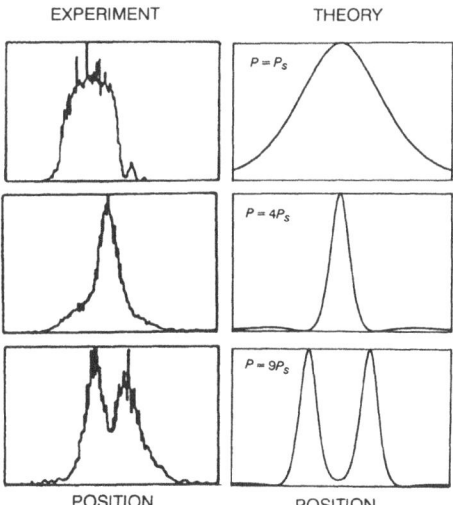

Fig. 3. Spatial beam shape at the output of a CS_2 waveguide for increased input power, corresponding to the three lowest order solitons. The propagation distance is about half the soliton period. (After Ref. 9).

because of two-photon absorption in these waveguides[19]. The effects of two-photon absorption on soliton propagation were analyzed theoretically[32]. It was found that in the presence of two-photon absorption higher order solitons break into individual fundamental solitons that move away from each other. The experimental observations agreed well with this prediction.

Attractive and repulsive interactions between two parallel soliton beams were also demonstrated in waveguides[20]. Solitons either attract or repel each other, depending on the relative phase between them, as is shown in a simulation in Fig. 4. As pointed out by a number of investigators[13,20], the interaction of solitons takes a simple interpretation in the spatial domain, as one can easily visualize the effect of one beam on the other. Since beams of light tend to bend towards higher index regions, one should only examine the index slope induced by

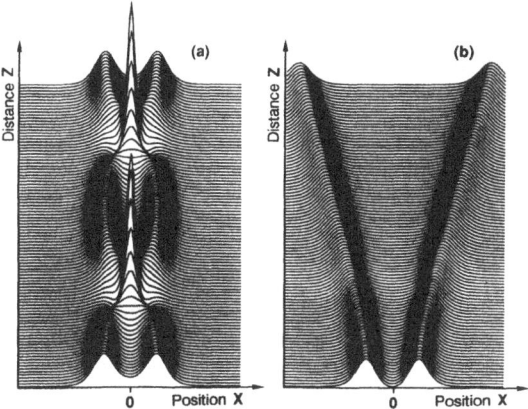

Fig. 4. Simulation of soliton interactions. Two fundamental solitons are launched parallel to each other along the z axis. In-phase solitons (a) attract each other, resulting in a bound pair. Out-of-phase solitons (b) repel each other.

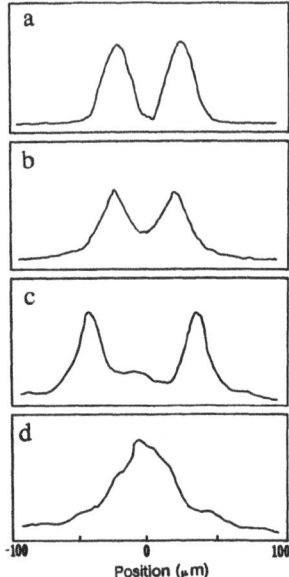

Fig. 5. Spatial soliton interactions - experimental results. (a). Light distribution at the input of the waveguide. (b) Output light distribution when the two beams do not overlap in time. (c). Interacting solitons with a π phase shift. (d). Interacting solitons with 0 phase shift. (After Ref. 20).

a second beam. Consider a soliton that without perturbation propagates along the z axis. A second in-phase soliton next to it perturbs the background index so that effectively the first soliton rides on an index gradient increasing towards the perturbing soliton. This gradient causes the soliton to bend towards the perturbing neighbor. When the two solitons are out of phase, they interfere to cancel the field at the center point between them. Now each soliton rides on an index gradient that increases away from the other soliton, causing them to move away from each other.

Fig. 5 shows experimental results demonstrating the interaction between two spatial solitons in a glass waveguide[20]. The two soliton beams are launched parallel to each other. When each beam propagates by itself, it propagates along the optical axis, and there is no change in the separation between the beams (2b). However, when the two beams are launched simultaneously, they repel (2c) or attract (2d) each other according to their relative phase.

3.2. *Bright Solitons in Bulk Samples*

Self-trapping of cw optical beams in a bulk medium has been demonstrated by Bjorkholm and Ashkin[10]. Their experiment showed the stabilizing effect of index saturation on self-focusing. In situations where saturation does not occur, the experiment must be designed to simulate two dimensional propagation in the bulk, three dimensional, nonlinear medium. To form the soliton feature, the experiment has to be designed so that the beam maintains a nearly constant shape in the y direction, normal to the propagation direction z and the diffraction direction x. In the case of bright solitons, one might try to use elliptical beams in order to assure that the diffraction length along the y axis is much longer then the propagation length in the medium. The soliton reshaping effect in the x direction could then be observed before substantial changes occur in the beam shape in the y direction. However, since the total power

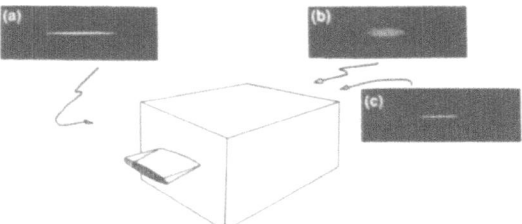

Fig. 6. Spatial soliton formation in bulk nonlinear medium. Two elliptical beams form a fringe pattern which stabilizes against filamentation. Insets show the input (a) and output beams at low (b) and high (c) powers. (after Ref. 24).

in the beam is then significantly larger than the self-focusing critical power, filamentation may occur. To prevent filamentation, the Limoge group demonstrated that instead of using a single elliptical beam, it is better to use two such beams interfering to form fringes across the long axis of the ellipse, as shown in Fig. 6. The fringe pattern divides the power into cells, the power in each cell being smaller than the critical power. This has been shown to greatly stabilize the beam against filamentation. This approach has been implemented successfully[21-23] to observe bright solitons and soliton interactions in CS_2 cells. Recently, Reitze et al.[24] used the same technique to stabilize against self-focusing and to obtain high power pulse compression of femtosecond pulses in bulk glasses.

4. Experiments with Dark Spatial Solitons

There are two difficulties that might arise in investigations of dark spatial solitons. First, dark solitons are waves with an extended background, while experiments must be done with finite size beams. In addition, there is the dimensionality problem, just as in the case of the bright solitons. In an experiment performed with a finite-extent beam, the beam diffracts; if the beam is large, the diffraction can be slow enough to observe the soliton features, although the soliton will change adiabatically as the background decays. This point has been tested in numerical simulations[25] and verified in experiments in the time domain[26].

The dimensionality issue turned out to be much less of a problem in dark soliton experiments compared with bright soliton experiments, because a defocusing nonlinearity does not lead to any catastrophic instability. In most experiments dark soliton-like features are observed as lines across the beam cross-section. In some experiments several such features were observed to coexist, and even cross each other, with minimal interference.

Dark spatial solitons have been investigated in a number of material systems with defocusing nonlinearities. The first experiments, by Swartzlander et al., were conducted in sodium vapor at wavelength just above 589 nm[27], where the absorption saturation is accompanied by a decrease in refractive index. A single wire or a wire mesh was placed at the input face of the medium, and the output field distributions were recorded photographically. The wires impose dark features in the beam. These features did not have the proper phase distribution of a dark soliton, but at the proper intensity they were observed to evolved into patterns that were composed of pairs of dark solitons.

Another study was conducted in a thermal defocusing nonlinear medium[28]. This is a liquid with a small amount of absorbing dye that enabled experiments with a low power cw laser. In this experiment it was verified that the solitons formed by a dark even initial field distribution

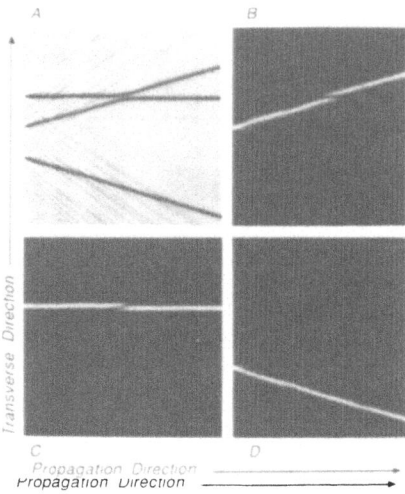

Fig. 7. Simulation of guiding light by light: An initial field distribution in a two-dimensional nonlinear medium generates three dark solitons in (a), one of which is black and two are gray. In (b), (c) and (d) low intensity (linear) waves are launched through the three waveguides defined by the solitons. Note how those waves cross the interaction region without scattering into the crossing waveguide. (After Ref. 36)

indeed evolve as predicted by the Zakharov and Shabat theory[29]. This theory predicts that a symmetric depression in an infinite field will indeed evolve into pairs of spatial solitons, and it derives the number of such pairs and the angle between them.

To observe a single black soliton, a phase jump has to be introduced into the input field distribution in order to generate the required antisymmetric field distribution. In a recent experiment[30], a single soliton was generated by introducing a phase shift by means of a glass slide that was placed to cover half of the input field. This experiment was conducted in ZnSe with picosecond pulses at 532 nm. The formation of a single black soliton at the peak of the temporal pulse was verified using a streak camera.

More recently, a series of experiments by Luther-Davis and Yang at the Australian National University[36] clearly demonstrated the waveguiding properties of dark solitons. In these experiments a properly shaped cw beam from an Argon laser generated dark solitons in a weakly absorbing medium through a defocusing thermal nonlinearity. Simultaneously, a weak red beam from a helium-neon laser was launched throuth the dark soliton waveguide. By controlling the phase profile of the green beam the angle of the gray soliton can be varied, thereby steering the red beam. Fig. 7 shows a simulation by the same authors of such a process, where three dark solitons are formed, propagating from left to right in (a). In (b), (c) and (d) light is launched through the three individual waveguides induced by these solitons. The figure demonstrates not only that these are good waveguides, but also that the guided light follows the individual solitons even when it inteacts with other solitons. This concept of guiding light-by-light could be an important utilization of spatial solitons.

5. Spatial-Temporal Solitons?

Most of the experiments with spatial solitons were performed with pulsed lasers. This usually complicates the interpretation of the experimental results, since the spatial shapes are changing within the temporal pulse. Some investigators used pulse shaping techniques to generate square pulses that eliminate this problem[17], and others used streak-cameras to resolve the evolving pattern[29]. These techniques can be applied as long as dispersive effects can be neglected, and no significant temporal reshaping occurs. This assumption might not be valid in experiments that are performed with ultrashort pulses. One should then expect significant temporal reshaping of the pulses. Particularly, if the dispersion is anomalous, we should consider the possibility of a nondiffracting, nondispersing temporal-spatial soliton, or 'light bullet'[6].

Consider the simultaneous occurrence of spatial and temporal self-focusing when a pulse propagates under the combined effect of self-focusing and dispersion in two dimensions. The nonlinear Schrodinger equation (2) should then be generalized to include the dispersion effects:

$$2ik(\frac{\partial}{\partial z}+\frac{1}{v_g}\frac{\partial}{\partial t})A+\frac{\partial^2 A}{\partial x^2}-k\frac{\partial^2 k}{\partial \omega^2}\frac{\partial^2 A}{\partial t^2}+2k^2\frac{n_2}{n_0}|A|^2 A = 0. \tag{13}$$

Using normalized units $\xi = kx$, $\zeta = kz$, $\tau = (-k\partial^2\omega/\partial k^2)(t-z/V_g)$, and $u = (n_2/n_0)^{\frac{1}{2}}A$, a dimensionless nonlinear Schrodinger equation for a pulse propagating along z can be derived:

$$i\frac{\partial u}{\partial \zeta}+\frac{1}{2}(\frac{\partial^2 u}{\partial \xi^2}+\frac{\partial^2 u}{\partial \tau^2})+|u|^2 u = 0, \tag{14}$$

where we assume that the dispersion is anomalous. Note the complete symmetry between the time (τ) and space (ξ) variables.

Since the temporal and the spatial parameters of a pulse undergoing self-focusing can be discussed on an equal footing, our good understanding of spatial self focusing[3] can immediately be applied to predict the evolution of pulse in a (anomalous) dispersive medium where it can diffract only along one spatial dimension, such as in a planar waveguide. We can predict then that such a pulse, if constructed properly so that it is symmetric in time and space, will undergo a symmetric collapse in both dimensions. Not less interesting is the extension to three dimensions; even faster collapse is predicted for a pulse which propagates in a bulk nonlinear material and undergoes simultaneous spatial and temporal self-focusing. For an observer traveling with the pulse, the pulse will appear to shrink symmetrically along the three spatial coordinates. This process could obviously be applied to pulse compression, in particular at high intensities, where the standard fiber-based compression technique fails. Consider, for example, nonlinear propagation at a wavelength of 1.55 µm in silica glass. From the anomalous dispersion of 16 ps/nm/km we can calculate that a space-time symmetric 100 fsec pulse requires a beam which is 350 µm wide. To collapse, such a pulse must carry an energy larger than 3 µJ.

The dynamics of the collapse process are not yet clear, although some insight can be obtain by analytical investigations of Eq. (18)[6,34]. Recently, Rothenberg has shown[35] that the paraxial approximation which is assumed in the derivation of the NLS equation may not be valid as the pulse approaches collapse. These and other approximations are currently investigated through numerical studies of the problem.

Finally, the question of the possiblity of the existence of stable temporal-spatial solitons, or light bullets, is still open. Such solitons are not possible in an ideal Kerr medium. They should, however, be possible in materials with different nonlinear responses, for example in a saturable nonlinear index medium. Such light bullets will be magical indeed: these are nondiffracting, nondispersing bunches of photons, held together only by the nonlinear interaction.

6. Conclusions

Spatial solitons are completely analogous to the more familiar temporal solitons; they are described by the same equation and they exhibit the same phenomena. However, the propagation and properties of spatial solitons can be explained in terms of diffraction, waveguides and modes - all concepts which appeal directly to physical intuition. For example, we have seen how the interaction of spatial solitons can have a simple intuitive meaning. For investigators with a good background in guided wave optics, the spatial soliton is a natural concept to grasp and a good introduction to the wide variety of soliton effects. The waveguide analogy has inspired the concept of guiding light-by-light. This is a novel and elegant way of manipulating light where waveguide structures are formed in an all-optical way.

Finally, it should be stressed that studies of spatial soliton phenomena are currently pursued by a number of groups, and new results are reported continuously. The experiments described in this Chapter should be viewed, then, as the beginning of a story that has not yet been completely told.

References

1. R.Y. Chiao, E. Garmire and C. H. Townes, Phys. Rev. Lett. 13, 479 (1964).
2. P.L. Kelley, Phys. Rev. Lett. 15, 1005 (1965).
3. For reviews of self-focusing, see Y. R. Shen, Prog. Quant. Electron. 4, 1 (1975); J. H. Marburger, Prog. Quant. Electron. 4, 35 (1975).
4. V.E. Zakharov and A. B. Shabat, Sov. Phys. JETP 34, 62 (1972).
5. A.Hasegawa and F. Tappert, Appl. Phys. Lett. 23, 142 and 171 (1973).
6. Y. Silberberg, Opt. Lett. 15, 1282 (1990).
7. A.W. Snyder and J.D. Love, "Optical Waveguide Theory" Chapman & Hall, London, (1983).
8. S. Trillo, S. Wabnitz, E.M. Wright and G.I. Stegeman, Opt. Lett. 13, 871 (1988).
9. V.V. Afanasev, Yu. S. Kivshar, V.V. Konotop and V.N. Serkin, Opt. Lett. 14, 805 (1989).
10. J.E. Bjorkholm and A. Ashkin, Phys. Rev. Lett. 32, 129 (1974).
11. A.W. Snyder, D.J. Mitchell, L. Poladian and F. Ladouceur, Opt. Lett. 16, 21 (1991).
12. Y. Chen, Opt. Lett. 16, 4 (1991).
13. Y. Chen and A.W. Snyder, Electron. Lett. 27, 565 (1991); Y. Chen, IEEE J. Quant. Electron. 27, 1236 (1991).
14. L. Poladian, A.W. Snyder and D.J. Mitchell, Opt. Commun. 85, 59 (1991).
15. Q.Y. Li, C. Pask and R.A. Sammut, Opt. Lett. 16, 1083 (1991).
16. S. Maneuf, R. Desailly and C. Froehly, Opt. Comm. 65, 193, (1988).
17. S. Maneuf and F. Reynaud, Opt. Comm. 66, 325, (1988).
18. J. S. Aitchison, A. M. Weiner, Y. Silberberg, M. K. Oliver, J. L. Jackel, D. E. Laeird, E. M. Vogel and P. W. E. Smith. Opt. Lett. 15, 471 (1990).
19. J. S. Aitchison, Y. Silberberg, A. M. Weiner, D. E. Laeird, M. K. Oliver, J. L. Jackel, E. M. Vogel and P. W. E. Smith. JOSA B, 8, 1290 (1991).
20. J. S. Aitchison, A. M. Weiner, Y. Silberberg, D. E. Laeird, M. K. Oliver, J. L. Jackel and P. W. E. Smith. Opt. Lett 16, 15 (1991).
21. A. Barthelemy, S. Maneuf and C. Froehly, Opt. Comm. 55, 201, (1985).
22. S. Maneuf, A. Barthelemy and C. Froehly, J. Optics (Paris) 17, 139 (1986).
23. F. Reynaud and A. Barthelemy, Europhys. Lett., 12, 401 (1990).
24. D.H. Reitze, A.M. Weiner and D.E. Leaird, Opt. Lett. 16, 1409 (1991).

25. W. J. Tomlinson, R. J. Hawkins, A. M. Weiner, J. P. Heritage and R. N. Thurston, J. Opt. Soc. Am. B 6, 329 (1989).
26. A. M. Weiner, J. P. Heritage, R. J. Hawkins, R. N. Thurston, E. M. Kirchner, D. E. Laeird and W. J. Tomlinson, Phys. Rev. Lett. 61, 2445 (1990).
27. G. A. Swartzlander, D. R. Andersen, J. J. Regan, H. Yin and A. E. Kaplan, Phys. Rev. Lett. 66, 1583 (1991).
28. D. R. Andersen, D. E. Hooton, G. A. Swartzlander and A. E. Kaplan, Opt. Lett. 15, 783 (1990).
29. V. E. Zakharov, and A. B. Shabat, Sov. Phys. JETP 37, 823 (1973).
30. G. R. Allan, S. R. Skinner, D. R. Andersen and A. L. Smirl, Opt. Lett. 16, 156 (1991).
31. V. Mizrahi, K. W. DeLong, G. I. Stegeman, M. A. Saifi and M. J. Andrejco, Opt. Lett. 14, 1140 (1989).
32. Y. Silberberg, Opt. Lett., 15, 1005 (1990).
33. Y. Chen, Electron. Lett. 27, 1985 (1991).
34. M. Desaix, D. Anderson and M. Lisak, J. Opt. Soc. Am. B, 8, 2082 (1991).
35. J. E. Rothenberg, Opt. Lett. 17, 134 (1992).
36. B. Luther-Davies, X. Yang, Opt. Lett. 17, 496 (1992); Opt. Lett. , 17, 1755 (1992).

Chapter 18

OPTICAL LOSSES CHARACTERIZATION OF CHANNEL WAVEGUIDES THROUGH A PHOTODEFLECTION METHOD

R. LI VOTI, M. BERTOLOTTI, L. FABBRI, G. LIAKHOU, A.MATERA, C. SIBILIA, and M.VALENTINO

1. Introduction

In order to measure the propagation losses in integrated optical waveguides several methods have been developed in the past, such as the sliding prism method[1], the calorimetric and pyroelectric methods[2,3], the method based on the measurement of out-of-plane scattered light[4] and the method using the polished endfaces of the waveguide as the reflecting surfaces of the optical resonator[5-7].

Recently a new method based on the photodeflection effect has been proposed[8]. Compared with the others, it is non contacting and applicable to a wide range of channel waveguides. In general, the technique analyzes the deflection of a test laser beam (probe) from its originary trajectory traveling in a medium with a thermally induced refractive index gradient[9]. In this particular case (see Fig.1) the refractive index gradient is due to the thermal gradient produced when the guided light, which is an endfire coupled time modulated pump laser beam, is absorbed during the propagation. The deflection angle of a probe ray is then given by the well known photodeflection formula

$$\vec{\Phi} = \int_{path} \frac{\nabla_t n}{n} ds = \int_{path} \frac{1}{n} \frac{\partial n}{\partial T} \nabla_t T ds \qquad (1)$$

with T, n, s, $\frac{\partial n}{\partial T}$, and ∇_t denoting the temperature rise, the refractive index, the probe path coordinate, the optothermal parameter and the gradient transverse to the ray path, respectively.

Depending on the probe beam path, two different transverse configurations can be distinguished and used (see Fig.1). The probe beam travels, in both paths, perpendicular to the channel direction, impinging finally on a position sensor connected to a lock-in amplifier.

M. Bertolotti, L. Fabbri, R. Li Voti, A. Matera, C. Sibilia and M. Valentino - Dipartimento di Energetica, Università di Roma " La Sapienza" and G.N.E.Q.P. of CNR, Via Scarpa 16, 00161 Roma, Italy.
G. Liakhou - Technical University of Moldova, 277012 Kishinev, Moldova

In the configuration in air, the probe skims through the layer of air close to the sample. Sometimes, if the sample has a low surface roughness, a stronger signal can be obtained experimentally by bouncing the probe laser beam near the channel zone with a small tilt angle and measuring the deflection angle of the reflected beam.

The configuration in air is usually used when a contactless detection system is needed (as in the case of photorefractive or probe absorbing waveguides). The deflection angle $\vec{\Phi}$ has two components, as usual: in the general case, the component directed along the z axis Φ_z is larger than the component along the channel direction Φ_y. In the case of a low propagation loss channel waveguide, because of the low value of the air's optothermal parameter with respect to that of a solid sample, an in situ configuration is more suitable in order to measure a stronger signal. The probe beam is now focused on the channel waveguide, crossing the whole sample. In this configuration the deflection angle has its only component along the channel direction.

2. Theoretical Analysis

In order to calculate propagation losses in a channel waveguide, the main idea of our approach is to measure, in any configuration, the amplitude of the deflection angle as a function of the distance from the input edge (y). Due to the absorption processes, the thermal power density in the channel is given by

$$w = \alpha \ I \ e^{-\alpha y} \tag{2}$$

where I and α denote the guided light intensity and the total loss coefficient respectively. Neglecting heat diffusion, a similar exponential behaviour with respect to y is also expected for the temperature rise and the deflection angle. On the other hand, taking into account the heat diffusion term, the temperature rise has a distribution given by the solution of the second order differential Fourier equation. In general, the 3-D solution given by Fourier analysis cannot be obtained in a closed form and is usually plotted by computer. Nevertheless neglecting the heat diffusion in the x-z plane, the 1-D solution is easily found, which corresponds to an average temperature rise in the x-z plane as a function of y given by

$$\begin{cases} \dfrac{T_{air}(y)}{T(0)} = \exp\left[\dfrac{1+i}{\ell_{air}}y\right] & y < 0 \\[2ex] \dfrac{T_{ch}(y)}{T(0)} = \dfrac{\alpha\ell_{ch}+e(1+i)}{\alpha\ell_{ch}-(1+i)}\exp\left[-\dfrac{1+i}{\ell_{ch}}y\right] - \dfrac{(1+i)(1+e)}{\alpha\ell_{ch}-(1+i)}\exp[-\alpha y] & y > 0 \end{cases} \tag{3}$$

where T(0) is the edge temperature, $\ell_{air/ch} = \sqrt{\dfrac{2D_{air/ch}}{\omega}}$, $D_{air/ch}$ is the thermal diffusivity of the air or of the channel, ω is the chopper angular frequency and e is the thermal effusivity ratio, defined as

$$e = \dfrac{e_{air}}{e_{ch}} = \sqrt{\dfrac{k_{air}(\rho c)_{air}}{k_{ch}(\rho c)_{ch}}} = \dfrac{k_{air}}{k_{ch}}\sqrt{\dfrac{D_{ch}}{D_{air}}} = \dfrac{k_{air}\ell_{ch}}{k_{ch}\ell_{air}} \tag{4}$$

where k is the thermal conductivity, ρ the density and c the heat capacity.

By a simple inspection of Eqs.3 the typical behaviour of the temperature rise in the channel waveguide (y > 0) is clearly seen. It depends on two exponential terms which have different decay lengths, connected respectively with thermal diffusion and optical absorption. However in the usual case, for low loss channel waveguides, the thermal diffusion exponential term has importance only for the first few microns ($y \ll \ell_{ch}$), being negligible for longer distances from the input edge. One way to calculate the propagation losses is therefore to study the deflection angle as a function of distance y of the probe beam by the waveguide edge. In other words, starting away from the edge, the amplitude of the photodeflection signal has only the exponential behaviour due to absorption processes, from which one obtain the propagation loss coefficient (see Fig. 2). However, this method is not useful for losses coefficient less than 1dB/cm. In fact, in order to minimize the noise one should perform measurements for very large values of y, sometimes larger than the available length of channel.

A more elegant way is suggested by the analysis of the deflection amplitude of the component along the channel direction (Φ_y) in the first few microns. The deflection angle is proportional to

$$\Phi \propto \frac{dT}{dy} \propto [\alpha \ell_{ch} + e(1+i)] \exp\left[-\frac{1+i}{\ell_{ch}} y\right] - (1+e)\alpha \ell_{ch} \exp[-\alpha y]. \tag{5}$$

In the analysis for short distance, the diffusion exponential term cannot be neglected with respect to the absorption term. Considering however the phase of the two terms, it is possible to find some length y_{min} for which the two exponentials have opposite sign, so that a minimum in the deflection signal is obtained. With the help of a computer one may find the distance y_{min}, which depends mainly on the effusivity ratio and on the loss coefficient, and if $\alpha \ell_{ch} \gg e$ the following empirical relation is found:

$$\alpha y_{min} = \ln(1+e) \approx e \quad \text{from which} \quad \alpha = \frac{e}{y_{min}}. \tag{6}$$

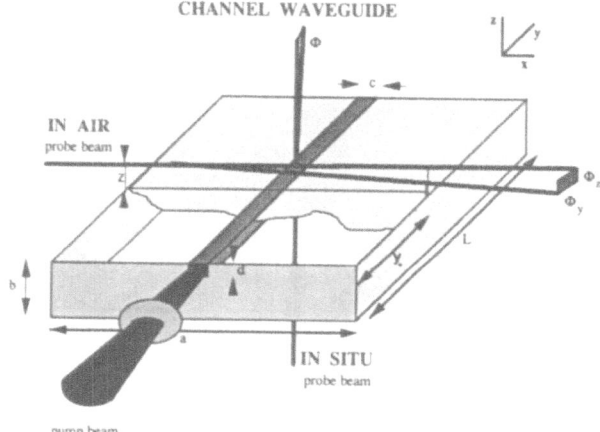

Fig. 1. Schematic representation of the configuration.

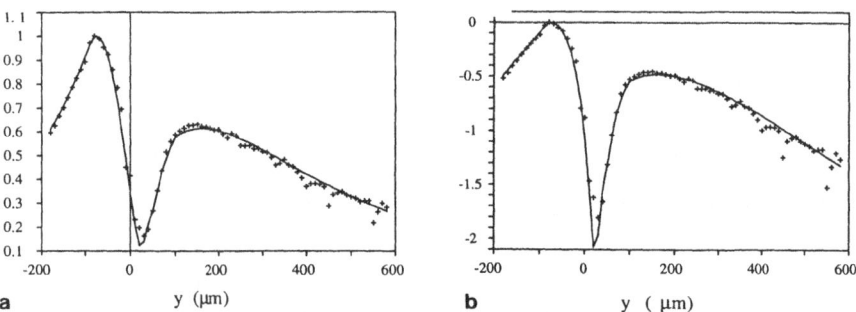

Fig. 2. The amplitude of the deflection signal for a glass on glass waveguide (the chopper frequency is f = 36 Hz, the measured loss is $\alpha \approx 50$ cm^{-1}: a) amplitude; and, b) logarithm of amplitude.

The minimum is directly connected to the existence of a maximum in temperature inside the waveguide. Due to the low value of the effusivity ratio in the general case of a solid sample, the existence condition $\alpha \ell_{ch} \gg e$ is in many cases verified.

3. Experimental Results

The channel waveguide which was used to test the first method, by measuring the exponential decay of the deflection signal, was a glass substrate with six channels 3 μm wide obtained by ion exchange. The refractive index increase in the channels is about $5*10^{-3}$ while the refractive index of the substrate is 1.55. In Fig. 2a the amplitude of Φ_y as a function of the offset y (μm) (configuration in air) is shown. In Fig. 2b the natural logarithm of deflection amplitude is reported. Note that starting from a certain distance, in the log plot, the experimental curve has a linear behaviour from which one obtains a loss coefficient $\alpha \approx 50$ cm^{-1} (±10%). The continuous lines in both Figs. 2 are the numerical simulations for a probe beam[10] of finite size w = 80 μm.

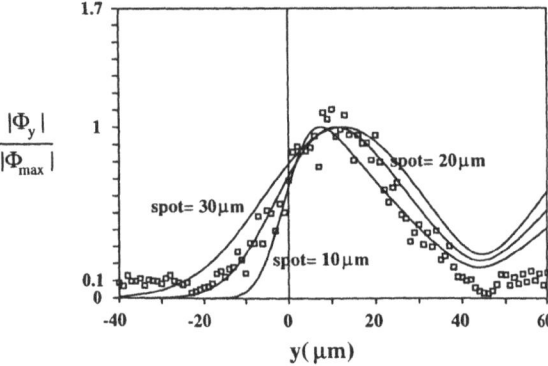

Fig. 3. The amplitude normalized of the deflection angle for a waveguide of Ti:LiNbO$_3$ (the chopper frequency is f = 18 Hz and the measured loss is $\alpha \approx 0.25$ cm^{-1}: (□) experimental points; and, (full line) numerical results for different beam size (10, 20, 30 μm)

The channel waveguide used to check the minimum method, was Ti:LiNbO$_3$.

The value of the effusivity ratio at room temperature was about $e \approx 0.0012$. The measurement reported in Fig. 3 was made at a chopper frequency f=18Hz so that the thermal length was about $\ell_{ch} \approx 130$ μm, which ensures that there will be a minimum.

In Fig.3 both theory and experimental data are shown for the normalized deflection signal in the in situ configuration. The theoretical curves take into account three different values of the probe beam size[10]. The fit is in good agreement with Eq. 6, giving a value of about $\alpha \approx 0.25$ cm^{-1} ($\pm 10\%$).

ACKNOWLEDGMENT. The authors would like to thank Prof. Arnone at the University of Palermo, which produced the tested Ti:LiNbO$_3$ channel waveguides, for help and discussion.

References

1. H.P.Weber, F.A.Dunn and W.W.Leibolt, Appl.Optics, 12,755 (1973)
2. K.H.Heagele and R.Ulrich, Opt.Lett. 4,60 (1980)
3. A.M.Glass, I.P. Kaminov, A.A.Aliman and D.H.Olson, Appl.Opt. 19,276 (1980)
4. Y.Okamura, S.Yoshinaka and S.Yamamoto, Appl.Opt. 22,3892 (1983)
5. I.P.Kaminov and L.W.Stulz, Appl.Phys.Lett. 33,62 (1978)
6. R.Reganer and W.Sohler, Appl.Phys.B36,143 (1985)
7. S.V.Bessonova, K.S. Buritskii, V.A.Chenykh and E.A.Shcherbakov, Sov.J. Quant. El. 19,559 (1989)
8. R.K.Hickeneil, D.R.Larson, R.J.Phelan and L.Larson, Appl.Opt. 27,1637 (1988)
9. W.B.Jackson, N.M.Amer and A.C.Boccara, Appl.Opt. 20,1333 (1981)
10. M.Bertolotti, L.Fabbri, E.Fazio, R.Li Voti, C.Sibilia, G.Ljakhou and A.Ferrari, J.Appl.Phys. 69,3421 (1991)

Chapter 19

INTEGRATED OPTICS APPLICATIONS FOR TELECOMMUNICATIONS

H.-P. NOLTING

1. Introduction

1.1. Historic Remarks

The situation of todays optical telecommunications is a curious contrast between the potential bandwidth and the actually utilized bandwidth of optical fibers. It is well known that the bandwidth of the fiber within the 1.3 and 1.55 µm window is in the order of 37 THz, a number which is nearly infinite compared with the bandwidth used in practice. Although it represents a great success that most of the terrestrial communication highways have been occupied by fiber connections, displacing analogue coaxial cable systems with a repeater spacing of 1.5 km and microwave links, the larger amount, consisting of customer access lines, is still monopolized by copper lines. For some time it has been widely agreed that the next big step for optical communications is within the local loop (fiber to the home). It is well known that actual bit rates on customer lines are very low, ranging from 64 kbit/s for plain old telephone services (POTs) to some Mbit/s for cable television (CATV). Therefore cost intensive point-to-point fiber connections have to compete with the very inexpensive copper lines. At several previous conferences [1], the break-even point for fiber and copper has been projected for the next 3 years but break-even has not been attained in any country.

1.2. Systems Devices and Technology

To discuss the advances in optical communication techniques we need to consider systems, devices (including OEICs and PICs) and material / technology aspects. There is a strong interaction between these three fields, leading to good system performance only for solutions which benefit from interdisciplinary work. We are in the process of developing modules of systems, devices and technology which have general significance beyond individual cases, in order to use them as a toolbox for other applications. The realization of this concept is very difficult and may sometimes be impossible. For example, it would be very helpful to establish a sequence of technology steps which are completely independent from all previous steps. The aim of this Chapter is to focus on a few examples which may be useful as building blocks in the future.

H.-P. Nolting - Heinrich-Hertz Institut für Nachrichtentechnik, Berlin GmbH, Germany

1.3. Organization of the Chapter

In Section 2 A we will discuss some system aspects, looking at the local loop and signal transparent applications, in particular optical frequency (wavelength) division (OFDM), optical time division (OTDM) and optical space division methods (OSDM). State of the art technology for the fabrication of optical devices will be reported on in Section 3, mainly focussing on monolithic integration technology and the optical motherboard approach for hybrid integration of larger units. Design aspects and critical issues for the fabrication of monolithic integrated polarization diversity receiver front-end OEICs will be treated in Section 4. Some investigation of new functional devices will be discussed in Section 5, mainly focussing on all-optical components. Afterwards some concluding remarks will be given in Section 6.

2. System Aspects

As pointed out in the Introduction, the effort in research and development is focussed on two targets, located at opposite extremes in bit rate: 1) reducing the cost of the fiber point-to-point link in the local loop, where the most cost-intensive modules today are the emitter and the receiver front-end, and, 2) exploiting the bandwidth of the fiber by frequency (OFDM) and/or time (OTDM) multiplexing techniques. In both cases monolithic integration of III-V devices and packaging to fiber pig-tailed modules play a significant role.

The competition between electronics and optics is now running at increasing speed. Nevertheless, both electrons and photons have their own advantages (+) and disvantages (-), as shown in the Table 1. Optical transmission lines - waveguides or 3D space [2] connections - can cross each other without interference problems. This is contradictory to electronic transmission lines, especially for higher frequencies. On the other hand, storing and switching is very easy for electrons, while photons have only weak interactions even in special optical materials (non linear optics).

2.1. Passive Optical Networks

There is a valuable basic idea which was introduced by British Telecom for the evolution from today's mixed copper and fiber telecommunication networks to the broadband fiber network of the future, where every customer can use several Gbit/s and has access to information all over the world with lots of new services[1]. A key step towards breaking out of the cost/application vicious circle will be optical networks where fiber infrastructure is shared between several end users via passive optical components [3]. This passive optical network (PON) architecture is an immensely flexible network platform capable of delivering narrowband and broadband interactive services. It also has an inherent capability for provision of distributed services.

Advantages of PON are the flexibility of adding any new and enhanced services by using frequency (OFDM) and/or time (OTDM) multiplexing techniques; easy upgradability; good flexibility; and savings in investment costs in the network itself. The network can be extended by optical fiber amplifiers or wavelength multiplexers, and only terminal equipment needs to be added[4].

Table 1. Comparison of basic properties for signal processing.

	electron	photon
switching	+	-
storing	+	-
transmission	-	+

2.1.1. OEICs in the Local Loop. Certainly there are other concepts and modifications of the PON concept, but in all cases we can identify two basic electro-optic converters, which are provided for each customer:

1) a bidirectional transceiver [5] for all personal data exchange such as telephone, picture phone, ISDN services and so on (Fig.1). The laser / driver as emitter and the detector / transistor stage as receiver are on the same chip. The isolation between forward and backward channels is accomplished by wavelength sensitive multiplexing / demultiplexing. The crosstalk between both channels needs to be very low and this has been attained in commercially available hybrid modules with crosstalk on the order of 50 dB.

2) a multi wavelength receiver (OFDM) for distributed services such as TV, HDTV etc. One possibility for realizing a multi wavelength receiver is a coherent optical receiver frontend [6]. The principle of a polarization diversity receiver is shown in Fig. 2. The incoming wave needs to be combined (via a 3 dB-coupler) with the wave from the local oscillator (tunable laser) and the superposition is converted to the electrical signal (detector) and amplified (JFET). In order to achieve optimum receiver efficiency, polarization matching schemes are necessary to overcome signal fading by the polarization fluctuations of the received signal within standard fibers, e. g. by using a polarization diversity reception scheme. Here TE and TM modes are separated by a TE / TM mode splitter, combined with the local laser light wave on separate paths and fed to balanced pairs of detectors.

An often discussed second solution is the combination of an amplifier, a tunable frequency selective filter and a receiver for direct detection. A comparison of these two approaches will be given later.

In this second approach we require photonic circuits with high complexity and high performance, which are candidates for monolithic integration, and much effort has been started at the HHI to investigate the integration process as an example for a large class of similar

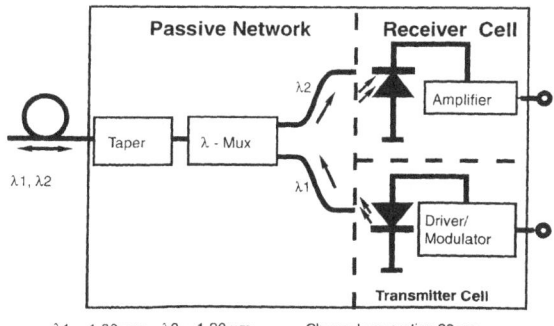

Fig 1. Principle of bidirectional optoelectronic converter.

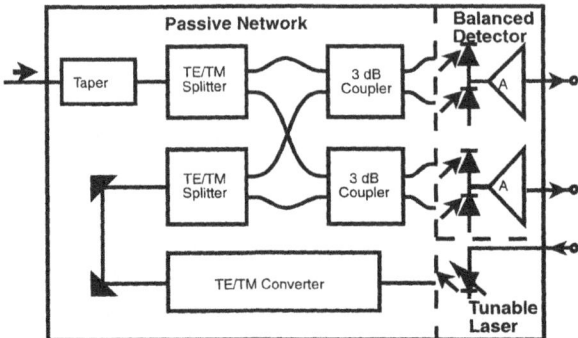

Fig. 2. Principle of polarization diversity receiver architecture.

devices. A schematic view of the local loop using OFDM techniques is shown in Fig. 3, demonstrating that each customer needs only one fiber connection to the central office via monomode fiber to have access to both narrow and broadband services.

2.2. Transparent Optical Switching and Routing Technologies

2.2.1. Switching Fabric. If the bandwidth increases in the local loop, tremendous growth is required in the switching nodes. A simple upgrade strategy based on the replacement of existing switching plant may not be the most cost-effective long-term approach, and advanced optical technology may play an important role [7]. Scale-up of the electronic switching plant is unlikely to offer the most cost-effective solution, particularly in the control layer.

Instead, the inherent transparency of the transmission fiber can be complemented by simple transparent [8] optical switching and routing technologies to provide a high degree of future flexibility to the network. The basic idea is that a transparent optical network carrying multi-Gbit/s data multiplexed in either the wavelength or time domains passes a network node. Optical switching technology (circuit or packet) routes the channels addressed to that node up and down between the all-optical layer and the electronically switched layer. Again, OFDM and OTDM techniques are favoured, accompanied by the inevitable space switching (OSDM).

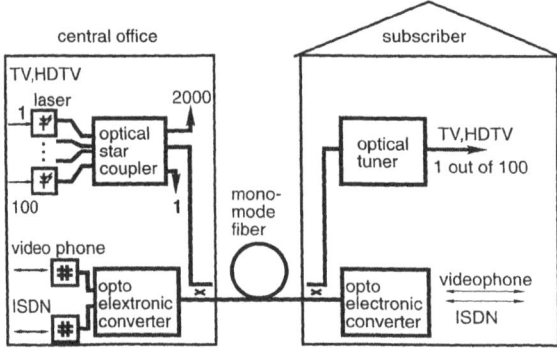

Fig. 3. Principle of broadband subscriber network.

At the far end repeated opto-electronic conversion needs to be avoided or, in other words, the switching node has to be transparent on the signal layer.

The basic idea is to make a transparent connection between two fibers in the same sense as the fiber is transparent itself. The simplest example is the mechanical fiber switch, which keeps all signal parameters constant. If the fiber carries more than one channel, as is in OFDM and OTDM, switching of each channel independently is a much more complex technique. Some functions such as frequency conversion, synchronization of data, and storing of data are difficult to realize in the optical domain and therefore investigation is focussed on these new functions. One main part of any transmission system is clock recovery to enable correct synchronization at the receiver or as part of an all-optical repeater. Examples of optical switching components (all-optical clock recovery) will be presented in Section 5.

2.2.2. Space Switching (OSDM). A broadband guided wave optical switch performs transparent switching in a natural way, if switch parameters such as low loss, low noise, low cross talk, digital switching behaviour etc. can be realized. An excellent review on space and time switches has been given by Midwinter[2] discussing the problems of large switching matrices, e. g. 128x128. It is obvious that guided wave optical switching matrices suffer from the large device length and the high cascading number of 2x2 switches. Therefore a large number of interconnections (fibers) and electrical control lines are neccessary. Hence completely new concepts are necessary.

If time switching is to be used, switching times less than 100 ps are mandatory. Electro-optical effects have the potential for this high speed regime, but unfortunately these effects are quite small, leading to long devices. Multi quantum well layers have the opportunity to enhance electro-optic effects by an order of magnitude compared with bulk semiconductor material. Bandgap engineering is a powerful tool to make semiconductors with physical properties better than natural materials. There is a hope for making materials with very low loss and very large changes of the refractive index as function of optical or electronical control. These effects are basic to high speed space switches. Under investigation are the quantum confined stark effect (QCSE)[9] and the bandfilling (BRAQWET)[10] effect. These effects have the potential for very high speed. These layers can also be used for tunable filters or tunable lasers.

2.2.3. Optical Frequency Multiplexing (OFDM). Increasing the line capacity by using different wavelengths on the same fiber is a natural means of exploitation in the optical domain. The neccessary components are passive wavelength sensitive multiplexers / demultiplexers and tunable laser sources. If transparent switching nodes are to be used, wavelength converters, discussed in Section 5, are indispensable. With these components wavelength routing will be possible, thus replacing space switching by frequency switching.

The maximum number of channels that can be used on one fiber is very high in principle, but is limited in reality by the non linearities of the fiber itself and the amplifiers in the line. BT has published a limit of 10 WDM channels in a net with a fiber span of 50 km [11], while on ECOC '92, there was a report on 100 FDM channels, 10 Gbit/s spaced using in-line amplifiers employing tunable gain equalizers for successful transmission [12].

2.2.4. Optical Time Domain Multiplexing (OTDM). Another way to exploit the capacity of the optical window is to raise the bit rate on the transmission fiber, of course, at the expense of the channel density. Today 10 Gbit/s components are commercially available and by all-optical multiplexing 100 Gbit/s, 50 km optical transmission [13] has been demonstrated. Synchronous and asynchronous nets are possible. The cornerstone underlying the implementation of OTDM

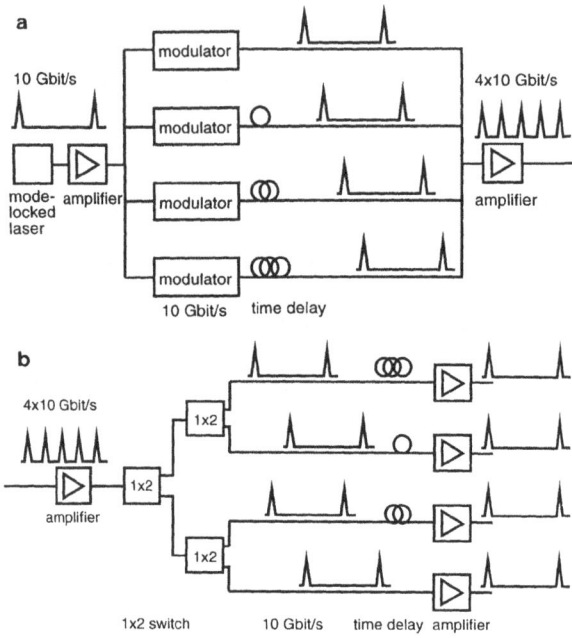

Fig. 4. a) 4×10 Gbit/s OTDM multiplexer; b) 4×10 Gbit/s OTDM demultiplexer.

networks is that the optical pulses produced by any transmitter are significantly shorter than the bit period at the highest multiplexed line rate in the network [14]. The implication of this is that modulated data streams may be interleaved and demultiplexed (see Fig. 4) with negligible crosstalk. Thus short pulses of a mode locked laser as source may be distributed to a number of modulators, working in parallel for each data stream. After an individual time delay they will be combined on one output fiber and amplified. Thus multiplexing in principle is simple with present devices. For demultiplexing different architectures are known, but a high speed switch is always neccessary.

2.3. Trunk Lines and Overseas Cables

The introduction of erbium-doped-fiber optical amplifiers (EDFA) has eleminated fiber loss as a fundamental limit to achievable transmission distance. This has been impressively demonstrated on OFC/IOOC '93 conference with a 10 Gbit/s, 9000 km intensity modulated, direct detection transmission experiment using 274 Er-doped fiber amplifier [15]. Mollenhauer et. al. [16] have demonstrated, using sliding-frequency guiding filters, error-free soliton transmission over more than 20,000 km at 10 Gbit/s, single-channel, and over more than 13,000 km at 20 Gbit/s in a two-channel WDM system. In this experiment the jitter in pulse arrival time, generated by spontaneous emission in the amplifiers, was reduced by sliding-frequency guiding filters. With such "sliding-frequency" devices the transmission is opaque to noise for all but a small final fraction of its length, yet remaining transparent to solitons.

2.4. Remarks and Conclusions on System Aspects

Most of the research effort today, where each new record surpasses the older ones, is directed towards searching for the limits of high capacity networks, with optical signal path and electronic control and management of the system. Looking at OFDM and OTDM techniques, devices with new functions have to be developed, especially wavelength converters, wavelength multiplexers / demultiplexers, high speed sources and detectors up to 100 Gbit/s, optical memories, all-optical repeaters, and optical logic elements.

Optical signal processing methods try to penetrate into the traditional domain of electronics. There is a competition between optical wavelength (frequency) division and time division methods. Today each method is investigated for its own ultimate limits, without looking for an optimized synthesis of both. The large bandwidth of the fiber can be used only once. Which domain or which combination is the best is an open question.

Present laboratory systems are limited not by the ability to generate short optical pulses, nor by the capacity of the optical fiber. They are limited by the electronics used for laser modulation and detection. As the trend towards still higher capacity systems continues, it will become more necessary to perform some of the signal processing operations with photonic components in order to eliminate the electronic bottleneck.

To bring the fiber all the way to the customer it is neccessary to reduce the cost of the fiber point-to-point link in the local loop. In both cases monolithic integration of III-V materials and packaging to fiber pig-tailed modules play a significant role.

3. Technology

3.1. Monolithic Integration

The basic idea of (monolithic) integrated optics started with early dreams in 1969 [17] and has grown in the last 5 years with the commercialization of MOVPE systems. Rapid progress and new approaches have led to new terms. Following Kogelnik [18], we will distinguish between opto-electronic integrated circuits (OEICs) [19], which include on-chip transistors and on-chip opto-electronic devices, and photonic integrated circuits (PICs) [20], which include on-chip lasers and on-chip optical waveguides.

The availability of OEICs and PICs is a prerequisite for the economic realization of optical communication systems. This applies especially for inserting the above mentioned electro-optical converters in the local loop. Outstanding qualities of OEICs and PICs are their compactness, reliability, functionality, reduced alignment and assembly effort, and robustness. Among all the possible materials for guided wave optical components, III-V semiconductors are the most suitable choice, because these are the only materials in the wavelength region of 0.9 ... 1.6 µm that allow a monolithic integration of optical (guided waves), opto-electronic and pure electronic functions. For telecommunications the wavelength region of 1.3 ... 1.6 µm is significant, because of the lowest loss and dispersion behaviour of the fiber. The potential applications of OEICs and PICs extend from the local loop and the long-haul transmission to the switching node. Long-haul trunk transmission requires a high level of performance and reliability, while high volume customer access demands both high performance and low cost. An additional push for developing monolithic integration techniques comes from optical interconnects between racks, boards, or even chips. Photonic switching, routing and signal processing also needs the increased functionality of OEICs and PICs.

Let us now look at the main problems of complex monolithic integrated optics. Some of them are severe, some are solvable and for some there are promising developments. Whenever monolithic integration of opto-electronic circuits is discussed, the Si ICs serve as a reference or competitior, respectively. As we will see in the following discussion and was pointed out in more detail by Erman [21], all problems are completely different than in Si technology. Nevertheless, it is implicitly assumed that the same story of success can be repeated for OEICs.

3.1.1. III-V-Material. III-V compound semiconductor physics, chemistry and technology are different from Si, and the duration of working and the amount of labor thus for put into the newer material system are much lower. Thus immaturity of the new material seems plausible. In addition, the material is much more complex than silicon. The III-V-crystal is composed of up to 4 elements: this leads to lower mechanical stability, higher diffusion coefficients, lower thermal conductivity etc. For optical devices different mixed crystal layers (ternary or quaternary) are neccessary for fundamental processes such as light generation, absorption and guiding. Each of the fundamental devices such as laser / optical amplifier, detector, passive waveguide, optical switch, transistor needs its own layer stack. Therefore the number of processes to fabricate PICs is high and the number of processes which are common to several devices of different types is low. These facts account for the very low fabrication yields today.

3.1.2. Epitaxy. The epitaxy process to grow layers of different compositions in the semiconductor material system InGaAsP lattice matched on InP is the key process for the fabrication of electro-optical and optical devices such as lasers, detectors, waveguides and transistors. In epitaxial growth technology for the fabrication of OEICs and PICs, both metalorganic vapor phase epitaxy (MOVPE) and molecular beam epitaxy (MBE) have become dominant techniques [22]. Substantial progress has been achieved for the production of highly uniform thickness, composition and doping of the heterostructures. At present, MOVPE dominates the device applications of phosphorus containing compound semiconductors, e. g. InGaAsP/InP and AlGaInP; and MBE dominates AlGaAs. On the horizon is the emergence of chemical beam epitaxy (CBE) or metalorganic beam epitaxy (MOMBE) which has been proven to be a very powerful combination of MOVPE and MBE.

For device application the relevant parameters are composition accuracy, uniformity and thickness, doping level and lattice mismatch (strain). The thickness can be as low as 2.5 nm (5 molecular layers) for MQW layers with high reproducibility [23]. 10 nm thick etch stop layers can be used without any problem. On large areas the thickness and composition homogeneity is about 5% (best results[22] reported 1%). The sequence of succeding layers is not limited in composition and the number of layers can be high. A background doping level on the order of 10^{15} cm^{-3} (and lower for InP) is sufficient for most applications. High p and n doping is possible, and forming semi-insulating InP layers by Fe doping results in a specific resistance value of 10^7 Wcm. A MOVPE process sequence combining AlInGaAs and InGaAsP on InP substrates for BRAQWET[10] structures for optical switches has been developed at HHI. One of the most needed processes for the fabrication of OEIC/PICs is selective area growth. Examples are the laser / waveguide coupling described in part C and the butt coupled waveguide / detector combination [24]. Additive process steps would reduce the fabrication effort to fabricate dissimilar devices with different sequences of layers by using precise epitaxy fill-in. At present, using MOVPE for selective epitaxy, the thickness and the composition depend on the dimension of the masked area. First experiments with MOMBE show favourable results.

A further important requirement on epitaxy is smooth and undisturbed growth over structured surfaces, e. g. gratings and embedded waveguides.

3.1.3. Structural Techniques. At present, device fabrication is based on the growth of plane epitaxy layers on the entire chip area and removing the not needed or disturbing parts of these layers by normal photolithographic definition of masks and dry or wet etching processes. This has to be repeated several times including other process steps such as diffusion, implantation and deposition of isolation and contact layers, depending on the complexity of the OEIC. Special care must be taken for devices with distinct spectral behaviour such as DFB or DBR lasers and two-/ multi-mode interference devices (optical switches, filters). Here the precise definition of the waveguide geometry is very important. The photolithographic process has only limited resolution, on the order of about 0,5 to 1 µm, thus to accommodate the likely fabrication tolerances waveguide widths of 3 to 5 µm are usual for passive waveguide devices such as directional couplers, resulting in laterally weak guiding waveguides. Electron-beam lithography may be helpful in the future to overcome this limit.

To define the height of a rib waveguide precisely, special dry etch techniques have been developed. One widely used method is to incorporate thin etch stop or etch marker layers, thereby transferring the control of the etch depth from the uncertain etch process to the very precise thickness control of the epitaxy process. Using etch marker layers, an endpoint detection method for CH_4/H_2 reactive ion etching of InGaAsP heterostructures have been developed with a depth resolution of 5 nm [25]. Also very useful is in situ etching depth monitoring for reactive ion etching of InGaAsP/InP heterostructures by ellipsometry. The instantaneous depth is determined with an accuracy of 10 nm [26], and this approach has been successfully used for the fabrication of asymmetric directional couplers with a periodic perturbation (meander couplers).

3.2. Opto-Electronic Multi Chip Module (OE MCM) Technique

Every OEIC needs a packaging system which includes electrical and optical interconnections to the "outside world". For the cost-effective employment of OEICs we have to consider the whole module, keeping in mind that todays costs are largely dominated by packaging (80% of the total price of a long wavelength laser module) . The development of the Opto-Electronic Multi-Chip-Module (OE-MCM) technique, on an optical "(Si) mother board", offers a new dimension in the design of optical units. This concept includes a number of approaches, all having however in common the use of Si as a mechanical support (see Fig. 5)

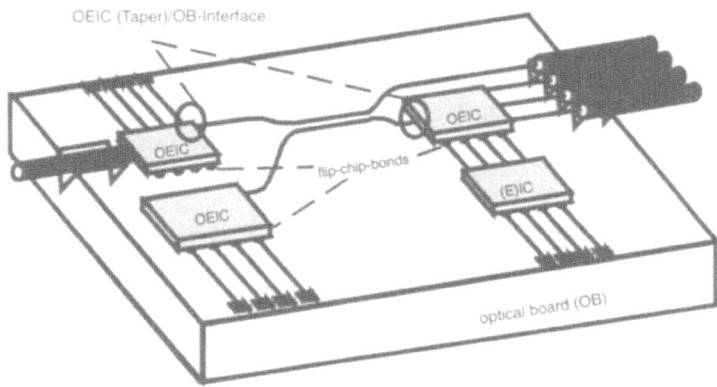

Fig. 5. Opto-electronic multi chip module showing: optical board (OB), OEICs, ICs, OB waveguides (e.g. Silica), optical and electronic connections by flip-chip techniques.

for the electronic chips, the PICs and the fiber and the use of standard micro-electronic processes (lithography, wet and dry etching for V-grooves, alignment pattern etc., flip-chip techniques[27,28] for electrical and fiber connections). Furthermore, it is possible to use the Si board as a substrate for low loss optical waveguides made of silica [29] or polymers, which provide the on-Si board optical interconnects with well defined polarization states of the waves, or can provide some passive functions (splitting, multiplexing, filtering,...). If we compare InP and silica based waveguides, we see that the silica waveguide has the advantage of very low loss (10 times lower than polymers and III-V-compounds), larger substrate area (8" to 2" today) and lower temperature coefficients of the refractive index. Thus the optical mother board, connecting many optical PICs as is neccessary for larger switching matrices, looks very promising, if it is possible to improve the mechanical alignment tolerances of the silica / InP waveguide connection. This can be done by including waveguide tapers in the InP chip, as described in part C.

3.3. Remarks and Conclusions on Technology

Rapid progress in MOVPE crystal growth techniques and structural techniques has led to promising results in monolithic integrated optics. Components for telecommunications require a high level of performance and reliability, and optical signal processing requires the increased functionality of OEICs and PICs.

4. OEICs

In the last 10 years monolithic integration technology has increased from two / three device combinations (for example, laser / driver, laser / modulator, waveguide / detector / transistor combinations) or electro-optical 4x4 switch matrices to more complex PICs such as a polarization diversity coherent receiver chip. In the beginning, complexity was limited, depending upon similar compositions of the layer stack for all devices on the chip. However, often people used a compromise for the same layer in different device types, as for instance the doping level in a detector and a JFET. This leads to non-optimized performance. One advantage of this approach is the low number of epitaxy runs, on the order of 1 to 3. The fabrication of state of the art tunable DFB, DBR or TTG lasers is much more complex, because an epitaxy sequence of up to 5 steps is neccessary.

However, the crucial test of a successful integration process is the combination of a laser, passive optical components forming a network (as in case of the heterodyne receiver) and a detector / transistor block. Keeping this in mind, it is easy to understand why the first realization of a coherent receiver chip uses a reverse biased laser[30] as a detector[20] and avoids any transistor on the chip.

4.1. Polarization Diversity Receiver OEIC

As an example of highly complex OEICs we will now discuss the problems of the monolithic integration of a bidirectional transceiver and especially of a polarization diversity receiver, whose waveguide architecture is sketched in Fig 6. The latter is a principal technological effort at HHI at the present time. The learning curve of HHI over the last decade can be seen in the sequence of the following papers: low loss waveguide in InP, phase modulator [31,32], meander coupler as a filter[33,34], waveguide / detector / JFET integration,

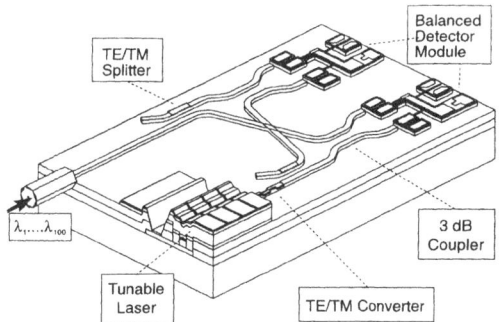

Fig. 6. Architecture of palarization diversity receiver chip.

balanced detector module, integrated wavelength demultiplexer-receiver on InP [5], multi quantum well (MQW) lasers, and polarization diversity waveguide network [35]. The evolutionary strategy of ongoing integration of the coherent receiver chip itself and actual experimental results on subcomponents such as tunable lasers, balanced detector units and passive networks is described in detail in the prospects and progress report of Heidrich[6].

4.2. Architecture of OEIC at HHI

4.2.1. Waveguide. The basic element of integrated optic devices is the dielectric waveguide [36]. It is the main part of most devices and serves for guiding, switching, polarization handling, filtering, splitting and combining of optical waves on a PIC. The concentration of optical power in a small cross-section (1 ... 10 mm^2) over long distances (100 ... 1000 ... 10000 µm) allows effective interaction of optical waves (gain, absorption, thermo-optical and electro-optical effects, etc.).

Optimization of a given device always leads to a distinct waveguide cross-section with its own field distribution. Integration of different devices on the same chip therefore needs mode transformation (tapered waveguides) or standardization of the waveguide to avoid coupling losses and generation of radiating waves, which may disturb the circuit function. Looking for a general solution, we have to connect the very compromised field distribution of a laser via a "standardized" passive waveguide with medium field distribution to the weakly guided fiber field.

4.2.2. Waveguide Concept of OEIC at HHI. After intensive studies of different possibilities, we have established the following waveguide sequence for the InGaAsP/InP material system at an operating wavelength of 1.55 µm, as shown in Fig. 7. The most important points are:
- Fe-doped semi-insulating substrates to minimize additional capacities and leakage currents for the detector unit;
- a passive waveguide of $\lambda_Q=1.3$ µm (p-doped, loss=10 cm^{-1}) as part of the DBR laser (high coupling efficiency of the grating);
- low loss, semi-insulating passive waveguide of $\lambda_Q=1.05$ µm (undoped, loss $\cong$ 1 cm^{-1}) with InP ribs for good fabrication tolerances of directional coupler structures as "standard" waveguides;
- waveguide taper for mode matching to the fiber as shown in Fig. 7;
- butt coupling built-in between both passive waveguides by masked, selective MOVPE[37]; and,

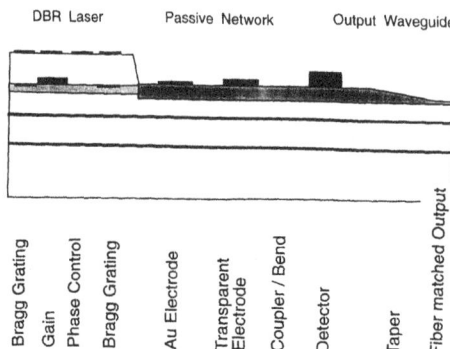

Fig. 7. Waveguide architecture of OEIC.

- waveguide detector coupling by evanescent coupling, detector and JFET grown together by MBE (lower growing temperature than MOVPE, very good thickness control) as "last" process.

4.2.3. Butt Coupling Built-in. The built-in butt coupling of the laser waveguide to the low loss semi-insulating network waveguide layer structure (see Fig.8) is one of the most crucial points for the OEIC. The semi-insulating waveguide layer stack is grown by selective MOVPE after

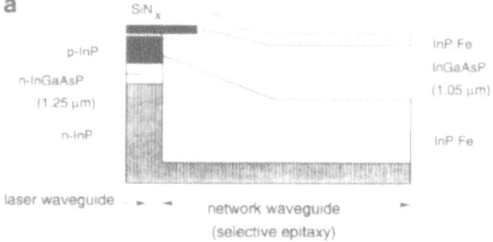

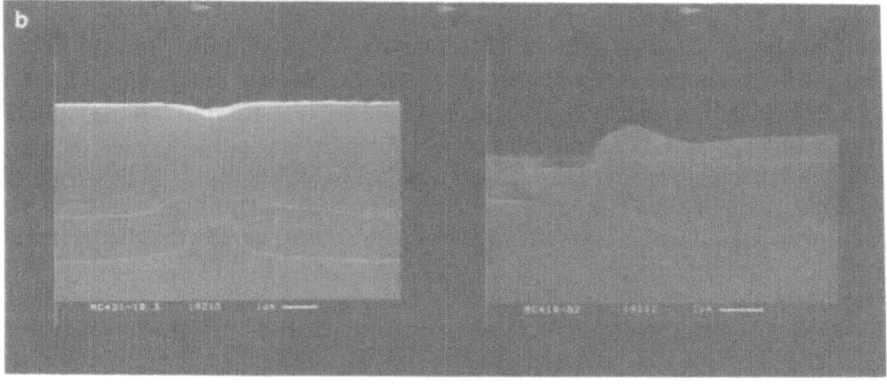

Fig. 8. Principle and SEM picture of laser-network-interface.

etching away the un-needed laser layers. In order to keep the increase of the layer thickness towards the mask edge within reasonable limits the masked areas have to be small and the mask has to be prepared with an undercut.[37] There is some hope that the butt coupling built-in connection shown in Fig.8 will be much smoother using MOMBE instead of MOVPE.

4.2.4. Standard Passive Waveguide. The cross section of the standard passive waveguide [38] is common to all passive devices (3 dB coupler, TE / TM mode splitter, TE / TM mode converter, bends, crossings), consisting of a 1.05 µm InGaAsP slab loaded with an InP-ridge with a height of 0.4 µm, a width of 3 µm and and a gap distance of 2 µm in a directional coupler. This waveguide meets all the requirements for the realization of process tolerant directional coupler devices such as 3 dB couplers and TE/TM-mode splitters [39] and of effective electro-optically controllable devices such as a TE/TM-mode converter[40]. To meet the neccessary fabrication tolerances laterally weak guiding waveguides are often used, resulting in large coupler length and bend radii. Therefore the chip area needed for the passive network is very large compared to the area of the laser or the detector. The only way to reduce the length of couplers and bend radii is to use high contrast (strong guiding) waveguides. This requires sub micron waveguide dimensions, which are hard to realize in the lateral geometry with common photolithographic techniques. One alternative is to use the high precision thickness control of epitaxy (MOVPE or MBE) to fabricate high contrast couplers with sub micron gaps in the vertical direction as has been demonstrated by Deri [41]. Unfortunately, this does not solve the bend problem. The use of carefully etched-down waveguides with interfaces to air gives too high contrast and leads to very small widths of the waveguide or to multimode waveguides [42]. Thus buried high contrast waveguides with nearly quadratic cross sections seem to be the best choice for the future. However e-beam lithography and dry etching processes need to be developed for good fabrication tolerances. With these submicron waveguides, length and radii reductions of more than a factor of 10 are possible.

With our present standard waveguide design, there is an imbalance between processing effort and the amount of area on the chip, if we compare laser and waveguide areas. Around 2/3 of the processing steps are necessary for the fabrication of the laser, which covers only about 5% of the chip area [6]. A much more compact waveguide design is mandatory, but lower fabrication tolerances are a prerequisite for stabilizing the yield of the process.

4.2.5. Waveguide Taper. Waveguide tapers are indispensable for mode transforming in complex PICs, if loss and radiating modes caused by abrupt waveguide transitions are to be avoided. The fact that tapers with no loss are possible[43] from a theoretical point of view has no effect on their practical realization. The ideal taper has to change the geometry and the refractive index of the waveguide simultanously and continously following special curved functions for transforming the mode. In III-V semiconductor technology specific and continous change of the refractive index is very difficult to realize. Fortunately, it is possible to use only continous change of waveguide dimensions for the fabrication of low loss tapers (see Fig. 9). This has been demonstrated theoretically[44] and experimentally [45,46]. The implementation of the taper into the OEIC/PIC chip and the increase of the output mode distribution of the wave has the tremendous advantage that the fiber / chip alignment tolerances increase (see Fig. 10) from submicron to a practicable value of at least 1 µm. This opens the way for a self-aligning optical assembly technology (OE MCM technique), which has been described above in Section 3.

4.2.6. Waveguide / Detector Coupling. The balanced detector unit including the transistor is grown by MBE in order to exploit the accurate thickness homogeneity which is mandatory for

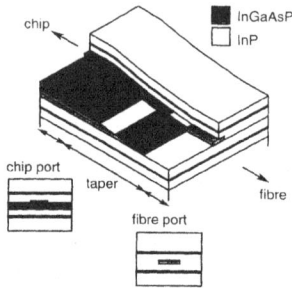

Fig. 9. Taper structure showing thickness tapering and guiding layers.

the fabrication of FETs. The layers are grown on the Fe-doped semi-insulating waveguide layers to minimize additional capacities and leakage currents and to be independent from the laser and waveguide design [47]. The coupling from the network output ports to the balanced detectors is accomplished by evanescent field coupling.

4. 3. *Process for Large Scale Photonic Integrated Circuits (PIC) at AT&T*

U. Koren and coworker from AT&T have described a process for large scale photonic integrated circuits [20,48], which has been used for a variety of laser structures, optical switches, passive waveguide structures and a coherent receiver chip. The key elements of this process are: using MOVPE, MQW for active layers, etch stop layers, wet etching, and carrier injection for electro-optical control of waveguides, and realizing passive waveguides with 1.3 μm bandgap material. It can be seen that the starting point of this development was the laser. A good example of this powerful process is the following device, which is a basic component of OTDM systems.

4.3.1. Laser for an OATM System. Mode locked lasers can be used as optical sources for OTDM systems, because they can generate very short pulses. At OFC 93 AT&T reported the monolithic integration of a soliton transmitter [49], using their PIC process. The soliton transmitter consists of a monolithic extended-cavity laser integrated with an electro-absorption

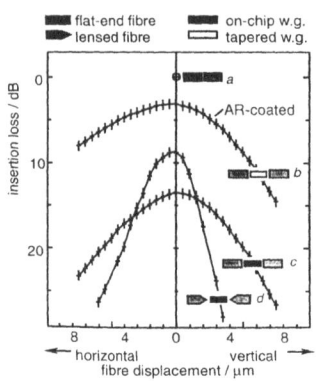

Fig. 10. Alignment tolerances: a) fiber- fiber, b) OEIC + taper-fiber, c) OEIC-fiber, d) OEIC-taper fiber.

modulator. Solitons are generated by active mode-locking of the extended-cavity laser, which is defined by one cleaved facet and the Bragg reflector grating segment. Data is encoded on the train of solitons by means of the electro-absorption modulator, which is monolithically integrated but external to the laser cavity. The minimization of the internal reflections between different segments of the backbone passive waveguide is a prerequisite to a good pulse shape of the device.

4.4. Comparison of Both PIC Processes

The AT&T process suffers from the following disadvantages: - the detector is not a "real", high efficiency, high speed detector composed of low doped ternary material: in fact, it is a reverse biased laser configuration; - the waveguide is not semi-insulating, therefore the fabrication of balanced detector pairs with high speed is not possible; and, - there is no transistor included.

An advantage of the AT&T process as compared with to the HHI process is the simple, low reflection connection between laser and passive waveguide.

4.5. Remarks and Conclusions on OEIC Technology

Process stability is a challenge with present state of the art technology. Allthough every step in the process sequence can be carried out successfully, some steps have low yield and are not completely stable. Further investigation and deeper insight in the physics of the process is neccessary.

For the future, the main goals of PIC technology must be to develop better selective epitaxy methods for butt joint built-in coupling of intra chip waveguide connections, taper technology for the chip / fiber or silica waveguide connection, and the reduction of the chip size by reduction of the waveguide network area (which requires high contrast waveguides).

5. Devices (single)

Here I will discuss the nature of some typical devices, independently of the fact that they may be integrated with other devices to obtain much more complex functionality.

5.1. Filters and lasers for OFDM systems

There is a great variety of different types of filters for optical frequency multiplexing, ranging from simple directional couplers, Mach-Zehnder type devices, Bragg gratings, meander couplers to spectrometers, arrayed grating couplers and passive and active tunable filters. A very compact device architecture for a passive wavelength sensitive multiplexer / demultiplexer is the arrayed grating coupler [50] shown in Fig. 11. It can be understood as an integrated version of cascaded Mach-Zehnder interferometers.

The most important parameters of filters are the number of channels that can be multiplexed, the loss, the wavelength spacing and the cross talk. All filters based on multi-wave-interference need very precise control of the waveguide parameters, especially if a specific wavelength has to be matched by a passive device. For instance Verdiell [51] has used weak guiding, diluted waveguides for a spectrometer and HHI has developed in situ etch monitoring techniques[26] for precise thickness control of the rib waveguides of meander-type

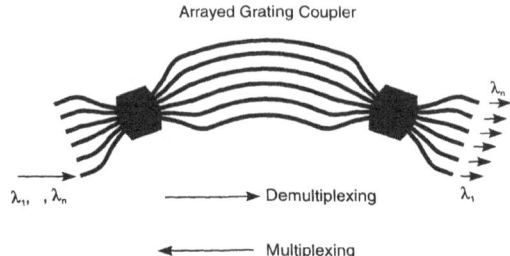

Fig. 11. Passive frequency de/multiplexer.

couplers. It is well known that the FWHM bandwidth of a filter is correlated inversely to the device length and directly to the dispersion of the waveguide. Thus filters with a very small channel separation are very long. They need precise waveguide homogenity over large areas.

Tunable filters are attractive, for example, as part of a multi-wavelength receiver. Unfortunately, the change in the refractive index of a waveguide is limited by the possible physical effects to on the order of 1% at best. This limits the tuning range of a single device to the order of 10 channels, independent of the device structure [52]. Larger tuning ranges are only possible by cascading different filter devices. One impressive but not very practical example is the combination of 7 tunable Mach-Zehnder interferometers [53] resulting in a resolution of 100 channels with a channel separation of 10 Gbit/s. A much more realistic approach would be the combination of a widely tunable meander coupler having tapered coupling for suppression of the side lobes, and a tunable, periodic filter as for example a Fabry-Perot resonator with much smaller FWHM. The broad filter suppresses all but one of the periodic wavelengths of the narrow filter (see Fig.12). This is the basic idea of the ACA laser[53] and the improved version the DFC laser [55], in which both tuning ranges are better matched. The laser as an active filter benefits from the gain suppression of the side lobes, which are inherent to all laser structures. Therefore, we may speculate that active filters will soon be superior to passive filters. When this occurs, the coherent receiver: will be superiour to the tunable direct detection scheme.

5.2. Compact space Switch Matrix Using Wavelength Routing

Let us consider a compact space switch matrix using wavelength routing, which could be realized in a hybrid integration of semiconductor wavelength converters and a silica-based

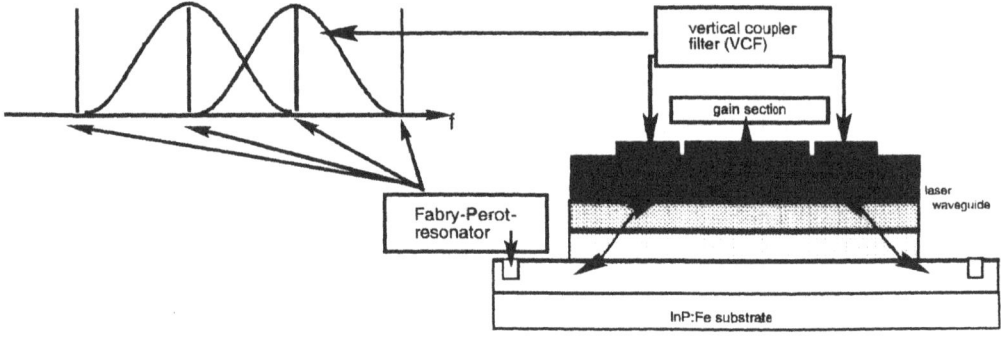

Fig. 12. Principle of large range tunable filter/laser.

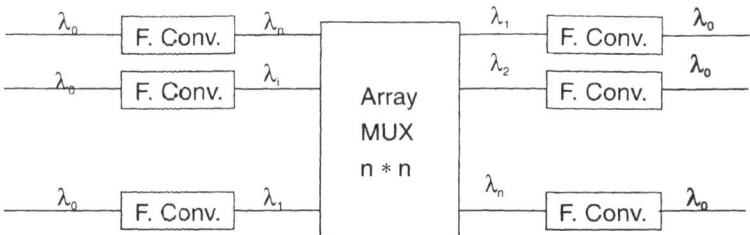

Fig. 13. n * n space switch matrix using wavelength routing.

passive array grating coupler, as sketched in Fig. 13. The passive array grating coupler provides a specified routing from each input to each output waveguide for a specific wavelength. Thus each input frequency converter can act as a 1xn space switch by tuning the frequency. In an OFDM system, frequency re-use is possible, with a multi-stage architecture. In this way, a very compact space switching arrangement is possible, because we are using 1xn switches, and the total number of cascaded switches is low. We need only one control line for each input waveguide to switch to the desired output waveguide. An additional advantage is the fact that the same components are used for multiplexing and demultiplexing in the frequency domain.

5.3. Low Cost, Wavelength Stable Laser for the Local Loop

Here I will describe a low cost, wavelength stable laser with improved temperature stability, which is suitable for the local loop. This device benefits from both PIC and OE-MCM techniques. As shown in Fig. 14, the device is a hybrid combination of a Fabry-Perot laser with a taper at one output and a silica waveguide with a Bragg grating for wavelength stabilization. The low temperature coefficient of 0.016 nm/K of the silica grating allows temperature stable devices. The taper enhances laser to silica waveguide coupling and thus the output power of the device. The laser / fiber alignment tolerances are large, thus are suited for batch process production. If we use a ridge guide laser, we need only one epitaxy process step. Thus this device is very promising, and may have low manifacturing coast.

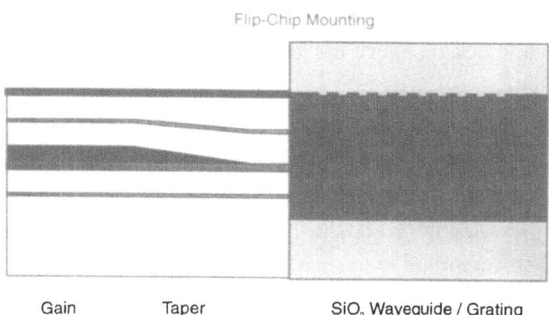

Fig. 14. Architecture of wavelength stable hybrid laser module.

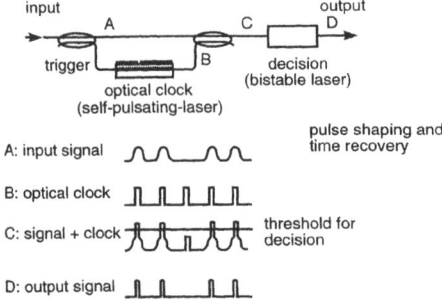

Fig. 15. Principle of all-optical signal regeneration.

5.4. New Switching Devices

This Section will focus on active semiconductor components (laser amplifiers) suitable for time and wavelength switching[56]. For all-optical signal processing, the non-linear effects of the laser amplifier are used. One key question is the upper speed limit of those components. It is now believed, that a speed limit on the order of several Gbit/s of absorptive non linear gain can be overcome by using dispersive nonlinear gain[57], with the potential of speed of several 10 Gbit/s even up to 1 Tbit/s. The absorptive nonlinear gain is limited by the electron recombination time, which is in the nanosecond range, and may be reduced by heavy Zn doping, but only by one order of magnitude. A new approach is possible, based on intraband relaxation mechanism, whose time constants may be on the order of picoseconds. This will be discussed in the next two examples.

Fig. 15 shows the principle of an all-optical repeater, which is of considerable interest for digital networks, especially in the nodes of the networks. It consists of a self-pulsating laser for clock recovery and a bistable laser for decision and pulse shaping.

5.4.1. Clock Recovery with Self-pulsating Two Section DFB Lasers.
Two-section DFB lasers show[58,59] current controlled self-pulsation with a tuning range between a few hundred MHz and 30 GHz. A first theoretical analysis [60] of this phenomenon shows that dispersive self-switching of the cavity build-up of both DFB gratings occurs. Thus small changes in the electron distribution may have very large effects on the laser behaviour.

Locking of self-pulsation to injected 18 GHz optical signals has been investigated for the two-section DFB laser[61]. Within a frequency detuning range of 150 MHz the pulsation locks to the injected modulation, and the RF linewidth in the locked state is below 10 Hz. More recently locking to a data stream of 18 Gbit/s has been demonstrated. In a 4x5 Gbit/s system experiment[62], all-optical clock recovery has been shown without receiver penalty compared with direct clock signals.

Initial results on decision and pulseforming show very promising results [63]. For example, the time jitter is similar to that of the best electronic repeater. Pulse shaping is poor, but there is the possibility of further improvement.

5.4.2. Wavelength Conversion.
Wavelength (frequency) conversion devices using all-optical processes are likely to be important elements in OFDM systems. These devices enable optical signals at one carrier wavelength to be transferred to another wavelength as shown in Fig. 16. Devices being tested generally use either of two different physical processes: four-wave mixing in nonlinear media, and DFB or DBR laser control by saturable absorbers.

Laser based wavelength conversion. Bistable lasers of different device architectures have been used for wavelength conversion experiments. The basic physical process is that one part of the cavity acts as a saturable absorber. Incoming light with amplitude modulation is able to switch the laser on and off, on its own wavelength, which may be tunable. Thus the signal is transferred to the new carrier. One very exciting device, which overcomes some known limitations, is a tunable DFB laser diode with a side-injection[64] light controlled absorber region. The two waveguides cross orthogonally: thus complete isolation between input and output signal, without any filter, is possible. The area of the saturable absorber is only 3x10 µm, due to the strong nonlinearity of the MQW structure of the active layers. The tuning range is over 100 nm and the bit rate 2.1 GHz. This seems to be a very practical approach, with its only disadvantages being that ASK is neccessary and the bit rate is limited by the electron lifetime.

Four-wave mixing based wavelength conversion. Four wave mixing in travelling wave semiconductor laser amplifiers has been shown to be very attractive for wavelength conversion of up to 30 nm (3.8 THz). This technique has the potential of operating at extremely high bit rates (30 ps [65]), it allows broadband continous tuning of input and output wavelengths and, in addition, the conversion is entirely transparent to the modulating format. Both bulk [66] and MQW [67] amplifiers have been used. In a system experiment at 622 Mbit/s, 10^{-9} BER with a power penalty as low as 1.1 dB has been achieved. The disadvantage is the strong decrease of the efficiency with rising wavelength conversion distance.

5.5. Remarks and Conclusions on Devices

A great variety of physical effects are under investigation for future photonic components. Different materials are competing today, e. g. , III-V semiconductors, polymers, glass and silica. For optical signal processing the nonlinear effects are of fundamental significance. Even during the last few years new high speed nonlinear effects have been discovered in materials with gain, thus raising the bandwidth of nonlinear semiconductor amplifier applications. Looking at the "tuning range" of tunable passive filters, lasers and optical switches by comparing the phase shift of each device for a fixed length, it is easy to see that up to now tunable lasers with a tuning range of 100 nm around 1.55 µm are the best. This, together with the compact wavelength routing scheme of arrayed grating couplers, seems to be the best approach for switching in the future transparent network.

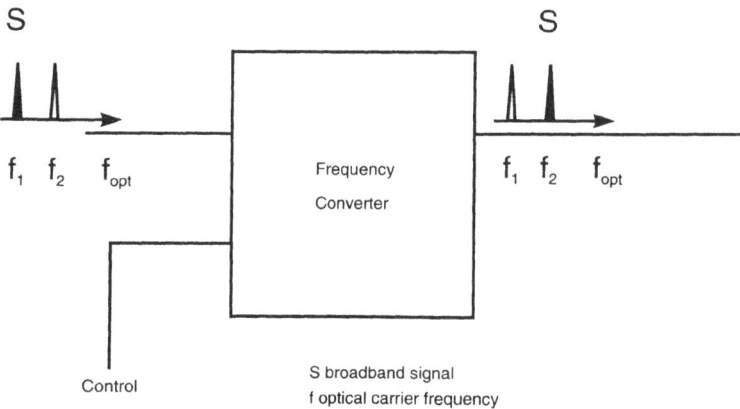

Fig. 16. Principle of frequency converter.

6. Conclusions

It is evident, that InP-OEIC technology is leaving its infancy. We have seen that monolithic integrated optics in combination with opto-electronic multi chip module techniques (OE-MCM), which include e. g. silica waveguide technology, are strong partners on the path towards future highly complex, high performance but low cost electro-optical modules for telecommunications. The basic element of optical signal processing is the waveguide itself. Efficient coupling and mode transformation between quite different waveguides is the biggest challenge of integrated optics. Starting with the laser as the light source, through passive or externally controlled waveguides with special functions such as switching or filtering, coupling to the silica waveguide or the fiber itself has to be accomplished. All these processes are possible in principle, but their maturity must await the future. Furthermore, we must look for new functions providing more complex functionality on easily fabricated chips.

We have discussed simple photonic devices which can perform complex functions, e. g. all-optical repeaters, switching in the space and frequency domain. With such devices we can exploit the bandwith of the fiber by raising the bit rate on the transmission lines (OTDM) and / or adding further wavelengths (OFDM). One of the most important driving forces of todays ATM systems is the possibility of decoupling the optimization of the network and of the services which are running on the network. In a transparent network this can be accomplished by using a new wavelength for a new service. Present electronic ATM signals can use their own individual wavelength channel, without changing the ATM network. Thus compatibility between new and old networks is guaranteed.

In this way, future telecommunication architectures may be strongly influenced by new photonic devices.

References

1. P. W. E. Smith, "Optics in Telecommunications: Beyond Transmission", in: "International Trends in Optics", J. W. Goodman, Ed. Academic Press (1991).
2. J. E. Midwinter, IEE Proceedings- J, Vol. 139, No 1, February 1992
3. D. B. Payne and J. R. Stern, IEEE, J. Lightwave Technology, LT-4, No 7, pp 864-869 (1986).
4. D. B. Payne, BT Technology, Vol.11, No. 2, pp 11-18 (1993).
5. C. Bornholdt, D. Trommer, G. Unterbörsch, H. G. Bach, F. Kappe, W. Passenberg, W. Rehbein, F. Reier, C. Schramm, R. Stenzel, A. Umbach, H. Venghaus and C. M. Weinert, Appl. Phys. Lett. 60 (8), 24.Feb. pp 971-973 (1992)
6. H. Heidrich, "Progress and Prospects towards Coherent receiver Frontend OEICs - Issues, Challenges and Perspectives- , ECIO 93,
7. P. E. Barnsley, BT Technology, Vol.11, No. 2, pp 19-34 (1993).
8. J. Bachus, "Optical tranparency", EOCLAN 93
9. H. Takeuchi, Y. Hasumi, S. Kondo and Y. Noguchi, Electron. Lett., Vol. 29, No. 6, pp 523-524, (1993).
10. N. Agrawal, D. Hoffmann, D. Franke, K. C. Li, U. Clemens, A. Witt and M. Wegener, Appl. Phys. Lett. 61, 249 (1992).
11. W. G. R. Walker, D. M. Spirirt, P. J. Chidgey, E. G. Bryant and C. R. Batchellor, Electron Lett., 28, pp 989-991 (1992).
12. H. Toba, K. Nakanishi, K. Oda, K. Inoue and T. Kominato, "A 100-channel Optical FDM In-Line Amplifier system employing Tunable Gain Equalizers", ECOC 1992, paper TuA4.2, p113

13. S. Kawanishi, H. Takara, K. Uchiyama, T. Kitoh and M. Saruwatare, "100 Gbit/s, 50 km Optical Transmission Employing All-Optical Multi/Demeultplexing and PLL Timing Extraction", OFC/IOOC 93, San José, California, Postdeadline paper PD2
14. D. M. Spirit and L. C. Blank, "Optical time division multiplexing for future figh-capacity network applications" BT Technology J, Vol. 11, No 2, 2 , p 35-45 (1993).
15. H. Taga, N. Edagawa, H. Tanaka, M. Suzuki, S. Yamamoto and H. Wakabashi,OFC/IOOC 93, San Jose, California, Postdeadline paper PD1.
16. L. F. Mollenhauer, E. Lichtman, M. J. Neubelt and G.T. Harvey, "Demonstration, using sliding-frequency guiding filters, of error-free soltiton transmission over more than 20,000 Km at 10 Gbit/s, single-channel, and over more than 13,000 km at 29 Gbit/s in two-channel WDM ", OFC/IOOC 93, San José, California, Postdeadline paper PD8.
17. S. E. Miller, Bell Syst. Techn J. 48, p 2059-2069, (1969)
18. H. Kogelnik, "Integrated Optics, OEICs, or PICs?" in: "International Trends in Optics", J. W. Goodman, Academic Press, (1991).
19. R. G. Hunsperger, "Integrated Optics: Theory and Technology", Springer Verlay. , Springer Series in Optical Sciences, Vol 33, (1991).
20. T.L. Koch and U. Koren: IEE Proceedings-J., Vol. 138, no. 2, pp. 139-147, (1991).
21. M. Erman, "Monolithic vs. Hybrid Approach for Photonic Circuits", ECIO 93, paper 2-1
22. W. T. Tsang, "Advances in MOVPE, MBE, and CBE", J. of Crystal Growth 120, p 1 - 24 (1992).
23. H. Röhle (HHI), private communication
24. A. Umbach, O. Kayser, D. Trommer and G. Unterbörsch, "Butt-Coupled PIN Photodiode on InP Using Selective Refill MOVPE Growth", Technical Digest on Integrated Photonic Research 1992, New Orleans, Louisiana, Paper TuA3-1, pp. 120-121.
25. H. Schmidt, F. Fidorra and D. Grützmacher, "Endpoint detection for CH_4/H_2 reactive ion etching of InGaAsP heterostructures by mass spectrometry", Inst. Phys. Conf. No 96, Chapter 6, Int. Symp. GaAs and Related Compounds, Atlanta, Georgia, p 431-434 (1988).
26. R. Müller, Appl. Phys. Lett. 57 (10), p 1020-1021, (1990).
27. M.J. Wale and C. Edge: "Self-aligned flip-chip assembly of photonic devices with electrical and optical connections", IEEE Trans. Comp. Hybrids, Manuf. technol., vol. 13, pp. 780-786, (1990).
28. K.P. Jackson, E.B. Flint, M.F. China, D. Lacay, J.M. Trewhella, R. Buchmann, Ch. Harder and P. Vettinger: "Flip-Chip, Self-Aligned, Optoelectronic Transceiver Module", in Proceedings of the 18th European Conference on Optical Communication (ECOC '92), 27.9-1.10.1992, Berlin (Germany), Vol. 1, pp. 329-332.
29. K. Okamoto and K. Kawachi, "Recent Progress in Silicon Based Integrated Waveguide Devices", in Proceedings, Topical Meeting on Integrated Photonics Research (IPR '93), March 22-24, 1993, Palm Springs, California, paper ITuE1, pp 280-283
30. H. Takeuchi, K. Kasaya, Y. Kondo, H. Yasaka, K. Oe and Y. Imamura: IEEE Photonics Technology Letters, Vol 1, no. 11, pp. 398-400, (1989).
31. H.G. Bach, J. Krauser, H.-P. Nolting, R. A. Logan, F. K. Reinhart, Appl. Phys. Lett., Vol. 42 (1983).
32. C. Bornholdt, W. Döldissen, D. Franke, J. Krauser, U. Niggebrügge, H.-P. Nolting and F. Schmitt, "High-Efficiency Phase Modulators in InGaAsP/InP", H.-P. Nolting, R. Ulrich: "Integrated Optics", Springer Series in Optical Sciences, Vol. 48 (1985), pp. 121-125.
33. H.-P. Nolting, D. Hoffmann and M. Schlichting, "Integrated Optical Wavelength Multiplexer/Demultiplexer for Optical Communications", in: "Integrated Optics", H.-P. Nolting, R. Ulrich, Springer Series in Optical Sciences, Vol. 48 (1985), pp. 215-220.
34. C. Bornholdt, F. Kappe, R. Müller, H.-P. Nolting, F. W. Reier, R. Stenzel, H. Venghaus and C. M. Weinert, Appl. Phys. Lett., Vol. 57, No. 24, pp. 2517-2519, (1990).

35. H. Heidrich, M. Hamacher, P. Albrecht, H. Engel, D. Hoffmann, D. Imhof, H.-P. Nolting, F. Reier and C.M. Weinert; "Polarisation Diversity Waveguide Network Integrated on InP for a Coherent Optical Receiver Front-End", in Proc. of the IOOC/ECOC '91, 9.-12.09.91/Paris (F), Part I, pp. 189-192.
36. H.-P. Nolting, Frequenz 41, pp. 21-25,.(1987)
37. P. J. Williams, P. M. Charles, I. Griffith, L. Considine and A. C. Carter, Electronics Lett., Vol 26, no. 2, pp. 142-143, (1990).
38. H. Heidrich, M. Hamacher, P. Albrecht, H. Engel, D. Hoffmann, D. Imhof, H.-P. Nolting, F. Reier and C.M. Weinert; "Polarisation Diversity Waveguide Network Integrated on InP for a Coherent Optical Receiver Front-End", in Proc. of the IOOC/ECOC '91, 9.-12.09.91/Paris (F), Part I, pp. 189-192.
39. P. Albrecht, M. Hamacher, H. Heidrich, D. Hoffmann, H.-P. Nolting and C. M. Weinert, IEEE Photon. Technol. Lett., Vol. 2 (1990), No. 2, pp. 114-115.
40. M. Schlak, P. Albrecht, H. Heidrich, H.-P. Nolting and F. Reier: "Electro-Optically Tunable General Polarization Transformer on GaInAsP/InP Using "Analog" und "Digital" Control Voltage Patterns", in Proceedings, Topical Meeting on Integrated Photonics Research (IPR '92), New Orleans, USA, 13.-15.04.1992, Technical Digest, pp. 44-45 (MB20-1).
41. R.J. Deri, E.C.M. Pennings, A. Scherer, A.S. Gozdz, C. Caneau, N.C. Andreadakis, V. Shah, L. Curtis, R.J. Hawkins, J.B.D. Soole and J.-I. Sing, IEEE Photonics Technology Letters, Vol. 4, no. 11, pp. 1238-1240, (1992).
42. E.C.M. Pennings, R.J. Deri, A. Scherer, R. Bhat, T.R. Hayes and N.C. Andreadakis, Appl. Phys. Lett., Vol. 59, pp. 1926-1928, (1991).
43. E. A. J. Marcatili, IEEE Journal of Quantum Electronics, Vol. QE-21, pp 307-314, (1985).
44. H.-P. Nolting, "Theoretical Investigations of Optical Waveguide Tapers on InGaAsP/InP", Proc., Integrated Photonics Research Topical Meeting (IPR '91), Monterey, California, USA, 09.-12.04.1991, pp. 111-113.
45. L. Moerl, L. Ahlers, P. Albrecht, H. Engel, H.-J. Hensel, H.-P. Nolting and F. Reier: "Efficient fiber-chip butt coupling using InGaAsP/InP waveguide tapers", in Proceedings Conf. on Optical Fiber Commun./Internat. Conf. on Integrated Optics and Optical Fiber Communsication (OFC/IOOC´93), San José (USA), February 21-26, paper ThK2.
46. T. Brenner, W. Hunziker, M. Smit, M. Bachmann, G. Guekos and H. Melchior, Electronics Lett., Vol. 28, no. 22, pp. 2040-2041, (1992).
47. D. Trommer, C. Bornholdt, R. Stenzel, A. Umbach and G. Unterbörsch, "Integrated balanced mixer receiver on InP", Proceedings of the Fourth Conf. on InP and Related Materials, Newport, RI, Apr. 1992, p 250-253
48. U. Koren, T. L. Koch, B. I. Miller and A. Shahar, "Process for large scale photonic integrated circuits", Integrated and Guided-Wave Optics, 1989, Houston, TX, Feb. 6-8, Paper MDD2-1
49. P. B. Hansen, G. Raybon, U. Koren, B. I. Miller, M. G. Young, M. A. Newkirk, M.-D. Chien, B. Tell and C. A. Burrus, "Monolithic semiconductor soliton transmitter", OFC/IOOC 93, San Jose, California, Postdeadline paper PD22
50. C. Dragone, IEEE Photonics Technology Letters, Vol 3. No. 9, p. 80-83, (1991).
51. J.-M. Verdiell, M. A. Newkiek, T. L. Koch, R. P. Gnall, U. Koren, B. I. Miller and B. Tell, "A Frequency Reference Photonic Integrated Circuit for WDM with Low Polarization Dependence", in Proceedings, Topical Meeting on Integrated Photonics Research (IPR '93), March 22-24, 1993, Palm Springs, California, paper IME2, pp 143-146
52. R. C. Alferness, "Electrically tuned optical waveguide filters", ECOC 92, paper WeB9.1 (invited), Vol.2
53. NTT, tunable MZ filter.
54. S. Illek, W. Thulke and M.-C. Amann, Electron. Lett., Vol 27, 2207-2208, (1991).
55. T. Wolf, S. Illek, J. Rieger, B. Borchert and W. Thulke, IEEE Photon. Technol. Lett. vol. PT-5, No. 3, (1993).

56. M. J. Adams, P. E. Barnsley, J. D. Burton, D. A. O. Davies, P. J. Fiddyment, M. A. Fisher, D. A. H. Mace, P. S. Mudhar, M. J. Robertson, J. Singh and H. J. Wickes, BT Technology J. Vol. 11 No 2, p. 89 - 97, (1993).
57. G. P. Bava, P. Debernardi and M. Tonetti, "Carrier Heating Effects on High Frequency Conversion in Semiconductor Laser Devices", Integrated Photonics Research Topical Meeting, March 22-24, 1993, Paper IMB6, p27-30
58. U. Feiste, J. Hörer, M. Möhrle, B. Sartorius and H. Venghaus, "Frequency Tuning Characteristics of Self-Pulsating Two-Section InGaAsP/InP DFB lasers"; Proc., 18th Conf. on Optical Communication (ECOC '92), Berlin 27.9.-1.10.1992, Vol. 1, pp. 245-248
59. M. Möhrle, U. Feiste, J. Hörer, R. Molt and B. Sartorius, IEEE Photon. Technol. Lett., Vol. 4, pp. 976-978, (1992).
60. U. Bandelow, H. J. Wünsche and H. Wenzel, "Theory of Selfpulsation in Two-Section DFB Laser Diodes"; to be published in IEEE Photonic Techn. Letters
61. B. Sartorius, M. Möhrle, D. J. As, J. Höhrer, H. Venghaus and U. Feiste, "High Frequency Locking at 18 GHz in a self pulsating DFB Laser", ECOC 93
62. P. E. Barnsley, G. E. Wickens, H. J. Wickes and D. M. Spirit, IEEE Photon Technol Letters, 4, No 1, pp 83-86 (1992).
63. E. Patzak, D.J. As, R. Eggemann and K. Weich, private communication.
64. H. Tsuda, K. Nonaka, K. Hirabayashi, H. Uenohara, H. Iwamura and T. Kurokawa, "2.1 GHz wavelength conversion over 100 nm input range with a side-injection light controlled bistable laser diode", Integrated Photonics Research Topical Meeting, March 22-24, 1993, Paper PDP-2
65. R. Schnabel, W. Pieper, R. Ludwig and H. G. Weber, Electronics Letters, Vol. 29, No. 9, p 821-822, (1992).
66. M. C. Tatham and G. Sherlock, "20 nm Wavelength Conversion using Ultrafast Highly Nondegenerate Four-Wave Mixing", Integrated Photonics Research Topical Meeting, March 22-24, 1993, Paper PDP-1
67. R. Schnabel, W. Pieper, R. Ludwig and H. G. Weber, "Ultrafast, Multi-THz Frequency Conversion of a Picosecond Pulse Train using a 1.5 mm MQW Semiconductor Laser Amplifier", Photonics in Switching 1993, Postdeadline paper.

Chapter 20

OPTICAL SIGNAL PROCESSORS AND APPLICATIONS

M.N. ARMENISE and V.M.N. PASSARO

1. Introduction

In the last few years, a considerable effort has been expended on the development of a number of both active and passive guided-wave devices, such as laser diodes, electrooptic modulators, acoustooptic transducers, photodetectors and microlens arrays, for optical signal processing applications including optical computing [1-3]. Many simple optical devices can be easily fabricated by using mature and reproducible technologies. The interest in optical processors involving simple devices is mainly due to a lower power consumption, reduced size, cost and weight and high throughput with respect to the corresponding electronic processors. A number of guided-wave optical circuits for communications, signal processing and optical computing have been demonstrated using the acoustooptic, magnetooptic and electrooptic effects [4]. Moreover, fields of application of particular interest are synthetic aperture radar (SAR), remote sensing and beam forming of linear array antennas, which are all well suited for optics-based processing because of the parallel processing allowed by optical architectures. Therefore, guided-wave optical processors can be successfully applied to the optical control of microwaves, to the reconstruction of two-dimensional images using both spatial and time integration, to data classification and identification in satellite applications and for performing a number of signal processing and optical computing functions.

The most studied materials for such guided-wave optical processing systems are lithium niobate [5] ($LiNbO_3$) which leads to a hybrid technology, and III-V semiconductor materials, such as gallium arsenide (GaAs) and related compounds, as potential candidates for the fabrication of monolithic integrated processors [6]. In fact, these two technologies have been successfully used in a number of applications such as optical communications, optical signal processing (such as SAR and remote sensing applications), optical beam forming networks and optical computing. In particular, where real-time and on-board SAR image formation with a side-looking focused radar is needed, the SAR data compression along both range and azimuth direction can be achieved with good resolution by using optical processors based on acoustooptic interaction cells and charge-coupled device (CCD) matrices, both in $LiNbO_3$ [7] and in GaAs [8] technology. The $LiNbO_3$ technology has been also applied to the development of an electro-optic comparator for reconstructing images in spacecraft SAR [9,10] and of a phase comparator for data identification and classification in remote sensing applications [11].

M. N. Armenise and V. M.N. Passaro - Politecnico di Bari, Dipartimento di Elettrotecnica ed Elettronica, Via E. Orabona 4, 70125 Bari, Italy

These guided-wave processors offer new technological and geometric solutions in terms of their optical components in order to solve some fundamental problems, such as long-distance SAR image reconstruction and the remote classification of large amount of data. Optical architectures and technologies have been also successfully applied to high-directionality far-field beam forming at high radio frequencies in linear active antenna arrays by using the compactness and the parallelism allowed by optical circuits [12]; and to the fabrication of multichannel modules formed by channel waveguide arrays, microlens arrays, acoustooptic transducers and electrooptic modulators for signal processing and optical computing functions (wavelength division multiplexing/demultiplexing, heterodyning detection, space-integrating correlation, spectral analysis, vectorial multiplication and so on).

2. Guided-Wave Processor for Real-Time SAR Applications

A SAR system provides the transverse position (range coordinate) and the longitudinal position (azimuth coordinate) of a ground target with respect to the flight line of the aircraft or satellite carrying the radar antenna [13-17]. A periodic pulse train (reference signal) is transmitted by the antenna and the sequence of radar echoes coming from the illuminated object is received by the radar (received signal) and collected according to the pulse repetition frequency (PRF). This imaging technique can form high resolution images at relatively long radar wavelengths with small antennas, when an appropriate processing of the received radar signal is performed with the transmitted signal, and it is very attractive for an optics-based implementation due to the high parallelism required. The reconstruction of the required two-dimensional image of the illuminated object can be carried out by using a guided-wave optical circuit in conjunction with an external CCD system, by which both spatial and temporal data compression is performed [18]. In this section we present some guided-wave solutions for the SAR image reconstruction, using both lithium niobate and gallium arsenide technology.

2.1. LiNbO3 Guided-Wave Processors

The first correlator scheme we have proposed [7] is sketched in Fig.1. A proton exchanged (PE) planar waveguide is fabricated on a Z-cut $LiNbO_3$ single crystal. A laser source (λ_0=0.84 µm), butt-coupled to the edge of the waveguide, is intensity modulated by the SAR reference signal and its beam is collimated by a geodesic lens. The following acousto-optic Bragg cell deflects the beam by applying the SAR received signal as modulating voltage and the undiffracted part of the light is shifted away by a filtering grating to the absorbing screen. The beam carrying the useful information leaves the waveguide, expanding in the three-dimensional space by natural diffraction, and it is vertically collimated and focused in the horizontal direction by an achromatic doublet on a CCD matrix. The light illumination of the CCD

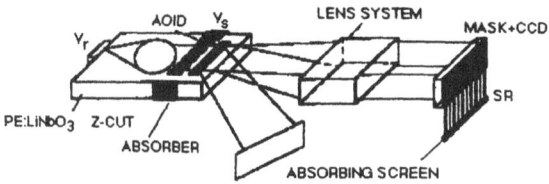

Fig. 1. Schematic of the $LiNbO_3$ acousto-optic correlator.

depends on a mask positioned in close proximity and modulated in transmissivity according to an appropriate chirp function.

Therefore, the circuit performs the range compression partially by the Bragg cell and partially by the cells of the CCD columns. The azimuthal compression is carried out by the CCD-mask system in the discrete-space domain. The crystal orientation has been chosen in order to have isotropic light propagation in the planar waveguide and no distortion of off-axis rays, while the appropriate PE process assures a purely monomodal condition (TM_o mode) and low optical damage.

The reference signal V_r drives the laser beam intensity, while the signal to be processed V_s modulates the amplitude of a carrier applied to the transducer, whose frequency determines the induced acousto-optic grating period. Because the spatial coordinate along the circuit axis is equivalent to the temporal coordinate of the correlation function between the reference and received signal, the interaction between the optical reference signal and received signal occurring in the Bragg cell actually performs a correlation function. In fact, each slice of the diffracted beam is integrated over the laser pulse period in the corresponding column of the CCD array. Finally, the weighting of the incident light involved by the mask, in connection with the PRF synchronized shift of the CCD row contents, acts as an azimuth compression and it is exactly a convolution, as it has been demonstrated by a detailed simulation [19].

Thus, this correlator can satisfy the main requirement for real-time processing, i.e., the ability to store all the azimuthal data for each range bin and to perform a multichannel correlation along the azimuthal axis. It is suitable for on-board application owing to its small dimensions (1.5x1.5x0.5 cm^3) and low power consumption ($\approx$ 100 mW). A range resolution of 3 m has been calculated assuming a range swath of 1 km for airborne SAR.

However, the reduced interaction length of the acoustooptic Bragg cell ($\approx$ 1 cm) due to the limited surface acoustic wave (SAW) propagation velocity (V_a=3.900 m/s for Z-cut PE:$LiNbO_3$ waveguides), does not processing of SAR data having very long processing time, as occurs in satellite applications.

For this reason, we have presented a new configuration of $LiNbO_3$ guided-wave optical correlator suitable for satellite applications, in which a serial operation between the reference and received signal allows the processing of very long data sequences [9]. It is based on a complex interferometric structure, involving four aperiodic phase-reversal traveling wave modulators (see Fig.2), which constitute two cascade Mach-Zehnder intensity modulators. These two Mach-Zehnder modulators are driven in order to reproduce the product signal

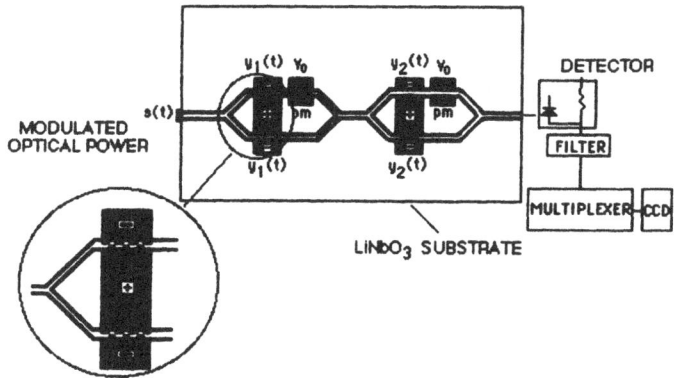

Fig. 2. Configuration of the Ti:$LiNbO_3$ interferometric device.

between the SAR received and reference signal, which is then time-integrated by a suitable photodetector. The filtered signal coming from the detector is proportional to the final correlation function, which can be electronically registered and multiplexed on a two-dimensional matrix by a sum-and-shift procedure. Thus, the processor performs the correlation function between the reference signal and received signal when they are applied to the laser diode and to the electrodes as driving voltage, respectively.

A titanium indiffused channel waveguide in an X-cut $LiNbO_3$ single crystal is assumed. A laser source is coupled into the waveguide and its beam ($\lambda_o=1.3$ μm, TE polarized) is driven by a microwave carrier at f_c frequency, whose amplitude-modulation reproduces the SAR reference chirp signal s(t). The beam suffers from the modulation induced by the Mach-Zehnder interferometers driven in push-pull linear mode by the base-band received $y_1(t)$ SAR signal at f_r frequency and by a bias $y_2(t)$ signal, respectively. The guided beam is then detected by a fast (>10 GHz) external photodetector by time-integrating the output optical intensity in a very short time period. The photogenerated current is filtered and the output signal, which is proportional to the integral product between the received and reference chirp signals, is multiplexed in order to reconstruct the correct correlation function by an electronic circuit in connection with a two-dimensional buffer array according to the electronic reference driving circuit [17]. Since the useful product between $y_1(t)$, $y_2(t)$ and s(t) is allocated around the f_r-f_c frequency, an appropriate filtering step having bandwidth 2xB is performed after the detection step, having f_r and f_c values depending on the chirp bandwidth of the particular application (airborne or spaceborne). The filtered signal drives an external electronic processing circuit formed by a high-frequency multiplexer and an external buffer in two-dimensional matrix form by an add-and-shift procedure on the same row for one pulse period by shifting the column content, thus reconstructing the temporal correlation function along the same row, appropriately weighted. In the design and simulation steps, we have considered the parameters of an airborne (AIR-C) and satellite (ERS1) SAR mission, as shown in Table I.

The processing time of this circuit, since it has small size (0.5x2.5x1 cm^3), is determined by the electronic circuits (a few ns for each received echo), while the maximum delay time needed to collect the whole data stream is given by the delay range (4.5 μs for airborne and ≈280 μs for satellite applications, respectively). The range and azimuth resolutions depend on the final multiplexing and post-processing performances (6.1 m and 37.2 m, respectively). This kind of architecture can be used also for many optical computing functions, such as fast cross-correlation, switching and matrix algebra.

2.2. GaAs Guided-wave Processors

Gallium arsenide technology allows a monolithic approach to the fabrication of guided-wave correlators in order to reduce circuit sizes and power consumption. The first scheme we present consists of a GaAs guided-wave integrated acousto-optic correlator, which can perform the range compression, and an external processing circuit based on a two-dimensional CCD array, which carries out the azimuthal compression [8,20]. This kind of processor is suitable mainly for airborne applications with side-looking focused radar.

The correlator scheme is sketched in Fig.3. We consider a planar optical waveguide formed by two $Al_xGa_{1-x}As$ epitaxial layers, a cladding layer and a buffer layer, grown on a GaAs substrate of very small size (3.7x1.7x0.1 cm^3).

A laser diode is butt-coupled into the crystal and emits TM-polarized light at the wavelength of $\lambda_o=0.85$ μm. Its beam is coupled into the waveguide and is expanded and collimated by two diffraction gratings having nonlinear profiles. An acoustooptic Bragg cell,

Table I. Simulation parameters of SAR missions.

PARAMETERS	SYMBOL	UNIT	AIR-C	ERS1
Radar carrier frequency	f_o	GHz	5.3	5.3
Chirp bandwidth	B	MHz	40	20
Chirp pulse time duration	T_o	µs	1.2	38
PRF	1/T	Hz	278	1720
Platform height	z	km	8	785
Medium range	R_o	km	10	854
Aircraft speed	v	m/s	210	7467
Minimum echo delay	t_1	µs	65.13	5583
Maximum echo delay	t_2	µs	68.50	5837
Delay range	$t_d = t_2 - t_1$	µs	3.37	254

fabricated on a piezoelectric ZnO layer, deflects the incident beams, giving at the output two diffracted and two undiffracted guided beams. The undiffracted light beams are shifted away by a filtering reflection grating, a TM-polarized diffracted light beam is coupled by a Fresnel lens array into a stripe waveguide array and a TE-polarized diffracted beam is uncoupled by the same lens array. Thus, the guided light into the stripe waveguide array can be out-coupled and focused along the transverse line by a grating, in order to illuminate the CCD matrix through a transmission mask.

The laser beam is modulated by the chirp reference signal at the PRF frequency while the received radar echoes, coming from the illuminated object, are applied to the electrodes of the AO transducer (AOT) as an amplitude-modulated signal. However, the role of reference and received signal can be interchanged. Range compression is carried out because each out-coupled elementary component of the useful beam, carrying one piece of information, is integrated over the laser pulse period in the corresponding column of the CCD matrix. Moreover, azimuthal compression is obtained again by the appropriate weighting of the CCD incident light, provided by the transmission mask and by the shift of the CCD row contents along the column direction, according to the PRF.

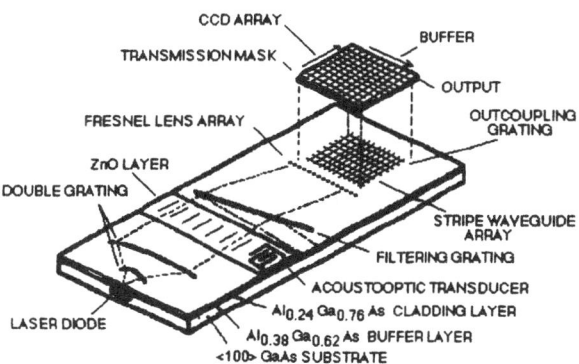

Fig. 3. Scheme of the GaAs optical processor for SAR applications.

In design and simulation, we have considered the parameters of the airborne SAR mission AIR-C. A delay time of 720 ms and a maximum processing time of ≈3.5 µs to detect the target presence in the swath area have been achieved by the design and simulation procedures. A sophisticated acoustic mode analysis of III/V structures [21] and a complete modeling of the guided-wave acoustooptic interaction [22] are required in order to take into account the frequency dispersion of SAW propagation which occurs in III/V multilayered structures. The proposed architecture allows a lower power consumption, smaller size and weight and lower computational cost with respect to analogous digital electronic processors and guided-wave optical processors based on LiNbO$_3$ technology. This kind of architecture can also be applied to a number of signal processing functions, such as time-integrating correlation and spectral analysis.

3. Guided-Wave Circuits for Remote Sensing Applications

Existing digital systems for data acquisition by satellite remote sensing involve very large and powerful computers, which are able to identify and classify a large amount of data, as occurs for example in continuous monitoring applications. In order to reduce the computational cost of such a system, it is desirable to transfer on-board as much as possible of the system hardware complexity. Then an on-board data screening system can be very useful, because it can select only a particular portion of transmitted data, depending on a set of user-controlled reference signals.

To this end, an optical processing system, based on an integrated optical circuit on a LiNbO$_3$ crystal, may be preferred to a digital electronic one because of its higher compactness and reduced weight. The input data, coming from sensors, can modulate an optical carrier by the electrooptic effect, which is recorded by the photorefractive effect in a holographic region and is compared with the optical carrier modulated by the reference signal, as proposed by Armenise et al. [11,23,24].

The configuration of this guided-wave phase comparator (see Fig.4), which has reduced size (0.1x1.5x0.25 cm^3), involves a Y-branch beam splitter, two stripe waveguides having appropriate final horns to reduce the background noise in the holographic region and a planar waveguide supporting the holographic region.

A laser beam (λ_o=0.6328 µm, TE-polarized) is split into two parallel arms by a Y-branch junction on X-cut Ti:LiNbO$_3$ channel waveguides. The two resulting guided waves, after suffering the phase changes due to a set of input modulators and to a reference modulator, interfere in the TIPE planar region, whose photorefractive sensitivity can be enhanced by Fe-doping, forming a transmission grating. The interference of the two waves creates a dark zone when the wave phase shifts are equal and, in that case, no output signal is detected. On the contrary, when the resulting phase changes are different, an illuminated zone is created in the hologram and a useful signal is detected, having an intensity proportional to the squared phase difference.

This operation allows us to identify and classify a number of input data by comparing them with the user-controlled reference signal. If the reference voltage is constant, the input phase changes can be continuously detected and the response time of the whole circuit is determined by the writing time t_g of each grating. This working mode may be called the "basic operation" mode. When the reference signal changes, the adaptation of the preprocessor to the new signal is determined by a time constant T_c, because the circuit can "see" a new reference only if it persists longer than this characteristic time: this is the "dynamic operation" mode.

Table II. Performance parameters of LiNbO$_3$ circuit for remote sensing.

PARAMETERS	SYMBOL	UNIT	VALUE
Free-space optical wavelength	λ_o	nm	633
Input optical power	P_O	mW	1
Circuit time constant	T_C	µs	1.1
Hologram writing time	t_g	µs	1.22
Equivalent noise phase change	ϕ_n	rad	3×10^{-3}
DC voltage minimum change	V_{min}	mV	11

The parameters characterizing the whole circuit performance are shown in Table II. The optical noise in the hologram, for λ_o=0.6328 µm and P_o=1 mW, is about 0.1% of the useful signal power. Continuous changes in the reference signal are allowed in the dynamic mode operation, within about 0.01 MHz. These values have been obtained by considering titanium indiffused channel waveguides and compared with proton exchanged guides formed with benzoic acid. However, better results could be achieved by using other proton sources [25].

4. Guided-Wave Circuits for Beam Forming Networks

The optical control of microwaves has been receiving considerable attention in recent years because of the progress made in the field of optical communications. In particular, optical beam forming of phased-array radar antennas has increasingly been considered to improve the performance in terms of phase/amplitude signal stability and reduced size and complexity of the Beam Forming Network (BFN). In these applications, the required phase encoding is applied to an optical frequency carrier, which is then transformed to the desired microwave frequency by a heterodyne mixing process. Each element is fed by a coherent microwave signal with different phase information so that a highly directional far-field lobe pattern is generated. The far-field pattern orientation is steered by phase variation of the control signal of each element. Integrated optics devices can be successfully utilized in a number of applications with substantial reduction of mass, volume and power consumption with respect to traditional solutions.

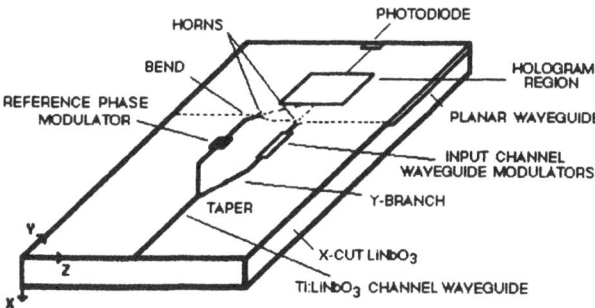

Fig. 4. Conceptual scheme of the Ti:LiNbO$_3$ preprocessor.

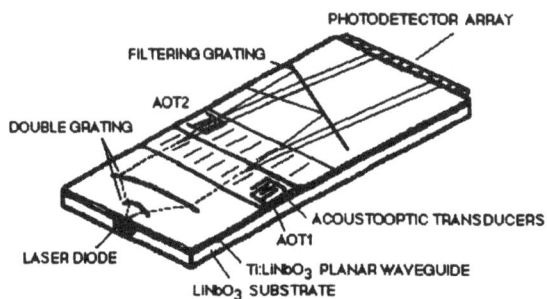

Fig. 5. Architecture of guided-wave processor for beam forming network.

The processor we propose is shown in Fig.5 and consists of three different sections. In the first section, a TM-polarized laser beam, at the free-space wavelength $\lambda_o=0.85$ μm, is coupled into a titanium-indiffused lithium niobate (Y-cut, X-propagating Ti:LiNbO$_3$) planar waveguide, expanded and collimated by two different grating lenses. In the second section, two acousto-optic (AO) transducers (AOT1 and AOT2), driven at (f_c-f_o) and (f_c+f_o) radiofrequency (RF), respectively, deflect and frequency-shift the incident beams coming from the gratings, both TM and TE modes due to the polarization conversion. f_c is the center frequency of both the transducers and f_o is a changeable control frequency. In the third section, a photodetector array detects the beams along the transverse direction, after an appropriate filtering of the undeflected beams at the output of AOT2 operated by another grating. The phase shift between each pair of array elements can be controlled linearly by changing the frequency f_o. Therefore, the photocurrent generated by each photodetector, after appropriate electronic filtering, can feed at $2 \times f_c$ the corresponding element of the linear phased-array antenna in order to obtain a highly directional far-field pattern.

The processor we have investigated, designed and simulated has small size (4x0.1x1.5 cm^3) and is suitable for transmitted beam forming of a linear array of phased active antennas. The design procedure has allowed us to determine the optimized parameters of the circuit, i.e. the RF center frequency of the two transducers $f_c=400$ MHz, the maximum control frequency $f_{omax}=8.74$ MHz, and the number of driven antennas $N_o=100$ with a signal-to-noise ratio on each detector of about S/N≈20 dB, assuming the propagation of fundamental optical modes (TE$_o$ and TM$_o$) into the Ti:LiNbO$_3$ planar waveguide. The photodiode array spacing is d=100 μm and the size of each photodetector is $d_z=30$ μm.

5. Multichannel Guided-Wave Devices

In the last few years, an increasing research effort has also been devoted to the development and the fabrication of laser diode arrays, microlens arrays and channel waveguide arrays. The aim is basically the realization of a number of different modules composed of multichannel integrated optical components to be used in RF spectral analysis and optical computing applications [4]. Fig. 6 shows a configuration of a multichannel integrated module, fabricated on a Y-cut LiNbO$_3$ crystal having sizes 2x0.1x1 cm^3. This circuit includes a planar waveguide, a channel waveguide array, a microlens array and a large aperture lens. Twenty channel waveguides, having a width of 6 μm and a spacing of 200 μm, are considered. The microlens array consists of 20 identical lenses, each of them having an aperture of 200 µm and a focal length of 2000 µm.

A number of multichannel devices can be built by inserting in the region between the microlens array and the large aperture lens different components, such as a single acoustooptic Bragg modulator, an electrooptic modulator array, or a combination of both, for performing a number of applications in the communications (e.g., switching, Wavelength Division Multiplexing, demultiplexing), optical computing and RF signal processing areas. In the scheme of Fig.6, the Bragg cell diffracts and frequency-shifts the optical power coming from the channel waveguide array, performing the product of the incident power and the acoustic power depending on the transducer voltage: this operation acts as a matrix multiplier when a large aperture lens is provided for adding the single contributions by focusing the optical power in a photodiode. When an electrooptic modulator is also considered (see Fig.6), the processor can execute multiplications and relative additions between the pulse train applied on the transducer and the pulse train on the phase modulator, thus performing cross-correlation of the two signals.

6. Conclusions

A number of guided-wave processors to be used in communications, signal processing and optical computing, have been presented. In particular, two guided-wave circuits based on lithium niobate technology have been proposed for SAR data compression in airborne and satellite systems, and one based on gallium arsenide technology for airborne system. Whereas gallium arsenide offers the possibility of monolithic integration on a single crystal, lithium niobate technology seems yet today more mature and versatile. Interesting results have been obtained for optical data preprocessing in remote sensing applications using a $LiNbO_3$ guided-wave phase comparator, and for the steering of linearly-spaced phased array antennas by forming the microwave beam on a $Ti:LiNbO_3$ planar waveguide. A configuration of multichannel guided-wave modules for performing RF signal processing, WDM and computing operations has been also described.

ACKNOWLEDGMENTS - The authors wish to thank Mr. Elio Cantatore for his valuable collaboration. This work has been partially supported by the CNR - T.E.O. Finalized Project.

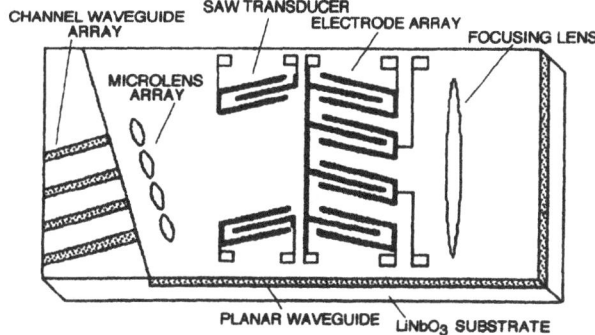

Fig. 6. Functional scheme of guided-wave multichannel module.

References

1. A.S. Semenov, V.L. Smirnov and A.V. Shmal'ko, Sov. J. Quantum Electron. 17:836 (1987).
2. M.N. Armenise and V.M.N. Passaro, "Recent advances in lithium niobate signal processors" in: "Proc. of 5° Congr. Naz. di Elettron. Quantistica e Plasmi," G.C. Righini, ed., Compositori, Bologna (1989).
3. C.S. Tsai, IEEE Trans. on Circuit and Systems 26:1072 (1979).
4. C.S. Tsai, IEEE Trans. on Ultrason., Ferroelect. and Fr. Control 39:529 (1992).
5. M.N. Armenise, IEE Proc. Pt.J 135:85 (1988).
6. T.L. Koch and U. Koren, IEEE J. Quantum Electron. **27**:641 (1991).
7. M.N. Armenise, E. Pansini and A. Fioretti, Proc. SPIE 993:225 (1988).
8. M.N. Armenise, F. Impagnatiello, V.M.N. Passaro and E. Pansini, Proc. SPIE 1562:160 (1991).
9. M.N. Armenise, V.M.N. Passaro, T.Conese and A.M. Matteo, Proc. SPIE 1704:125 (1992).
10. M.N. Armenise, V.M.N. Passaro, T.Conese and A.M. Matteo, Proc. SPIE 1904 (1993).
11. M.N. Armenise and V.M.N. Passaro, Bulgarian J. of Physics 16:563 (1989).
12. M.N. Armenise, V.M.N. Passaro, T.Conese and A.M. Matteo, Proc. SPIE 1794:489 (1992).
13. C. Elachi, T. Bicknell, R.L. Jordan and C. Wu, Proc. IEEE 70:1174 (1982).
14. K. Tomiyasu, Proc. IEEE 66:563 (1978).
15. C.D. Daniel, IEE Proc. Pt.J 133:7 (1986).
16. D. Psaltis and K. Wagner, Optical Engineering 21:822 (1982).
17. P. Kellman, Proc. SPIE 185:130 (1979).
18. T. Bicknell, D. Psaltis and A.R. Tanguay, J. Opt. Soc. Am. A 2:P8 (1985).
19. M.N. Armenise, F. Impagnatiello and V.M.N. Passaro, Proc. SPIE 1230:582 (1990).
20. M.N. Armenise, F. Impagnatiello and V.M.N. Passaro, Optical Comp. and Process. 2:79 (1992).
21. M.N. Armenise, V.M.N. Passaro and F. Impagnatiello, J. Opt. Soc. Am. B 8:443 (1991).
22. M.N. Armenise, A.M. Matteo and V.M.N. Passaro, Proc. SPIE 1583:256 (1991).
23. M.N. Armenise and V.M.N. Passaro, Proc. SPIE 1177:123 (1989).
24. M.N. Armenise and V.M.N. Passaro, IEE Proc. Pt.J 137:347 (1990).
25. C. Ziling, L. Pokrovskii, N. Terpugov, I. Savatinova, M. Kuneva, S. Tonchev, M.N. Armenise and V.M.N. Passaro, J. Appl. Physics, 73:3125 (1993).
26. N. A. Riza and D. Psaltis, Applied Optics 30:3294 (1991).
27. N. Riza, IEEE Phot. Technol. Lett. 4:177 (1992).

Chapter 21

PROGRESS OF HIGH-SPEED OPTOELECTRONICS

HUAN-WUN YEN

1. Introduction

During the past decade, there has been a great deal of progress in the development of high-speed optoelectronic components. Semiconductor lasers, photodetectors, and electro-optic modulators with bandwidth in the tens of gigahertz range have been reported. This, coupled with the maturity of fiber optic communications technology, has created an active field of research and development where photonic technologies are used to accomplish a variety of microwave signal processing functions[1,2]. These functions include: generation, modulation, switching, filtering, and transmission of microwave signals; injection locking, time-shifting, phase-shifting, and frequency modulation of microwave oscillators; RF spectrum analysis and signal identification; and circuit and material characterization.

There are many reasons for using optical techniques rather than the conventional RF approaches to perform these functions. In some cases a high degree of electrical isolation is desired and optics offers the best solution. In other cases, there is a need for high-speed switching or gating of microwave signals but there are no microwave switches with short enough rise and fall times to do the job. There is also a large class of applications where microwave signals need to be transmitted over a long distance or delayed for a relatively long period of time. This is traditionally accomplished by passing microwave signals through either copper waveguides or coaxial cables. The copper waveguides are heavy and bulky while the coaxial cables are relatively lossy. Therefore, using conventional approaches there is a practical limit on how far one can go with guided RF transmission. On the other hand, with high-speed lasers (or modulators) and photodetectors it is possible to transmit microwave signals over optical fiber cables with negligible propagation loss. As long as the electrical-to-optical and optical-to-electrical conversion losses can be compensated for with appropriate electronic amplification, the optical-microwave technique offers many new options to microwave system designers.

In general, optical techniques offer very good electrical isolation (50 - 70 dB) and fast response (on the order of a nanosecond) for microwave switching applications. They provide the capability for wide bandwidth, low crosstalk, low dispersion, minimal electromagnetic interference, and very long delays (up to milliseconds).

Huan-Wun Yen - Hughes Research Laboratories, 3011 Malibu Canyon Road, Malibu, California, 90265 USA

These features not only can help solve the difficult problems facing microwave engineers but also offer other potential benefits such as smaller size, lighter weight, and lower power consumption. In addition, they free up system design constraints by allowing subsystems to be configured and located at more convenient locations without suffering from severe performance degradation.

2. Background

Integrated optics was originally pursued with the vision that someday complex optical processing functions can be carried out in a compact, rugged optical chip similar to electronic integrated circuits. Therefore, a number of optical signal processing ideas were proposed and investigated during mid and late seventies. Some examples are: A/D converters, integrated optic spectrum analyzers, tapped delay lines, and transversal filters. Although integrated optic chips as originally perceived did not materialize the research effort did lead to the development of several important devices based on optical waveguide technology; most notably, $LiNbO_3$ modulators and switches. There have also been substantial improvements in diode laser performance, both in lower threshold and in increased modulation bandwidth. High-speed photodetectors are also being developed to satisfy the ever increasing data rate needs of optical communications. With all the high-speed components in place it is only a matter of time before one can begin to put together functional microwave-optical systems.

There were several major milestones that propelled the field of microwave-optics. One is the realization and understanding of multi-gigahertz direct modulation of semiconductor lasers. The functional dependence of laser resonance frequency on device parameters was determined using a rate equation formulation and the results were used to guide the design of high-frequency lasers. The power spectrum of laser relative intensity noise and the optical feedback induced noise behavior of lasers were extensively studied. As a result, very low noise lasers have been developed and optical isolators are engineered into laser transmitters. Another major event is the successful development of traveling wave electro-optic modulators with modulation bandwidth greater than 20 GHz. These integrated optic modulators also have acceptable drive power level (10 to 100 mW) compared to their bulk counterparts. Today, devices with 40, 60, and even 94 GHz modulation capability are being pursued for millimeter wave applications. A third major event is the development of mode-locked semiconductor lasers that can generate extremely short optical pulses. These pulses in turn are used to excite "optoelectronic switches" for microwave generation, switching, gating, and sampling. This technique has also been used to characterize high-speed microwave circuits, semiconductor lasers, detectors and modulators. The merit of using optical control signals is that there is minimal electrical loading to the devices being controlled and there is no mutual coupling effect that could affect the characteristics of the control signal.

3. Status of Optoelectronic Components

Key components that are required for carrying out optical-microwave processing functions are: optical sources, optical modulators, optical switches, high-speed photodetectors, optical fiber and waveguide devices, and optoelectronic switches. These components will be discussed below.

3.1. Optical Sources

Depending on whether a direct modulation scheme or an external modulation scheme is used, we will employ either a high-speed semiconductor laser with decent noise characteristics (measured by RIN - relative intensity noise) or a CW high-power, low-noise solid-state laser. If the application involves propagation over a long length of optical fiber then the wavelength selected will be either 1.55 or 1.32 μm. For applications that do not involve long lengths of fiber, wavelength selection is dictated by cost, availability, and convenience.

To be used in a link where the linear dynamic range requirement is high (100-110 dB/Hz$^{2/3}$) a distributed feedback (DFB) laser with RIN in the neighborhood of -155 dB/Hz is necessary. The optical power measured from a single mode fiber typically ranges from a few mW to slightly above 10 mW. Semiconductor lasers with modulation bandwidth between 10 and 20 GHz are available commercially. The expression for the laser resonance frequency f_r (usually a good measure of the laser modulation bandwidth) is given by:

$$f_r \propto (AP/\tau)^{1/2} \tag{1}$$

where A is the differential gain of the laser, P is the photon density in the laser cavity, and τ is the photon lifetime of the laser cavity. Therefore, for lasers to have a large modulation bandwidth, the device must be designed to have large differential optical gain, good optical confinement, and be able to be driven to several times above threshold (and build up high optical intensity in the cavity)[3,4,5]. Direct modulation works well in practice up to about 10 GHz. For applications beyond 10 GHz external modulation is more effective.

In an external modulation link the linear dynamic range increases with the input laser power. Therefore, it is desirable to have a relatively high power CW laser as the source. The source is typically a diode-pumped YAG laser lasing at 1.32 μm wavelength, whose output power (taken from a polarization preserving fiber) can be as high as 100 mW. The RIN figure for these lasers is very good, around -165 dB/Hz, although one needs to beware of low frequency noise peaks. With the proper feedback circuit one can, in principle, remove these noise peaks.

Regardless of the types of laser used, it is critical to provide a proper level of optical isolation (30 to 60 dB) between the highly coherent laser source and the rest of the system. Without proper isolation the feedback induced noise in the laser can easily raise the system noise floor by more than 30 dB. Commercially available high performance lasers are equipped with one to two stages of optical isolators.

3.2. Optical Modulators

The modulator is the device that puts the microwave signal onto an optical carrier. It is used in combination with a cw laser source. The external modulators available today are primarily based on the electro-optic crystal LiNbO$_3$. Typically, optical intensity modulation is used but in some cases phase or frequency modulation is used.

The most common intensity modulator is of the Mach-Zehnder interferometer construction. It has a well-known modulation transfer function of the form

$$I = I_o \cos^2(\pi V/2V_\pi) = (I_o/2)[1 + \cos(\pi V/V_\pi)] \tag{2}$$

where I_o is the output optical intensity when the device is not biased, V is the voltage applied to the device, and V_π is the voltage at which the modulator output goes to zero. For linear operations, the device is biased at $V_\pi/2$ so that the optical output is at $I_o/2$. The modulation signal is then applied to the electrodes to cause the output intensity to vary between 0 and I_o. If the electrode is constructed and driven as a capacitor ("lumped element"), then the bandwidth is RC time constant limited (typically around 1 GHz). If the device is constructed to have a traveling-wave electrode then the bandwidth is limited by the relative velocity mismatch between the optical carrier propagating down the waveguide and the modulation signal traveling down the transmission line (electrodes). However, if the electrode propagation loss is sufficiently high it could limit the modulator bandwidth as well. Devices with bandwidths of up to 20 GHz are available. There are also laboratory devices that can be operated at frequencies up to 94 GHz with 5 to 10% bandwidth. The modulation efficiency and the bandwidth can be traded off,[6] but generally this has to be done at the device design stage.

3.3. Optical Switches

Low voltage, low loss, high-speed, reliable, and affordable optical switches have been pursued by researchers for over two decades. Despite this effort, it is still difficult to have a device that offers all the features listed above.

There are two basic types of switch: one contains moving parts and the other contains no moving parts. The types with no moving parts are perceived to have better reliability. Switches based on directional couplers fabricated on electro-optic crystals ($LiNbO_3$, e.g.) give low voltage (TTL compatible) and fast switching time (sub-nanosecond), but the extinction ratio is a very sensitive function of the voltage settings. The device operation is also sensitive to the optical field polarization. In addition, the transition from single-mode fiber to $LiNbO_3$ waveguide and back to single-mode fiber involves critical optical alignments. Therefore, it is difficult to construct fiber-pigtailed switches with insertion loss of less than 1 dB.

On the other hand, mechanical switches that use electromagnetic force to physically move a mirror or prism to effect spatial switching of fiber output ports show many advantages such as low loss and polarization independent operation. However, the response time of these devices is from milliseconds to tens of milliseconds.

3.4. High-Speed Photodetectors

To demodulate microwave signals from the optical carrier we need photodetectors with low capacitance and short transit time. This means high-speed detectors must be inherently small in area, with short electrode spacing, and must be properly packaged to minimize parasitic capacitances and inductances.[7] Because of their small dimensions, high-speed photodetectors can be saturated by moderate levels of input optical power. This saturation effect is manifested by an increased level of nonlinearity in the detector response. At low optical power levels, the photodetector is extremely linear. As the optical power is increased towards the saturation level, the harmonics and intermodulation products begin to rise and eventually become the limiting factors for system dynamic range. There are research efforts on going to develop detector structures such as waveguide-coupled detectors and traveling-wave detectors to mitigate this problem, and thereby extend the optical system dynamic range.

3.5. Optical Fiber and Waveguide Components

Because of the enormous efforts by the telecommunications industry, low loss optical fibers and cables are widely available. Optical fiber loss at 1.32 μm is about 0.3 dB/km and at 1.55 μm is about 0.2 dB/km. For regular single-mode fibers, there is a zero dispersion wavelength at around 1.32 μm. For dispersion shifted fibers, the zero dispersion point can be moved to around 1.55 μm at the expense of a slightly higher loss. For all but a handful of microwave applications the fiber dispersion is not an issue if a single frequency laser is used and the fiber is chosen properly to match the optical source wavelength.

Another optical component that is used extensively in constructing an optical system is the coupler. It can be constructed using fused fibers or with silica waveguides formed on a silicon substrate. Typically the fiber couplers have lower insertion loss while the silica waveguide couplers give more precise control of the coupling ratio. The silica waveguide loss has been brought down to less than 0.1 dB/cm and the substrate size has grown to over 6 inches in diameter. This technology is still evolving and could well replace fiber devices in many short distance applications because of its ease of handling.

3.6. Optoelectronic Switches

Opto-electronic switches are fabricated by forming gaps in microwave striplines deposited on high resistivity semiconductor substrates. When an optical beam illuminates the exposed semiconductor in the gap, the photocarriers generated in this region lower the resistivity of the semiconductor by several orders of magnitude, thereby making the stripline continuous. When the optical beam is removed, the photo-generated carriers dissipate rapidly through various recombination processes. As a result, the electrical resistance at the gap reverts back to its original state and the conduction path is again blocked. Therefore, the optoelectronic switch can be used to control or modulate a microwave oscillator output. If one end of the stripline is connected to a dc power supply and the switch is illuminated by the output of a mode-locked laser, then the switch will produce a train of very short electrical pulses which can be used for characterizing the frequency response of devices, circuits, and materials.

4. Microwave Fiber Optic Links

The most elementary subsystem for microwave-optics is an analog fiber optic link or delay line. A generic fiber optic link, shown schematically in Fig. 1(a), consists of a transmitter,

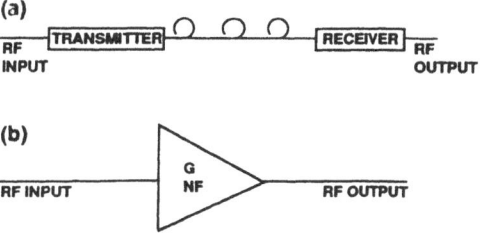

Fig. 1. (a) Block diagram of a microwave fiber optic link; and, (b) Microwave equivalent circuit of the link.

a length of interconnecting fiber, and a receiver. The primary function of the transmitter is to modulate the intensity of the optical source and to launch it into the fiber. The receiver, conversely, takes the arriving optical signal and extracts the modulation information. The fiber optic link can be regarded as just another microwave component (or black box) that is characterized by an equivalent microwave gain, noise figure, dynamic range, and time delay as illustrated in Fig. 1(b). As we discussed earlier, the transmitter can be realized with either a directly modulated semiconductor laser or a combination of a CW laser and an electro-optic modulator. The receiver usually consists of a high-speed photodetector and a low noise linear amplifier.

The link loss (or gain) is determined by three main terms: transmitter efficiency, fiber splicing and propagation losses, and receiver responsivity. In a direct modulation transmitter, the efficiency is determined by the product of the laser slope efficiency and the laser-to-fiber coupling efficiency. For DFB lasers with built-in optical isolator and single-mode fiber output, this number is around 0.2 mW/mA. The single-mode fiber coupled receiver responsivity with no amplifier is around 0.9 mA/mW. Thus, for a short link (neglecting fiber losses) the direct modulation link will have an RF link loss of about 15 dB. This number can be made smaller for narrow band applications where impedance matching with resonance enhancement is possible. For an external modulation link, the transmitter efficiency is determined by the CW laser power, the modulator insertion loss, and V_π of the modulator. Using the modulator transfer function given earlier, it is straightforward to calculate the slope around the bias point ($V=V_\pi/2$) to be $-(I_o\pi/2V_\pi)$. For a realistic wideband modulator with V_π = 14 Volts, optical insertion loss = 5 dB, and I_o = 50 mW, the RF insertion loss is on the order of 25 dB. In narrow band applications one can use a modulator with a small V_π in addition to using a higher power laser to further reduce the insertion loss.

The spurious-free dynamic range (SFDR) is a commonly used term to describe over what signal range a microwave fiber optic link can faithfully transmit the RF signal. It is expressed in dB, and is the ratio of the largest undistorted signal to the smallest measurable signal (S/N = =1). Fig. 2 shows a typical plot of the output microwave power versus input power. Spurious responses of the link come from the second harmonic, the third harmonic, and the third order

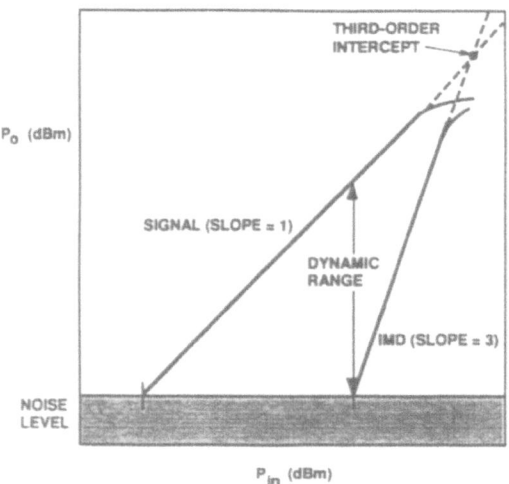

Fig. 2. Output fundamental and third order intermodulation power versus drive level of a microwave system.

intermodulation products. If the bandwidth coverage is less than an octave, then neither 2nd nor 3rd harmonics contribute to spurions responses. However, the third order intermodulation distortion (IMD) which is generated by the mixing of co-existing signals will remain inband (e.g., the 3rd order mixing of f_1 and f_2 will generate $2f_1 - f_2$ and $2f_2 - f_1$). In Fig. 2, the output signal versus input signal (log-log) plot follows a straight line with a slope of 1. On the same chart, the IMD product versus input signal is also a straight line except the slope is 3. Signal distortion is contributed by each and every element in the link. Under most circumstances, the modulator or the semiconductor laser is the main contributor to link nonlinearities. Several feedback and feed-forward schemes have been implemented by the cable TV industry to improve the overall linearity of the link. However, the servo electronics will limit the applicable frequency range. There are also on-going efforts in the development of linearized modulators[8] that could improve the SFDR by at least 5-8 dB. However, these schemes are generally quite sensitive to manufacturing and operational parameters.

It is evident from Fig. 2 that keeping the noise floor low will improve SFDR also. Aside from the kT thermal noise and the added noise from the amplifiers, laser noise and detector shot noise are two major contributors. The detector shot noise is of course proportional to the DC detector current (I_d) and the laser noise can be expressed in terms of the product of RIN and I_d^2. It is important to select a laser with low intrinsic RIN and use a proper amount of optical isolation to prevent further degradation of the noise floor.

5. Applications

Ther are a number of potential applications for analog fiber optic links: for example, cable television, fiber-in-the-loop, cellular phone and PCN (personal communication network) antenna remoting, satellite antenna remoting, phased array antenna steering, delayline-based test equipment, and microwave signal processing.[9-14] Two of these applications are discussed below.

5.1. Fiber Optic Delaylines for Radar Testing

Testing and servicing of modern sophisticated radar systems require state-of-the-art technology that can handle diverse waveforms without introducing spurious responses. Testing the proper operation of a radar can best be conducted by using a copy of the transmitted pulse which has been stored for an interval corresponding to the round trip transit time to the target and to which a Doppler frequency shift is added. The superior delay-bandwidth product of fiber optics makes it an uncontested choice for such a delayline application compared with competing technologies. Two versions of fiber optic radar test target have been developed. One is a compact, 30 µs fixed delay unit for built-in-test applications. It operates over the wide frequency range of UHF through X-band microwaves. The other is a programmable fiber optic delayline capable of providing a total delay of 1 ms in increments of 83.6 ns for RF signals up to 2.5 GHz. The programmable feature of this unit permits the simulation of continuous tracking (to within a range cell of 42 ft) of a target moving at Mach-2 speed over a range of 80 nautical miles. To compensate for losses in the 208 km long delayline three erbium doped fiber optical amplifiers excited by a single 1480 nm pump laser are used. Both test target devices were constructed using direct-modulated semiconductor lasers and have been successfully tested for functionality.

5.2. *Optical Control of Phased Arrays*

Phased array antennas will play a major role in future advanced radar and communication systems. The aperture of a phased array is composed of spatially distributed transceiver modules. These modules are controlled by a central computer that feeds the phase and amplitude information to each radiating element of the array. Because of this precise control of the radiation field at each module, the phased array antenna beam can be shaped as required, pointed accurately, and scanned without physical movement. It can also form multiple, independent beams simultaneously.

In an active array there can be several thousand transceiver modules. The interconnects required to address each module are indeed very complex. Optical interconnects offer several performance advantages over conventional microwave distribution networks for advanced phased array antennas. Besides handling a variety of digital control signals, they can also transmit and process microwave analog signals to achieve true time delay beam forming functions over a wide frequency band. The flexible form factor of optical fibers and the light weight and compact size of integrated photonic modules offer great potential for the antenna system to be partitioned and packaged differently. Although initial demonstration results are encouraging much more development effort is still needed.

6. Summary and Future Outlook

High-speed optoelectronic components and subsystems have the potential to perform a variety of functions in the transmission and processing of high frequency analog signals. This has been made possible by the development of high speed, low noise, and high linearity semiconductor lasers and photodetectors, as well as wideband electro-optic modulators and switches. Although these components have reached a high degree of sophistication, they are by-and-large discrete components. They need to be carefully fiber-pigtailed and interconnected via fibers or dielectric waveguides to form functional subsystems. As a result, the components and subsystems tend to be high cost, not suitable for volume production, and perceived to be susceptible to environmental factors. We need advanced optoelectronic device designs, a higher level of device integration, and advanced packaging techniques to eradicate this barrier of inserting optoelectronics into real systems.

Fortunately, there is also a large class of applications that requires only moderate performance specifications. These can be served by existing components and subsystems. These relaxed specifications will allow OEICs to be fabricated at reasonable cost, which in turn, will spur even more application opportunities. Therefore, the future for integrated optics and, in particular, high-speed optoelectronics is indeed bright. We will see a lot of analog fiber optic links for cable TV/fiber-in-the-loop, and cellular/PCN antenna remoting applications in the near term. As R&D efforts mature, the applications for large antenna remoting, satellite ground station signal distribution, and wideband phased array antennas will follow.

References

1. W.Platte, IEE Proceedings, vol. 132, pt J, pp. 126-132 (1985)
2. D.D. Hall, M.J. Wale and N.J. Parsons, The GEC Journal of Research, vol. 10, pp. 80-84 (1993)
3. K.Y. Lau and A. Yariv, IEEE J. Quantum Electron., vol. 21, pp. 121-138 (1985)

4. J.E. Bowers, B.R. Hemenway, A.H. Gnauck and D.P. Wilt, IEEE J. Quantum Electron., vol. 22, pp. 833-844 (1986)
5. K.Y. Lau and A. Yariv, Appl. Phys. Lett., vol 46, pp. 326-328 (1985)
6. D.W. Dolfi and T.R. Ranganath, Electron. Lett., vol 28, pp. 1197-1198 (1992)
7. J.E. Bowers and C.A. Burrus, Jr., J. Lightwave Technol., vol. 5, pp. 1339-1350 (1987)
8. M.L. Farwell, Z. Lin, E. Wooten and W.S.C. Chang, IEEE Photo. Technol. Lett., vol. 3, pp. 792-795 (1991)
9. H.W. Yen, Appl. Phys. Lett., vol. 36, pp. 680-682 (1980)
10. H.A. Hung, P. Polak-Dingels, K.J. Webb, T. Smith, H.Huang and C.H. Lee, IEEE Trans.Microwave Theory Tech., vol. 37, pp. 1223-1231 (1989)
11. A.Daryoush, IEEE Trans. Microwave Theory Tech., vol. 38, pp. 467-476 (1990)
12. I.L. Newberg, C.M. Gee, G.D. Thurmond and H.W. Yen, IEEE trans.Microwave Theory Tech., vol. 38, pp. 664-666 (1990)
13. J.H. Schaffner, R.R. Hayes, R.L. Joyce, J.B. Lewis, J.L. Pikulski, H.W.Yen and C.M. Gee, Opt.Engineering, vol. 33, pp. 187-193 (1994)
14. W. Ng, A. Walston, G. Tangonan, J.J. Lee, I. Newberg and N. Bernstein, IEEE J. Lightwave Technol., vol. 9, pp. 1124-1131 (1991)

Chapter 22

OPTOELECTRONIC SWITCHING APPLIED TO RADAR STEERING

G. L. TANGONAN, R.Y. LOO and W. W. NG

1. Motivation

The present trend in radar design is towards wide instantaneous bandwidth and sharing of multiple functions (search, tracking, and electronic warfare) into a single aperture that operates over multiple bands. This redesign necessitates a rethinking of present day beam steering/control approaches. Specifically, researchers are now actively pursuing microwave optical delay lines as replacement for coax lines.

Microwave signals can be impressed onto a lightwave carrier that propagate with low loss (< 0.75 dB/ km - electrical) over fiber optic waveguides. The microwave modulated lightwave can be switched using optical spatial switches or optoelectronic switches, thereby effecting time delay and beam control/switching functions.

A common optical feed for all frequencies of operation can simplify the beam-forming architecture of a phased array antenna and provide true time delay steering. As in other lightwave systems, fiber-optics offer high isolation against electromagnetic interference and ground loops. The use of fiber-optic delay lines for microwave phased array antenna steering is especially attractive in airborne applications where size and weight are of prime concern, and for large surveillance arrays.

In this Chapter we report on the development of new optical integrated circuits aimed a performing the specific task of true time delay steering of phased arrays. These new circuits combine fiber or waveguide delays with switchable arrays of detectors and lasers[1,2]. We focus on the device aspects related to the switched arrays and the incorporation of delay line with these arrays.

Several unique problems arise in applying optoelectronic switching to radar steering. These problems result from the fact that radar engineers require matching of the delay lines in amplitude and phase. Thus the building of delay lines in a box using photonics takes on the added difficulty of ensuring a reasonable match between optical components in RF response. We begin with a short review of true time delay steering. We present results of true time delay module development for L and S-band using laser and detector switching, L and X band steering using laser switching, and an integrated optical chip for X band systems.

G.L. Tangonan, R. Y. Loo and W. W. Ng - Hughes Research Laboratories, 3011 Malibu Canyon Road, Malibu, California 90265 USA

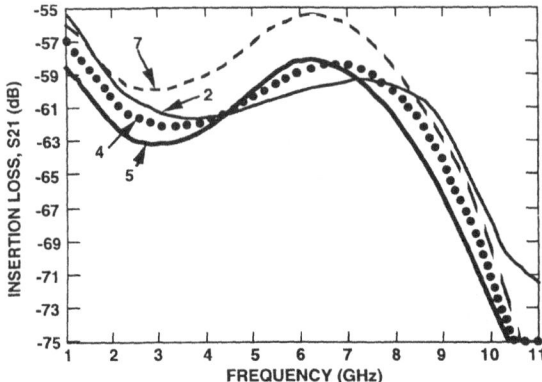

Fig. 1. Laser and detector switching combined for true time delay switching.

2. True Time Delay Steering of a Phased Array

To steer the radiated beam, the microwave signal (of frequency ω_m) exciting each radiating element is first routed through a RF "phasing unit". If the ratio of the RF output to the RF input in these units is given by $a_n \exp(i\psi_n)$, the far field of the antenna beam at an angle θ is given by the following expression:

$$E(\theta,t)=\Sigma a_n \exp(i\omega t)\exp i(\psi_n+nk_m l \sin\theta) \qquad (1)$$

where l is the distance between the radiating elements and k_m is the wave number ($=\omega_m/c$) of the radiated beam. The time independent part of $E(\theta,t)$ is proportional to the array factor of the antenna. The phase front and therefore direction of the radiated beam, is steered electronically by controlling the relative phase between successive radiating elements of the array. For example, to point the radiated beam at an angle θ_o, ψ_n is set to the value:

$$\psi_n=-nk_m l \sin\theta_o \qquad (2)$$

In many applications, this is satisfactorily accomplished with microwave "phase-shifters" designed to provide specific phase shifts (mod. 2π) over fixed frequency ranges. If we take the differential of Eq.2, we see immediately that for a fixed set of ϕ_n's, the radiated beam will drift by an amount $\Delta\theta_o$ given by:

$$\Delta\theta_o=-\tan\theta_o(\Delta\omega_m/\omega_m) \qquad (3)$$

for an instantaneous change of the microwave frequency equal to $\Delta\omega_m$. This undesirable shifting of the radiated beam leads to a drop of the antenna gain in the θ_o direction, giving rise to a phenomenon commonly called "beam squint".

To achieve a wide instantaneous bandwidth, "time-shifters", as distinguished from the "phase-shifters" described above, must be used to create the relative phase shifts between the radiating elements. In the so-called "true-time-delay" approach, compensation for the path difference between two radiators is provided by lengthening the microwave feed to the radiating element with a shorter path to the microwave phase front. Thus, a fixed set of delay lines compensates for the path differences between the radiating elements at all frequencies. Specifically, the microwave exciting the (n+1) th antenna element is propagated through an additional delay length $nL(q_o)$, where $L(q_o)=\Lambda \sin q_o$. For all frequencies ω_m, ψ_n is then given by:

$$\psi_n = -n\omega_m t_d(\theta_o) \tag{4}$$

where $t_d(\theta_o)=L(\theta_o)/$(group velocity in delay medium). Since the time delays pertaining to a particular steering angle are independent of frequency, we need not switch the delay lines to accommodate an instantaneous change of ω_m.

3. Optoelectronic Switching for Phased Array Steering

Radar signal steering using optoelectronic switching utilizes laser and detector arrays, optical splitting and RF switching, as shown in Fig. 1 for a 5 bit delay line switch. The overall function of the switch is to time delay an RF input; optics is used "within the box" because of the ease of optical splitting and the high isolation attainable with detector and laser switching. Each laser in the source array is hard-spliced to a fixed length of fiber delay. The lengths of fiber are selected to form differential delays with 0 δl, 1 δl, 2 δl, and 3 δl. The four different delays form a 2 bit laser switched delay line. The input signal is directed by an RF switch to one of the lasers in the array. This particular laser is the only laser biased on, so only that laser performs efficient microwave to optical conversion. A fiber combiner (4 x 8) distributes the 4 input optical beams to eight outputs. Each of the outputs has a different delay attached in sequence 0 δl', 1 δl', 2 δl', ... 7 δl', to form a 3 bit delay line array. The outputs of the fiber lines are coupled to an 8 detector array. By biasing "on" one detector of the array, one of the output delays is selected. The combined array thus performs as a 5 bit delay line requiring only a single laser and detector operating at one time.

4. Laser Switching

Laser switching was first implemented for radar applications by Ng et al in a 3 bit delay line demonstration operating over 2 and 9 GHz[2]. The lasers used in the demonstration were high speed 1.3 μm GaInAsP/InP buried crescent lasers, with typical threshold currents(I_{th}) of 10-14 mA and differential quantum efficiencies of ~30%/facet. Their relative intensity noise(RIN) at $6I_{th}$ were typically better than -144 dB/Hz away from the resonance peak. Since the modulation bandwidth of the laser increases as $(I/I_{th}-1)^{1/2}$, laser "bias-switching" enables the RF input at 2 or 9 GHz to be modulated into the optical output of the selected laser. Excellent RF on/off ratios (>60 dB) were obtained by biasing the quiescent lasers of the module below threshold.

Ideally the lasers should be biased so their resonance frequencies are much higher than the maximum transmission frequency of interest. Because the resonance frequencies of some of the lasers in the demonstration system saturated at ~9 GHz, Ng reduced the bias currents of all lasers so that their resonance peaks occurred at ~7 GHz to achieve a reasonable S/N ratio of ~107 dB/Hz at 9 GHz. The reason for the careful alignment of their resonance frequencies is as follows. The phase part of the laser modulation response undergoes a sharp transition at its resonance frequency, as predicted for a second order transfer function By biasing the lasers so that their resonance frequencies coincided with each other, and selecting lasers with

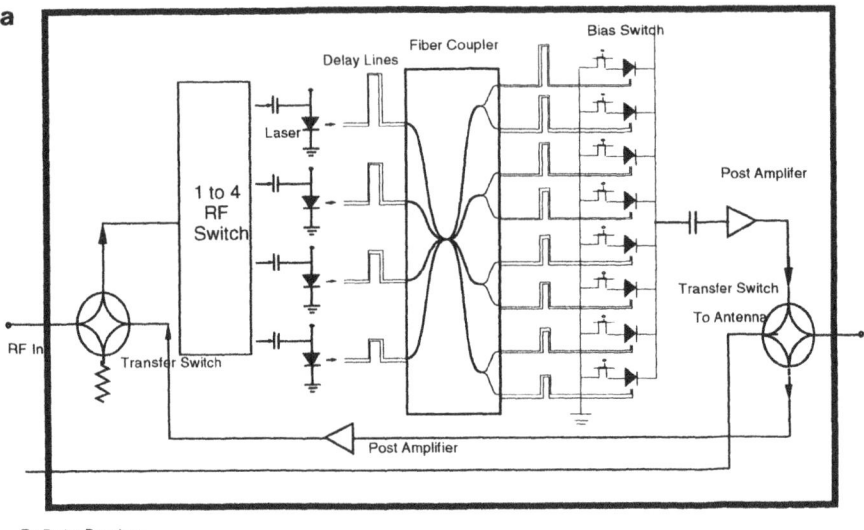

Fig. 2a. Frequency dependence of RF output (S_{21} magnitude).

"matching" modulation responses, intrinsic nonlinearities of the lasers phase responses cancelled each other. Thus the differential insertion phase measured between two optical links is due entirely to differences in propagation times between the two optical fibers. Fig. 2 shows the amplitude and phase response of the laser array used in the L and X band demonstration. The insertion phase obtained for subsequent channels represents the differential phase($\Delta \phi =$ =ϕ_i-ϕ_1, i=3...8) measured with respect to the shortest delay line. Specifically, the difference in time delay Δt_d between a particular delay line and the shortest line is calculated by the equation: $\Delta t_d = (\Delta \phi / \Delta f)/360°$, where Δf is the corresponding frequency sweep. The RF output S/N and the differential phase delay exhibited by the time delay lines proved sufficient for the L and X band array demonstration.

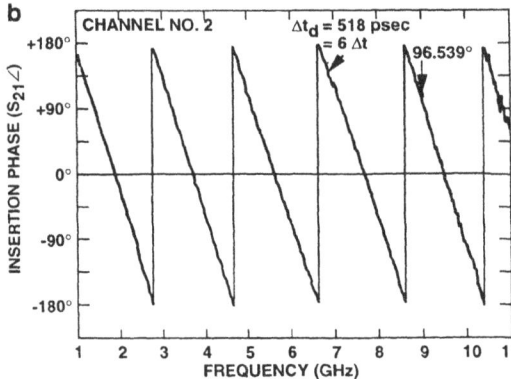

Fig. 2b. Differential RF insertion phases (angular part of S_{21}) between delay lines.

In the first demonstration of optical true time delay steering, we steered an L and X band common aperture antenna, in 4° increments, from -28° to +28° at 2 and 9 GHz. We demonstrated no "beam squint" in the radiated beam as the microwave frequency was switched from 1.9 to 9 GHz. These results demonstrated the wide instantaneous bandwidth - of almost one frequency decade - in an optically steered microwave phased array antenna.

5. Detector Switching

Switchable detector arrays with high speed performance are an important addition to the menu of optoelectronic integrated circuits (OEICs). The particular configuration which has received most study is that shown in Fig.1, with a detector array and switching FETs. In this array the detectors have a common cathode, which forms an RF combiner on chip. The FETs provide the detector bias switching. Different detector types have been used in these arrays - metal-semiconductor-metal (MSM) photodiodes and PIN diodes.[3,4,5] In both cases the FET serves to open and close the detector -bias connection. MSM photodiodes have a symmetric I-V characteristic so that zero volts implies zero photocurrent. The PIN diode behaves quite differently however in the FET switched mode. Open circuited from the bias, the PIN is actually driven to forward bias with zero current. In both cases experimental results have shown that excellent RF isolation from this configuration is attainable at frequencies up to 5 GHz.

An 1x4 InGaAs PIN chip (0.5 mm x 1.6 mm, with spacing 400 μm) with 30 μm diameter was fabricated and tested for true time delay applications. The photodetectors are connected in a common cathode configuration with the GaAs FET switches. These 4 detectors have very uniform electrical and optical characteristics. The detector leakage current is very low: a typical value of 65 pA at -5.0 v. The capacitance is 65 fF at -5.0 v for each detector and the quantum efficiency for the non-AR coated diode is 60%. In the bias switch the detector array is flipchip bonded onto a ceramic submount. Multiple input beams from four fibers are incident and focused on each of the four mesa detectors through the backside of the transparent InP substrate.

We used an optical heterodyne technique to measure the frequency response and on/off ratio of each individual detector in the array. The detector under test was switched on and off by applying zero and negative voltage, respectively, to the gate of the FET switch. Within the frequency range from dc to 6 GHz, these packaged detectors are nearly identical to within 0.5 dB, with flat frequency performance response to about 4 GHz. The 3 dB bandwidth is 5 GHz and we obtained a 70 dB (electrical) isolation on/off ratio. The average photocurrent for these measurements was 7 mA.

We also measured the crosstalk isolation between two neighboring detectors using different optical signals on each detector. In this experiment, we used two similar pigtailed lasers (1.5 mW output power), with one laser modulated at 820 MHz and the other at 840 MHz. We focused the two input beams on the two separate neighboring detectors. We measured the RF power output from the bias switch (closing one FET switch corresponding to 820 MHz, and opening the FET switch corresponding to 840 MHz, respectively). The difference in output power between the RF signals at 840 MHz, (On) and at 820 MHz, (Off) is the crosstalk isolation between the two neighboring detectors. The crosstalk isolation is 43 dB (electrical) with the average current being 1.0 mA. Similar results obtained out to 4 GHz indicates that the same level of isolation is obtained to S-band.

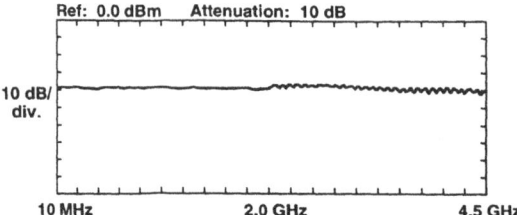

Fig. 3. Frequency response for the 5 bit time delay module with laser and detector switching.

6. Combined Laser and Detector Switching Module

Both laser and detector switching have recently been combined into a single module. With proper selection of the laser type we have been able to demonstrate excellent performance to 4 GHz. This will enable the insertion of this module into S - band radars. The size of the module is 8 cm x 11 cm x 6 cm and it weights about 400 gm. Inside the module there are 4 semiconductor lasers, a 4x8 optical coupler, an array of 8 photodetectors, and a decision circuit to select a particular delay time. This true time delay scheme achieves 32 combinations of delay lines - 5 bit operation, which is required of optical control of the phased array. Excellent frequency and phase results have been obtained with this module. Shown in Fig. 3 is the measured frequency response for the module. As shown, the frequency response is extremely flat, indicating that both the lasers and detectors are operating with > S band response. The overall insertion loss for the module is 37 dB.

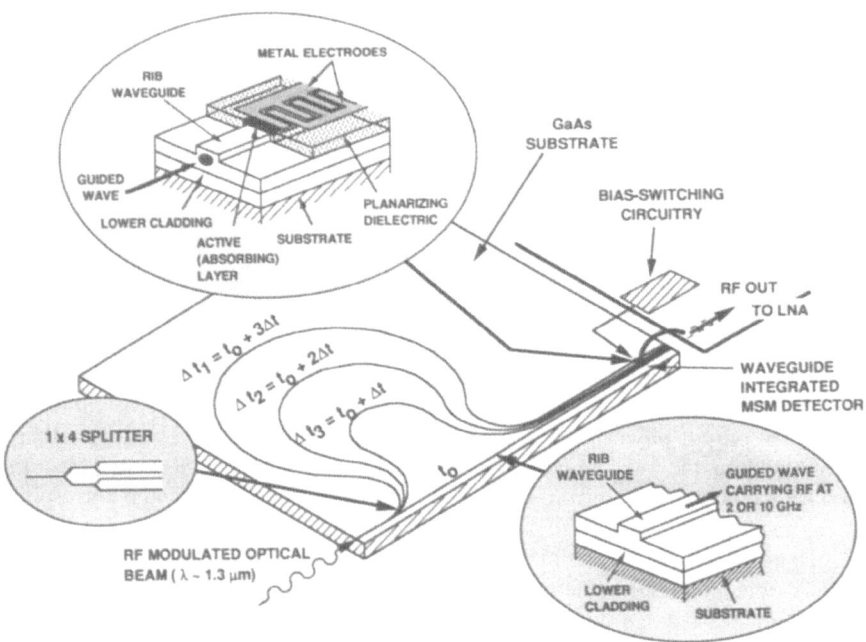

Fig. 4. Integrated optical chip with waveguides and detectors.

7. Future Directions

The next generation of OEICs for optoelectronic switching is being developed today. Laser and detector array based componentry with 10 GHz bandwidths are being developed in several laboratories for applications in optical digital networks. We can anticipate that these new array components will be directly applicable to true time architectures. A circuit being fabricated at Hughes with this dual use in mind is an InP OEIC including a 1x8 InGaAs PIN array with InP based HBT transimpedance amplifier and a selector switch. On a second front, a new generation of integrated optics chips is being developed with integrated delay lines and photodetector switches. Ng et al[6,7] recently reported the first integrated true time delay network. Specifically, the monolithic time-delay network integrates the delay lines (curved GaAlAs/GaAs rib-waveguides) and $In_{0.35}Ga_{0.65}As$/GaAs MSM detectors on a GaAs substrate. The chip design is shown in Fig. 4. For 100 μm long waveguide detectors a responsivity of 0.58 mA/mW was obtained and ON/OFF switching ratios > 40 dB were demonstrated at L and X band. Aside from compactness, the GaAs time-shift network offers more precision for the RF phase than fiber-delay-lines because the lengths of the waveguides on the GaAs wafer, as defined by photolithography, can be fabricated to micrometers of accuracy. Finally, the integration of detectors to the optical waveguides eliminates a delicate fiber-device interface between the delay-lines and the optoelectronic switches that control the delay times.

References

1. G.L. Tangonan, W. Ng, A. Walston and J.J. Lee, "Optoelectronic Switching for Radar Steering", in Photonic Switching II, K.Tada and H.S. Hinton, Ed., 391 (1990).
2. W. Ng, A. Walston, G. Tangonan, J.J. Lee, I. Newberg and N. Bernstein, J. Lightwave Tech. vol. LT-9, 1124-1131 (1991).
3. G.L. Tangonan, V. Jones, J. Pikulski, D. Jackson, D. Persechini and S.R. Forrest, Electron Lett. 24, 275(1988).
4. S. R. Forrest, G. L. Tangonan and V. Jones, J. Lightwave Tech. vol. LT-7, 607 -614 (1989).
5. R.Y. Loo, G.L. Tangonan, V. Jones, M. LeBeau, Y. Liu and S.R. Forrest, "High Speed InGaAs Photodetector Array for Optoelectronic Switching", DOD Fiber Optics Conference '92, McLean, Virginia, March 24-27 '92
6. W. Ng, D. Yap, A. Narayanan, R. Hayes and A. Walston, SPIE V.1703, 379 - 383 (1992).
7. W. Ng, A. Narayanan, R. Hayes, D. Persechini and D. Yap, "High-Efficiency Waveguide-Coupled l = 1.3 mm In_xGaAs_{1-x} /GaAs MSM Detector Exhibiting Large Extinction Ratios at L and X Band", IEEE Photonics Tech. Lett., May 1993.

Chapter 23

APPLICATION OF OPTICAL LINKS IN HIGH ENERGY PHYSICS EXPERIMENTS

C. DA VIA, M. GLICK and J. SÖDERQVIST

1. High Energy Physics Experiments at Future Colliders

During the last century a series of experiments at charged particle beam accelerators and colliders has allowed matter to be analysed with greater and greater precision. Present knowledge is summarised in the "Standard Model". This theory lists the known basic constituents of matter and the forces which govern how they behave, together with the carrier particles, which communicate the forces. The "Standard Model" has so far been very successful in predicting the outcome of experiments, but it is not an exhaustive description of how matter and nature behave. The model contains more than 20 "arbitrary parameters". To understand the missing pieces of the particle puzzle there is a need to perform experiments at higher energy. One hopes to discover the top quark and the Higgs particle, which could explain the interesting and mysterious problem of the origin of mass. Another theory predicts the existence of the Super Symmetric (SUSY) particles which interact very weakly with normal matter, thus placing high performance demands on the detectors.

To this end, the European Organisation for Nuclear Research (CERN) is planning to build within the next decade the Large Hadron Collider (LHC). A similar collider (SSC) is being considered in the United States. The LHC will collide protons on protons with a centre of mass energy of 15 TeV, at a repetition rate of 40 MHz. In each of these particle bunch interactions more than 1000 particles are produced. This high event rate is a prerequisite for studying interesting rare decays. To be able to untangle the complex events several million detector channels covering the full solid angle are required - from one to two orders of magnitude larger than any current high energy physics experiment. The radiation produced in particle collisions as depicted in Fig. 1 adds a new difficulty for the construction of detectors.

In Fig. 1 the barrel region configuration of a typical detector for the LHC is shown, along with the approximate number of channels per subdetector and the expected yearly doses of neutron and gamma radiation[1,2]. The largest number of channels are for particle tracking around the beam axis, in a nearly hermetic volume. The ends of the barrel will be closed with endcap detectors with similar layouts.

The rest of this Chapter closely follows Ref. 4.

C. Da Via and M. Glick - CERN, Geneva, Switzerland; J. Söderqvist - Manne Siegbahn Institute of Physics, Stockholm, Sweden

Advances in Integrated Optics
Edited by S. Martellucci *et al.*, Plenum Press, New York, 1994

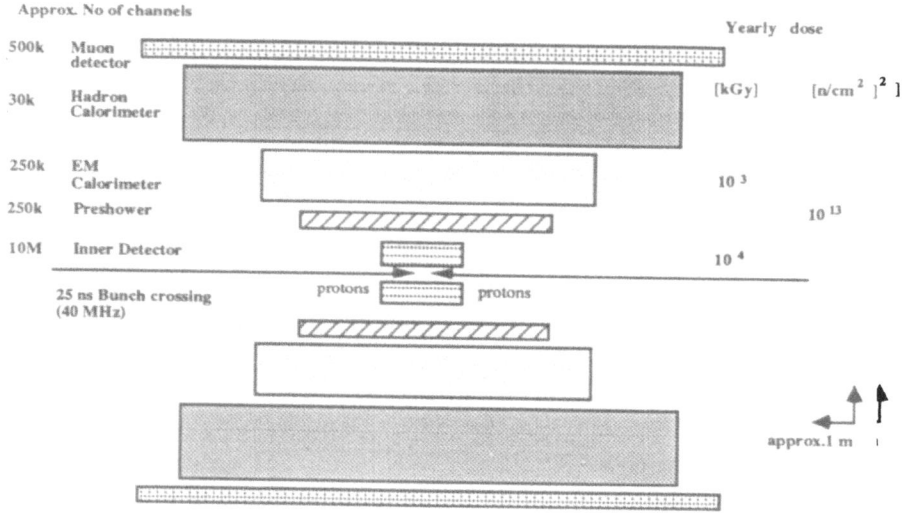

Fig. 1. A typical detector for hadron collider experiments.

2. Front-end Electronics

The large number of detector channels severely limits the method generally used for electronics in the current generation of experiments, where the functions of the front-end electronics are kept to a minimum and most detector signals are transferred outside via conventional copper cables (coaxial and twisted pairs). In future collider experiments, the front-end electronics will include signal processing, buffering and multiplexing. The number of cables must be kept to a minimum to limit the uninstrumented detector volume. Optical links will indeed minimise the cabling volume. The design of the front-end electronics, its actual implementation and the readouts, present major challenges[3], and may also offer the opportunity for a novel application of advanced integrated optics technology on a large scale[4].

The requirements for front-end electronics can be summarised as follows: minimise the amount of materials in cables; low power dissipation; radiation hardness; multiplexing capability; reliability; and, low cost. Optical links meet these requirements and, in addition, are immune to electromagnetic interference and ground loop problems. The links will bring signals to remote electronics situated approximately 50 m from the detector. The development of technologies for producing radiation hard, low-power optical links in volume quantities (several hundred thousands), at an affordable cost, constitutes a novel and interesting challenge for optoelectronics.

The calorimeter will need to measure particle energies ranging from a few hundred MeV to hundreds of GeV, with great precision. The requirements for calorimeter channels are a dynamic range of 15 bits and a linearity of 12 bits, making complex local digital signal processing a necessity[5]. Data transfer will require several thousand digital optical links at rates up to 1 Gb/s if serial transmission is used[6].

Tracking requires a dynamic range of only a few bits, but uses the largest number of links. For the preshower the dynamic range should be around 9 bits. For these kind of links deviations from linearity should be within 1%. The front-end electronics architecture for tracking/preshower detectors presently under investigation is schematically shown in Fig. 2

together with digital high speed links for timing and control signals[7]. It should be pointed out that in all the practical cases considered so far for tracking detectors, the required bandwidth would not exceed about 100 Mb/s, equivalent to about 35 MHz. The detector signals are buffered in analogue pipelines, multiplexed, digitized locally by fast ADCs, stored in digital pipelines, and read out over digital optical links, whose overall number may exceed hundred thousand links. This form of signal processing and data reduction implies a considerable complexity of the front-end electronics. Complexity increases with the "depth" of the analogue pipeline; at present the first level trigger latency is about 2 µs. The power dissipation for the preshower must be less than a few mW/channel since it is contained in liquid argon (LAr).

The scheme would fail in cases in which CMOS radiation hard technology would not be safely applicable, as may be the case at LAr temperature (87 K), and an alternative solution must be found. A link transferring prompt analogue signals would then be needed.

One possibility for multiplexing would be to use subcarriers (SCM) that would allow one to multiplex a number of baseband detector signals onto one fibre, using a 1 GHz bandwidth[4]. However, the preferred scheme at present is one in which signals are transferred after on-chip buffering and multiplexing.

These requirements, together with the short link length, call for a special type of optical interconnect. The requirements for components used in the above described links will be different from those of components demanded by telecommunication industry, but the technology and the production processes have much in common. Therefore the development of these links should be of broad interest in the integrated optics community.

Two different schemes for optical links have been considered: direct modulation of LEDs or lasers, or external modulation using electro-optical devices, as shown in Fig. 2. In order to determine the most effective type of optical links an R&D project (called RD23) has been started[4] by a collaboration involving CERN, HEP research institutes and industry. The direct modulation scheme is being investigated by various subdetector groups at CERN.

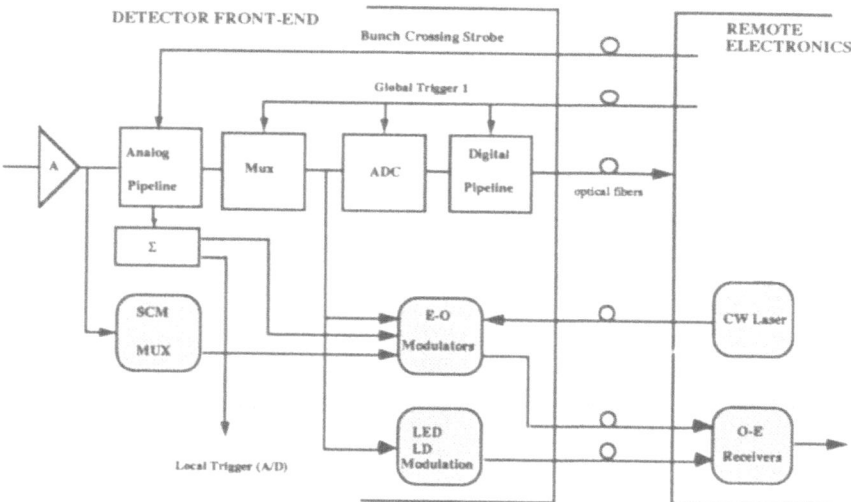

Fig. 2. Tracker/preshower front end electronics.

3. Direct Modulation

Direct LED modulation is a widely used technique for short links, often with emitter wavelength in the first transmission window, i.e. around 850 nm. However, the conversion efficiency is low and LEDs require multimode fibres. The multimode fibres are more sensitive to radiation damage than single-mode fibres. Furthermore, radiation induced losses are enhanced at shorter wavelengths[8]. Laser diodes offer higher conversion efficiency and are easily coupled to single-mode fibres. However, very low threshold currents (< 1 mA) would be required to comply with the power budget. In both types of emitters, radiation induces non-radiative recombination centres, which decrease minority carrier lifetime and increase threshold current density[9]. To maintain optical output power the devices need to be operated at high current density, which might not be feasible within the tight power budget.

4. The RD23 Project: LiNbO3 and MQW III-V Semiconductor Modulator Arrays

Two main alternative technologies for electro-optic intensity modulators are being investigated within the RD23 project [4]: (i) single-mode waveguide structures on LiNbO$_3$ (Mach-Zehnder interferometers), and (ii) multiple quantum well (MQW) III-V semiconductor reflective modulators.

These devices translate electrical signals into optical ones by external modulation of CW laser beams. The lasers are located outside the particle detector; optical fibres convey the incoming laser power to the detector front-end, and the outgoing optical signals to the readout electronics. The operating wavelengths are 1.3 µm or 1.55 µm, so that low-cost telecommunication type fibres and detectors can be used. Electro-optic modulators are basically passive devices which present a circuit load equivalent to a capacitor of a few pF, requiring negligible electrical drive power. Optical power losses can be reduced to a few hundred µW per link. Optical power is, according to a feasibility study, a key factor in determining the overall cost of the system. Preliminary measurements show that the radiation resistance of materials, technologies and devices can meet the most severe constraints[4].

RD23 has previously tested commercially available single channel devices in LiNbO$_3$. An overall linearity of about 1% can be obtained by restricting operation to the region $V_{signal} < 0.12 V_\pi$ around the quadrature point[4].

The interest of RD23 in semiconductor MQW is focused on asymmetric Fabry-Perot cavity modulators. These devices work with unpolarised light and are compact, but require laser sources which are well matched. The operating wavelengths shift with temperature, so that temperature control of sources and modulators is required. Initial measurements were carried out on a channel in an 8-channel linear array by GEC-Marconi[10], an array which was originally developed for digital interconnects. It has proven to meet analogue link requirements.

For optical input power levels <1mW, the limiting noise source may turn out to be the relative intensity noise (RIN) of the laser, which should be lower than -135dB/Hz to match the requirements for the dynamic range[4].

The RD23 collaboration has now developed demonstrators which are now being tested. A chip with a 16 LiNbO$_3$ modulator array is now being evaluated[11]. A MQW device with an array of 4 modulators[12] has also been manufactured and will soon be tested. With these new demonstrators the work is well on the way. The outcome of the tests of these devices will guide the future development of this work with new optical interconnects.

5. Conclusions

Light wave analogue and digital links can find a large volume application in future High Energy Physics experiments. Links using direct modulation as well as external electro-optic intensity modulators are being considered. A collaboration between HEP laboratories and industry has been established to develop optimised integrated optics devices. Demonstrators for modulators in lithium niobate and in III-V semiconductors have been designed and built. Within the HEP community all aspects of optical readout links will be investigated with the aim of designing and building detectors using optical interconnects.

ACKNOWLEDGEMENTS. We wish to thank Dr. G. Stefanini for many useful discussions. Prof. P. Carlson and Dr. L-O. Eek are acknowledged for their comments and suggestions.

References

1. CMS Letter of intent CERN/LHCC 92-3 (1992)
2. ATLAS Letter of intentCERN/LHCC 92-4 (1992)
3. Anghinofi et al., CERN/ECP 91-11 (1991)
4. G. Stefanini, RD23 collaboration, "Optical links in front-ends of high-energy physics experiments", European Conference on Integrated Optics, Neuchatel ,12-22 (1993)
5. B. Löfstedt et al., "A digital Front-End and Readout Microsystem for Calorimetry at LHC", CERN/DRDC/90-74 (1990)
6. N. Ellis et al., "First level trigger systems for LHC detectors", CERN/DRDC 92-17 (1992), now project RD27
7. B.G. Taylor, "Multichannel Optical Fibre Distribution System for LHC Detector - Timing and Control signals", CERN/ECP 92-16 (1992)
8. Y. Morita and W. Kawakami, IEEE Trans. on Nucl. Sci. Vol. 36, 584(1989)
9. B. Leskovar, IEEE Trans. on Nucl. Sci., Vol. 36, 543(1989)
10. A.J. Moseley et al., Electronic Letters 26 913(1990)
11. T.P. Young et al., "Design and initial evaluation of a compact optical splitter and 16-channel modulator array on Ti:LiNbO$_3$", European Conference on Integrated Optics, Neuchatel , 10-10 (1993)
12. N.P. Green et al., "Optoelectronic modulator technology for analogue data transfer", European Conference on Integrated Optics, Neuchatel , 12-30 (1993)

Index

Absorption, 6, 9, 18, 21, 37, 40, 44, 61, 69, 71, 74, 84, 111, 113, 116, 134, 146, 152, 158, 166, 173, 175, 182, 185, 213, 214, 216, 218, 221, 222, 225, 229, 235, 248, 249, 250, 265, 267, 275, 286, 289, 292
   losses, 6, 59, 70, 76
   spectra, 171, 195, 196
Acoustooptic, 59, 79, 90, 92, 303, 304, 305, 306, 308, 311
Active semiconductors, 222
Additivity, 63, 64, 66, 69
Adiabatic tapers, 138
Aging, 59, 65, 66, 67
AlGaAs, 44, 111, 213, 214, 218, 221, 222, 225, 249, 257, 286
Anisotropic etching, 138, 156, 236, 238
Anomalous dispersion, 259, 269
Asymmetric coupler, 141

Band filling, 71
   renormalization, 71
Beam forming of linear array antennas, 303
Bending, 138, 229
Bidirectional transceiver, 281, 288
Birefringence, 24, 79, 80, 130, 131, 153, 156, 228, 231, 232, 233, 236
   waveguides, 152
Bistability, 34, 39, 59, 111, 183, 296, 297
Black solitons, 261, 262
Blazed gratings, 155
Blue films, 189
Bragg grating, 143, 293, 295
Bright solitons, 98, 259, 260, 262, 263, 264, 266, 267
Buried channel guides, 113, 121, 131, 133, 134, 135, 234, 291, 325
Butt coupling, 3, 153, 155, 239, 289, 290, 291

C-H bond, 185
Cascaded second order nonlinearities, 214, 223, 225
Cascading, 70, 74, 76, 283, 294
Cascading effect, 49
CdS, 39, 40, 63, 75, 167, 170
CdS_xSe_{1-x} doped glass, 37
CdS_xSe_{1-x} in a glass, 72
Chalcogenides, 170
Channel glass waveguides, 25
Chemical sensors, 158, 234, 240
Cherenkov-scheme, 24
Chiang method, 16
Chiral molecules, 65
Chirality, 66
Co-directional guided waves, 23
Co-doping, 146
Coherence relaxation, 60
Cohesive field, 60
Collapse, 95, 100, 168, 269
Composites, 60, 70, 71, 72, 76, 172
Concentration clusters, 146
Conjugated polymers, 69, 70, 73, 74, 186, 195
Conjugation defects, 74
Contra-directional guided wave modes, 42
Corona poling, 67
Coupled-mode theory, 23, 40, 96, 97, 178
Coupling efficiency, 3, 37, 39, 40, 137, 138, 140, 141, 153, 179, 181, 183, 239, 289, 318
   length, 27, 40, 116, 140, 218, 231
Covalent crystals, 59, 63
Critical switching power, 28
Crossbar switches, 117
CS_2, 264, 267

Dark line, 3
Dark solitons, 259, 260, 261, 267, 268
Demultiplexing, 218, 219, 281, 284, 295, 304, 311

337

Detectors, 116, 244, 246, 248, 257, 281, 285, 286, 292, 316, 323, 327, 328, 329, 331, 332, 333, 334, 335
Detuning parameter, 42
Dielectric confinement, 70, 71
Dielectric susceptibilities, 166
Diffraction, 45, 46, 47, 95, 96, 100, 103, 104, 141, 159, 222, 259, 260, 261, 266, 267, 270, 304, 306
Dipole-dipole, 65, 66
Direct laser writing, 210
Direct LED modulation, 334
Directional couplers, 5, 8, 21, 43, 44, 61, 79, 82, 84, 85, 91, 140, 141, 143, 144, 173, 199, 200, 201, 203, 204, 205, 213, 214, 216, 217, 218, 220, 221, 225, 231, 232, 287, 289, 291, 293, 316
Dispersion, 34, 35, 36, 72, 86, 95, 96, 98, 146, 167, 191, 196, 214, 215, 216, 259
  curves method, 14
  relation, 31
Distributed-Bragg-reflector, 113
Doping, 59, 66, 71, 73, 90, 111, 112, 122, 146, 156, 170, 240, 252, 286, 288, 296, 308

Effective index, 4, 12, 14, 15, 16, 35, 36, 84, 86, 131, 142, 152, 262
Electric field domain, 248, 252, 255, 256, 257
Electroabsorption modulators, 113, 114, 115
Electrode poling, 67
Electron beam irradiation, 122, 131, 134, 135
Electron delocalization, 69, 186, 187, 196
Electrooptics, 59, 61, 64, 68, 75, 79, 80, 82, 90, 91, 92, 104, 160, 199, 201, 234, 237, 303, 304, 308, 311
Epitaxial-lift-off, 155
Er-doped fiber amplifier, 284
Erbium, 146, 156, 284, 319

Fermi exclusion, 70
Ferroelectric glassy films, 169
Fiber loop, 157
  optic probe, 8
Fibre-waveguide coupling, 154
Figure of merit, 44, 59, 61, 62, 173, 216

Filamentation, 46, 267
Film losses, 185, 192
Films, 6, 25, 32, 123, 124, 125, 126, 127, 128, 129, 130, 131, 134, 169, 185, 186, 189, 191, 194, 195, 196, 197, 235, 236, 238
Flame hydrolysis deposition, 124, 126, 138, 152
Flip-chip technique, 155, 288
Fluorescence, 171, 185, 235
Fock-Leontovich equation, 97
Four-layer tapered structure, 192
Franz-Keldysh free electron model,113, 186
Frequency conversion, 21, 24, 79, 85, 90, 283
  generation, 26, 51, 59, 89, 201, 224
  preserving, 59, 61, 68, 75
Front-end electronics, 332, 333
Fundamental soliton, 47, 100, 260, 264, 265

GaAs, 25, 44, 63, 70, 73, 85, 110, 111, 121, 213, 216, 243, 244, 248, 249, 252, 253, 303, 306, 327, 329
Gate matrix switch array, 145
Glass waveguides, 11, 25, 44, 264
Glasses, 4, 12, 58, 59, 67, 68, 71, 72, 126, 131, 134, 147, 165, 166, 167, 170, 171, 172, 189, 238, 267
Graded index, 46, 80, 192, 262
Grating, 3, 39, 87, 89, 91, 116, 140, 143, 152, 155, 159, 178, 179, 180, 185, 223, 228, 229, 234, 235, 236, 237, 239, 286, 289, 293, 295, 296, 304, 305, 306, 307, 308, 310
  couplers, 37, 173, 178, 231, 237, 293, 297
  coupling, 39, 174, 180, 183
Gray solitons, 262, 263
Grün range, 135
Guest-host system, 67
Guided-wave components, 11
Gyro, 157, 232, 233, 238
Gyrometry, 157
Gyrotropy, 75

Helmholtz equation, 96, 97
High energy physics, 331, 335
  linear refractive index, 167
Higher-order solitons, 100
Hydrogen bonding, 65

In-plane scattering, 3, 9, 152
Index saturation, 266
Indium phosphide, 151
Infrared detectors, 244
    imaging, 155
Inhomogeneous artificial solids, 60
InP, 64, 109, 110, 111, 113, 118, 121, 253, 286, 287, 288, 289, 291, 298, 325, 327, 329
Integrated mechanical alignment features, 138
Integrated optics, 1, 21, 37, 57, 65, 66, 76, 79, 84, 85, 89, 92, 109, 114, 117, 121, 123, 124, 127, 135, 137, 139, 145, 151, 152, 155, 157, 158, 160, 185, 186, 197, 199, 207, 227, 228, 229, 234, 237, 239, 244, 279, 285, 286, 288, 298, 309, 314, 320, 329, 332, 333, 335
Intensity-dependent effects, 50
Interaction of solitons, 265
Interconnection, 110, 151, 238, 283, 287
Interfacing, 57, 59, 66, 67, 70, 73, 237, 238
Intersubband transitions, 244, 245
Inverse Raman scattering, 74
Ion exchange, 4, 12, 16, 121
    waveguides, 156
Ionic crystals, 59, 64, 65, 66

Kerr, 29, 31, 33, 35, 39, 42, 46, 48, 75, 95, 103, 181, 224, 260
    effect, 42, 45, 46, 59, 61, 63, 68, 74, 75, 98, 104
    medium, 95, 96, 259, 263, 269
    like media, 21
Kramers-Kronig, 69
$KTiOPO_4$, 25
KTP, 25, 51

Langmuir-Blodgett, 186, 189
    films, 189, 192, 194, 196, 197
Laser densification, 170
Laser diodes, 85, 146, 154, 157, 158, 222, 303, 334
    irradiation, 6, 134
Leakage, 39, 178, 179, 229, 289, 292, 327
Light bullet, 49, 96, 100, 269
$LiNbO_3$, 25, 64, 67, 68, 76, 79, 80, 81, 84, 85, 86, 87, 89, 90, 91, 92, 121, 138, 140, 145, 166, 199, 204, 207, 208, 210, 227, 233, 234, 239, 277, 303, 305, 306, 308, 310,

$LiNbO_3$ (cont'd)
    311, 314, 315, 316, 334
Linear glass waveguide, 37
Linear-nonlinear interface, 48
Liquid crystal, 37
Lithium niobate, 4, 6, 9, 12, 24, 79, 83, 166, 239, 303, 304, 310, 311, 335
Local densification, 122, 134
Longitudinal, 3, 39, 60, 64, 90, 210, 304
Losses, 5, 6, 8, 9, 11, 18, 39, 49, 59, 70, 71, 76, 80, 81, 84, 85, 90, 92, 111, 117, 133, 134, 137, 145, 152, 153, 156, 158, 170, 173, 185, 186, 192, 194, 196, 197, 208, 210, 216, 217, 232, 273, 274, 275, 289, 313, 318, 319, 334
Lumped electrodes, 83

M-lines, 3
Mach-Zehnder, 5, 61, 79, 82, 84, 140, 142, 145, 199, 204, 207, 208, 214, 217, 219, 220, 223, 293, 294, 305, 306, 315, 334
Magnetooptic, 59, 75, 303
Main-chain systems, 67
MAP, 65
Material growth, 59
Metallic particles, 168
Micro-optic, 155, 159, 228
Mixing optical fields, 28
Mode number, 3, 22
    structure, 22
    locking, 90, 91, 92, 293
Modelling, 1, 11, 12, 19
Modulators, 79, 81, 82, 83, 84, 90, 92, 110, 111, 113, 114, 115, 118, 140, 143, 145, 155, 185, 199, 201, 203, 204, 205, 207, 208, 210, 211, 227, 244, 284, 288, 293, 303, 304, 305, 308, 311, 313, 314, 315, 318, 319, 320, 334, 335
Molecular beam epitaxy, 111, 243, 248, 286
    crystals, 59, 65, 66
    engineering, 63, 185
Monolithic integration, 110, 117, 280, 281, 285, 288, 292, 311
Morphological resonances, 70
Multi-chip modules, 155, 287, 298
Multi-quantum well structures, 44
Multimode fibres, 334
MUX/DEMUX, 140

Nernst-Planck equation, 12
Nonlinear cladding, 31, 34, 35
Nonlinear coupled mode equations, 42
    coupling, 21, 26, 27, 28, 37, 42, 43, 44
    directional coupler, 43, 61, 214, 216, 217, 218, 220, 225
    dispersion, 31, 34
    film, 33
    grating coupling, 39, 174, 180, 183
    materials, 21, 49, 50, 213
    optics, 49, 57, 62, 65, 66, 73, 74, 75, 95, 165, 213, 246, 259
    phase modulation, 51, 259
    phase shifts, 50, 51
    polarization, 22, 23, 25, 28, 29, 60, 62, 86
    propagation, 21, 100, 269
    refractive index, 21, 47, 167, 174, 181, 182, 183, 215, 222, 261
    saturation, 44
    surface-guided wave, 31
    wave propagation, 28
    waves, 32
    X-junction, 61, 214, 217, 221
Nonlocal, 39, 70, 74, 96, 104
NPP, 65, 66
Nucleation, 72, 167

Optical absorption, 229, 275
    amplifier, 113, 117, 145, 222, 284, 286, 319
    breakdown, 59, 64
    computing, 303, 304, 306, 310, 311
    delocalization length, 73
    fibers, 1, 5, 25, 109, 110, 114, 154, 159, 166, 197, 259, 263, 264, 279, 317, 320, 326
    Kerr effect, 59, 61, 68, 69, 74, 75, 98, 104
    links, 326, 331, 332, 333
    memories, 155, 156, 159, 285
    nonlinear materials, 21
    switching, 117, 183, 213, 214, 215, 216, 218, 219, 221, 223, 224, 225, 282, 283
Optoelectronics, 1, 109, 113, 151, 153, 154, 155, 156, 157, 159, 165, 228, 237, 238, 244, 248, 257, 313, 314, 317, 320, 323, 325, 327, 329, 332
Optomechanical devices, 158
Organic compounds, 185
    dyes, 166, 168, 171

Organic compounds (cont'd)
    integrated optics, 185
    waveguides, 37
Orientation parameter, 68
Overlap integral, 24, 27, 87, 88, 143
Overtones, 152, 158
Oxygen-polyedra, 59

Packaging, 59, 110, 159, 236, 280, 285, 287, 320
Parametric amplification, 26
    generation, 21
    oscillation, 26, 89
Patterning, 111, 185, 195, 197
Periodic stacked structures, 64
Phase matching, 3, 21, 22, 25, 39, 40, 42, 63, 64, 67, 68, 70, 79, 80, 85, 86, 87, 89, 218, 223
Photocurrent spectroscopy, 250, 257
Photodeflection method, 273
Photodetectors, 115, 151, 155, 238, 303, 313, 314, 316, 320, 327, 328
Photoelastic effect, 130, 233, 236
Photoinduced bleaching, 195, 197
Photonic integrated circuits, 110, 285, 292
Photorefractive effect, 49, 68, 96, 104, 308
Planar waveguide, 2, 21, 22, 143, 170, 178, 233, 235, 239, 263, 269, 304, 305, 308, 310
Plasmon polaritons, 29
Pockels effect, 61, 63, 69, 74, 208
Polaritons, 29, 31
Polarization, 22, 23, 24, 25, 28, 40, 50, 51, 60, 61, 62, 68, 69, 70, 74, 75, 80, 82, 85, 86, 87, 89, 92, 104, 144, 152, 156, 159, 166, 228, 229, 231, 232, 233, 234, 235, 245, 280, 281, 288, 289, 310, 315, 316
Poled nonlinear organic polymers, 24
    polymers, 67, 68, 76
Poling efficiency, 67
Poly-α-methyl-styrene, 173, 174, 183
Poly-phenyl-acetylene, 173, 183
Polydiacetylene, 73, 173, 186, 187, 189
Polymers, 24, 59, 65, 67, 68, 69, 70, 71, 72, 74, 76, 166, 171, 173, 174, 176, 177, 181, 185, 186, 189, 192, 194, 195, 196, 197, 288, 297
Polymeric materials, 160, 173
Polymerization, 128, 174
Polythiophenes, 186

POM, 65, 66
Population relaxation, 60, 61
Power splitters, 139, 140, 145, 227
Prism-coupling, 3
Processability, 57, 59, 67, 73, 173
Propagation losses, 1, 6, 9, 11, 80, 81, 84, 90, 111, 133, 134, 137, 152, 173, 179, 192, 208, 210, 215, 216, 273, 274, 275, 313, 316, 318
    wavevector, 28
Proton exchange, 80, 81, 89, 90, 91, 207, 304, 309
Pulse compression, 267, 269

Q-switching, 91, 92
Quadratic nonlinear materials, 49
Quantum confinement, 71, 72
Quantum well, 44, 70, 72, 111, 213, 243, 244, 245, 246, 248, 249, 250, 253, 255, 256, 283, 289, 334
    confined-Stark-effect, 113
Quasi-phase-matching, 25, 92

Radiation damage, 334
    hardness, 332
    induced losses, 334
    losses, 9
    mode, 97, 220, 221, 262, 263
Raman effect, 214, 215
Rare earth, 146, 147, 174
Reactive ion etching, 131, 136, 217, 287
Refractive index, 2, 4, 5, 6, 11, 12, 14, 16, 18, 19, 21, 22, 24, 26, 28, 29, 31, 34, 37, 39, 40, 42, 43, 46, 47, 50, 61, 80, 95, 96, 104, 110, 111, 122, 123, 127, 130, 131, 134, 135, 137, 143, 146, 152, 165, 166, 167, 170, 171, 173, 174, 175, 181, 182, 183, 192, 194, 195, 197, 208, 213, 214, 215, 217, 222, 223, 232, 235, 243, 259, 260, 261, 263, 267, 273, 276, 283, 288, 291, 294
    profile, 12, 13, 47
Relaxation time, 39, 182, 187
Remote sensing, 303, 308, 311
Retardation effect, 74
Rhodamine 6G, 171

Rhodamine B, 171, 185
Ring resonator, 152, 157, 229, 232
RRS, 189, 191, 196

SAR, 303, 304, 305, 306, 308, 311
Saturable nonlinear coupler, 43
    nonlinear media, 263
    nonlinearity, 42
Saturation, 21, 33, 34, 42, 43, 44, 49, 51, 71, 72, 111, 174, 177, 195, 263, 266, 267, 316
    effect, 31, 70, 316
Scaling laws, 69, 214
Scattering losses, 9, 18, 133, 186, 192, 196, 197, 216, 217
    detection techniques, 8
Second harmonic generation, 21, 22, 25, 26, 49, 50, 62, 66, 67, 68, 79, 80, 91, 92, 166, 223, 224, 244
Second order effects, 67
    nonlinear interaction, 89
    nonlinear phenomena, 22
    polarizability, 63, 64
    susceptibility, 62, 63
Self-defocusing cladding, 33, 34
    guide, 44
    Kerr-like nonlinearities, 29
Self-focusing, 29, 33, 34, 35, 44, 45, 48, 51, 95, 222, 259, 260, 263, 266, 269
    bounding media, 34
    cladding, 33, 34
    Kerr-like nonlinearities, 29
    nonlinearity, 30, 48
Self-trapped beams, 95, 259, 261, 263
    solutions, 100, 106, 263
Self-trapping, 102, 259, 263, 266
Semiconductor containing glasses, 167
    doped glass waveguides, 44
    waveguides, 110
Semimagnetic semiconductors, 75
Sensors, 155, 156, 157, 158, 159, 227, 228, 229, 232, 233, 234, 236, 240, 308
SHG, 24, 25, 51, 223
Signal processing, 151, 227, 229, 233, 236, 237, 238, 239, 248, 285, 288, 296, 297, 298, 303, 304, 308, 311, 313, 314, 319, 332, 333

Silica-on-silicon, 121, 122, 124, 126, 137, 138
Silicon, 121, 123, 124, 125, 126, 128, 131, 137, 139, 151, 152, 153, 154, 155, 156, 157, 158, 159, 168, 170, 227, 228, 232, 233, 236, 238, 240, 244, 286, 317
   integrated optics, 152
Sol-gel, 127, 128, 130, 132, 133, 134, 139, 147, 168, 169, 170, 172, 235
   deposition, 124, 127
Soliton, 21, 33, 45, 47, 48, 49, 59, 74, 95, 96, 98, 100, 103
   emission, 48, 49
   pairs, 263
   period, 261, 264
Spatial nonlocal nature, 105
Spatial solitons, 45, 48, 222, 259, 260, 261, 262, 263, 266, 267, 268, 269, 270
   temporal solitons, 268
Spatially dispersive, 96
Spin coated films, 189, 191, 195, 196
   coating, 67, 127, 128, 130, 133, 173, 174, 175, 178, 183, 185, 186
   exchange interaction, 75
Sputtering, 82, 124, 125, 157, 204, 210, 235
Stability of solutions, 33
Stark effect, 213, 214, 215, 216, 283
Stars, 140
Step-index waveguide, 4, 12, 13, 80
Stress-optical tensor, 130
Striking, 72, 73
Superlattices, 248, 252, 253, 254, 255, 256, 257
Surface plasmon, 29, 37, 72
Switch, 44, 82, 84, 85, 91, 109, 110, 116, 140, 145, 215, 217, 222, 223, 283, 284, 286, 288, 294, 297, 316, 317, 325, 327, 329
Switches and switch array, 140
Switching matrices, 79, 84, 92, 207, 227, 283, 288
Synthetic aperture radar, 303
System SiO_2-TiO_2, 171

Tapering, 112, 192
Tapers, 138, 154, 288, 291
Temporal solitons, 259, 268
Thermal conductivity, 151, 154, 275, 286
   effects, 21, 129
   oxidation and nitridation, 124
Thin-film heater, 143

Third-order nonlinearities, 28, 248
   susceptibility, 28
III-V semiconductors, 109, 110, 285, 297, 335
Three-prism setup, 7
Three-wave mixing effects, 49
Threshold devices, 34
Ti:$LiNbO_3$ technology, 207
Time division multiplexing, 109
Titanium indiffusion, 80, 87, 89, 207, 210
TM modes, 111, 156, 179, 180, 194, 197, 233, 235, 281
Topographic guides, 122, 131
TPA, 213, 214, 215, 216, 225
Transferability, 63
Transverse, 3, 8, 31, 45, 60, 64, 73, 95, 97, 100, 102, 104, 107, 137, 140, 175, 181, 210, 228, 259, 264, 273, 304, 307, 310
Trapped beams, 95, 259, 261, 263
Tunable detector, 248
Tuning, 24, 113, 143, 179, 240, 294, 295, 296, 297
Two-photon, 70, 213, 215, 224
   absorption, 214, 218, 265
Two-prism technique, 7

Unstable region, 48
Urea, 65, 66

Van der Waals, 65
Vectorial, 96, 263, 304

Wannier-Stark, 113
Waveguide lasers, 79, 85, 90, 91, 92, 147
Waveguide lasers in rare-earth, 90
Waveguides, 1, 4, 6, 8, 15, 16, 19, 21, 36, 37, 39, 44, 49, 79, 80, 82, 89, 90, 91, 110, 111, 112, 121, 122, 123, 128, 131, 133, 134, 137, 138, 140, 152, 153, 154, 156, 158, 170, 174, 173, 175, 178, 179, 183, 185, 186, 189, 192, 194, 195, 196, 197, 201, 208, 210, 216, 217, 221, 225, 228, 229, 231, 233, 236, 259, 263, 264, 265, 268, 270, 273, 274, 275, 277, 280, 285, 286, 288, 289, 291, 292, 293, 297, 298, 305, 308, 309, 310, 313, 317, 320, 323, 329
WKB method, 13

Y-junction, 140, 141, 145, 207, 208, 210, 211, 220, 221

ZnO, 37, 307
ZnS, 37, 39

The manufacturer's authorised representative in the EU is Springer Nature Customer Service Centre GmbH, Europaplatz 3, 69115 Heidelberg, Germany. If you have any concerns regarding our products, please contact ProductSafety@springernature.com

Printed and bound by CPI Group (UK) Ltd, Croydon, CR0 4YY

10/11/2025

01994629-0018